Studies in Systems, Decision and Control

Volume 68

Series editor

Janusz Kacprzyk, Polish Academy of Sciences, Warsaw, Poland
e-mail: kacprzyk@ibspan.waw.pl

About this Series

The series "Studies in Systems, Decision and Control" (SSDC) covers both new developments and advances, as well as the state of the art, in the various areas of broadly perceived systems, decision making and control- quickly, up to date and with a high quality. The intent is to cover the theory, applications, and perspectives on the state of the art and future developments relevant to systems, decision making, control, complex processes and related areas, as embedded in the fields of engineering, computer science, physics, economics, social and life sciences, as well as the paradigms and methodologies behind them. The series contains monographs, textbooks, lecture notes and edited volumes in systems, decision making and control spanning the areas of Cyber-Physical Systems, Autonomous Systems, Sensor Networks, Control Systems, Energy Systems, Automotive Systems, Biological Systems, Vehicular Networking and Connected Vehicles, Aerospace Systems, Automation, Manufacturing, Smart Grids, Nonlinear Systems, Power Systems, Robotics, Social Systems, Economic Systems and other. Of particular value to both the contributors and the readership are the short publication timeframe and the world-wide distribution and exposure which enable both a wide and rapid dissemination of research output.

More information about this series at http://www.springer.com/series/13304

Arturo Locatelli

Optimal Control of a Double Integrator

A Primer on Maximum Principle

 Springer

Arturo Locatelli
Dipartimento di Elettronica, Informazione e
  Bioingegneria
Politecnico di Milano
Milan
Italy

ISSN 2198-4182   ISSN 2198-4190   (electronic)
Studies in Systems, Decision and Control
ISBN 978-3-319-82504-5   ISBN 978-3-319-42126-1   (eBook)
DOI 10.1007/978-3-319-42126-1

*To Franca, my parents, children
and grandchildren*

Preface

Optimal Control Theory has been very popular along a fairly large number of years, starting from the late 1950s when some fundamental results have been established. Among them there is no doubt that the *Maximum Principle* must be considered as a cornerstone. However, the (possibly) excessive enthusiasm for its reputed capability of solving any kind of problem, which has characterized the beginning of its story, has been followed by an (equally) unjustified rejection by considering it as purely abstract concepts with no real utility. In recent years it has been recognized that the truth lies somewhere between these two opinions, and optimal control has found its (appropriate yet limited) space within any curriculum where system and control theory plays a significant role. Consistently, this book is intended to supply an introductory but fairly comprehensive treatment of the founding issues of Pontryagin's Maximum Principle.

The book is suited for students who are already familiar with the basics of system and control theory and possess the calculus background usually taught in undergraduate engineering curricula. Furthermore its structure allows different ways of reading it and teaching its contents. A first level presentation of the Maximum Principle can be carried on by referring to Chap. 1, Sects. 2.1 and 2.2 (restraining the attention to Sects. 2.2.1 and 2.2.2) and to Chaps. 3–5. Subsequently, the material of Sect. 2.2.3 and Chap. 6 can be explored and a further in-depth study can be achieved through Sect. 2.2.4 and Chap. 7. A second level presentation which supplies a deeper insight makes reference to the more complex problems dealt with in Sect. 2.3 and Chaps. 8–10. Their treatment is almost self-consistent so that the fruition of this material needs Sects. 2.1–2.3, whereas it does not necessarily require considering Chaps. 3–7. Finally, the topics in Chaps. 11 and 12 can be added both to the first and the second level presentations.

Milan, Italy

Arturo Locatelli

Contents

1 Introduction 1

2 The Maximum Principle 3
  2.1   Statement of the Optimal Control Problems 3
  2.2   Necessary Conditions: Simple Constraints 6
      2.2.1 Purely Integral Performance Index 9
      2.2.2 Performance Index Function of the Final Event 11
      2.2.3 Non-standard Constraints on the Final State 13
      2.2.4 Minimum Time Problems 14
  2.3   Necessary Conditions: Complex Constraints 20
      2.3.1 Description of Complex Constraints 20
      2.3.2 Integral Constraints 21
      2.3.3 Punctual and Isolated Constraints 22
      2.3.4 Punctual and Global Constraints 23
  2.4   Necessary Conditions: Singular Arcs 26
  2.5   The Considered Problems 27

3 Simple Constraints: $J = \int$, $x(t_0) =$ Given 31
  3.1 $(x(t_f), t_0, t_f) =$ Given 32
  3.2 $(x(t_f), t_0) =$ Given, $t_f =$ Free 38
  3.3 $x(t_f) =$ Not Given, $(t_0, t_f) =$ Given 48
  3.4 $x(t_f) =$ Not Given, $t_0 =$ Given, $t_f =$ Free 62

4 Simple Constraints: $J = \int$, $x(t_0) =$ Not Given 77
  4.1 $(x(t_f), t_0, t_f) =$ Given 77
  4.2 $x(t_f) =$ Not Given, $(t_0, t_f) =$ Given 84
  4.3 $x(t_f) =$ Not Given, $(t_0, t_f) =$ Free 92

5 Simple Constraints: $J = \int + m$, $x(t_0) =$ Given,
$x(t_f) =$ Not Given 107
  5.1 $(t_0, t_f) =$ Given 107
  5.2 $t_0 =$ Given, $t_f =$ Free 116

6   Nonstandard Constraints on the Final State   127

7   Minimum Time Problems   149

8   Integral Constraints   161
    8.1   Integral Equality Constraints   161
    8.2   Integral Inequality Constraints   178

9   Punctual and Isolated Constrains   193

10   Punctual and Global Constraints   227
    10.1   Punctual and Global Equality Constraints   227
    10.2   Punctual and Global Inequality Constraints   240

11   Singular Arcs   271

12   Local Sufficient Conditions   297
    12.1 $x(t_f)$ = Partially Given, t_f = Free   299
    12.2 $x(t_f)$ = Given, t_f = Given   303
    12.3 $x(t_f)$ = Free, t_f = Free   304
    12.4 $x(t_f)$ = Free, t_f = Given   307

Appendix   309

References   311

Chapter 1
Introduction

Since its appearance in the early sixties, the Pontryagin's *Maximum Principle* is a fundamental part of Control Theory. Indeed it is the basic tool when we have to cope with *Optimal Control* problems, in particular those stated in terms of finite-dimensional, continuous-time dynamical systems.

As well known, the Maximum Principle is rooted in the Calculus of Variations and many people legitimately see it as a natural extension of this chapter of mathematics, specifically oriented to control problems. Thus it should not be surprising that the supplied results intrinsically have the nature and present the limits of necessary conditions. As a matter of facts these conditions stem from a customary and simple idea. Specifically, a solution of the problem must entail that any change of its free parameters does not cause an improvement of the value of the performance index (the index to be optimized) if the change is accomplished starting from the situation which is claimed to be optimal. The conditions which result by exploiting *first order variational methods*, namely by imposing that the improvement does not take place when first order variations occur, are necessary optimality conditions. However this simple idea is often developed in a fairly complex mathematical context, especially when the problem is featured with non trivial constraints on the state and/or control variables.

The aim of this book is to present in a simple and friendly manner the achievements resulting from first order variational methods. They are illustrated by a large number of problems which, almost without exceptions, refer to a particular second-order, linear and time-invariant dynamical system, the so-called *double integrator*. There is no doubt that it is the simplest system which supplies a fairly comprehensive overview of the topics under consideration. Furthermore a deeper insight on the significance of the solutions can frequently be gained by recalling that the double integrator constitutes the mathematical model of a body with unitary mass which is constrained to move along a straight line under the action of a force. Finally, the required computational burden is made as small as possible by the choice of this particular system, so that the presented material can easily be grasped. Consistently, we have chosen not to give a formal proof of the various necessary conditions and

© Springer International Publishing Switzerland 2017
A. Locatelli, *Optimal Control of a Double Integrator*, Studies in Systems, Decision and Control 68, DOI 10.1007/978-3-319-42126-1_1

leave to the specialized literature the task of satisfying the need for a more complete discussion.

The book is organized in the following way.

At the beginning of the second chapter (Sect. 2.1) we describe the optimal control problems to be later considered. They refer to a general finite-dimensional dynamic system, not necessarily a double integrator. The necessary conditions are presented in the subsequent Sects. 2.2 and 2.3 with reference to various classes of problems which, for convenience of presentation, are grouped according to the form of the performance index and the nature of the constraints imposed to the state and/or control variables. These constraints are classified as *simple constraints* and as *complex constraints*. Then we discuss the issues raised by the presence of the so-called *singular arcs* (Sect. 2.4). Problems related to this last topic constitute the content of Chap. 11, whereas those including only simple constraints are faced in Chaps. 3–6 and in Chap. 7 where reference is made to the particular, yet important, class of the *minimum time problems*. The subsequent Chaps. 8–10 consider problems where complex constraints must be satisfied. For the sake of completeness we mention and briefly illustrate in the last chapter of the book the important topic of *local sufficient conditions* resulting from *second order variational methods*. Indeed if a more precise evaluation of the changes of the performance index is performed by pushing the analysis up to second order terms and an actual deterioration is imposed, then local sufficient optimality conditions result. Space limitations have suggested restraining the attention to only few, yet significant, scenarios.

The way the material is collected and discussed in the present book closely mimic the one adopted in *Optimal control: an introduction* [15], a previous textbook of ours. This fact can not be surprising: indeed in both cases the underlying motivations strongly reflect the author's beliefs, come to maturity after many years of teaching activity, on how the basic theory of optimal control should be offered to students.

Before ending this short presentation, we point out that having (almost) always made reference to the same simple system and desired to allow the reader to concentrate on a particular class of problems without requiring an exhaustive reading of the preceding material, implies a certain degree of redundancy. In view of the objectives of this book, we believe this drawback to be fully acceptable.

Chapter 2
The Maximum Principle

Abstract We present here the most significant results which are customarily grouped under the name of Maximum Principle. It supplies a set of necessary conditions of optimality for a wide class of optimal control problems. Firstly, we give a formal, yet synthetic, description of these problems is given. Then we state the theorems which, in principle, allow one to find a solution of them which satisfies the necessary conditions of the Maximum Principle.

In this chapter we present the most significant results which are customarily grouped under the name of Maximum Principle. It supplies a set of necessary conditions of optimality for a wide class of optimal control problems. A formal, yet synthetic, description of these problems is given in the next Sect. 2.1. Then we state the theorems which, in principle, allow one to find a solution. For the sake of conciseness, we use this term to make reference to a solution of the control problem which satisfies the necessary conditions of the Maximum Principle. Similarly, NC is a shortened form for necessary conditions.

2.1 Statement of the Optimal Control Problems

The optimal control problems to be dealt with in the following are characterized by:

(a) The controlled system. It is a continuous-time, finite-dimensional dynamic system
(b) The set S_0 to which the system initial state x_0 must belong
(c) The set S_f to which the system final state x_f must belong
(d) The set of functions to which the system control variable must belong
(e) The performance index J
(f) A set of constraints on the state and/or control variables.

© Springer International Publishing Switzerland 2017

A. Locatelli, *Optimal Control of a Double Integrator*, Studies in Systems,
Decision and Control 68, DOI 10.1007/978-3-319-42126-1_2

More in detail, the system to be controlled is defined by the differential equation

$$\dot{x}(t) := f(x(t), u(t), t) \tag{2.1}$$

where *dot* stands for the derivative of $x(\cdot)$ with respect to the time t and $x(t) \in R^n$, $u(t) \in R^m$ are the vectors of the state and control variables, respectively. The functions $f(\cdot, \cdot, \cdot)$, $\partial f(\cdot, \cdot, \cdot)/\partial x$, $\partial f(\cdot, \cdot, \cdot)/\partial u$, $\partial f(\cdot, \cdot, \cdot)/\partial t$ are continuous with respect to all their arguments.

Together with this equation we have the boundary conditions

$$x_0 := x(t_0) \in S_0 \tag{2.2a}$$
$$x_f := x(t_f) \in S_f \tag{2.2b}$$

which account to saying that the state at the initial time t_0 and at the final time t_f belongs to specified sets S_0 and S_f. These sets are often referred to as the sets of the admissible initial and final states, respectively. The initial time as well as the final one may or may not be given.

The set of functions among which the control can be chosen is denoted by $\overline{\Omega}$. It is the set of the m-tuple of piecewise continuous functions which take on values in a given closed subset $\overline{U} \subseteq R^m$. Thus $u(\cdot)$ must be piecewise continuous and $u(t) \in \overline{U}$, $\forall t \in [t_0, t_f]$.

The performance index J must be minimized and is constituted by the sum of two terms which need not to be both present. One of them is the integral of a function of the state, control and time, while the second one is a function of the final event (the couple (x_f, t_f)). Thus the performance index takes on the form

$$J := \int_{t_0}^{t_f} l(x(t), u(t), t)dt + m(x(t_f), t_f) \tag{2.3}$$

In Eq. (2.3) the functions $l(\cdot, \cdot, \cdot)$, $m(\cdot, \cdot)$ and their partial derivatives with respect to x and t are continuous with respect to all their arguments.

Finally, the constraints of point (f) above have various forms and can account for a fairly wide class of limitations imposed to the state and control variables. The description of these further constraints, referred to as complex constraints, is presented in Sect. 2.3.1.

Let denote with $\varphi(\cdot; t_0, x_0, u(\cdot))$ the system *state motion*, that is the solution of Eq. (2.1) when $x(t_0) = x_0$ and the input is the function $u(\cdot)$. Then the optimal control problem can be stated in the following way.

Problem 2.1 (*Optimal control problem*)
Find a vector of functions $u^o(\cdot) \in \overline{\Omega}$, defined over the interval $[t_0^o, t_f^o]$, in such a way that Eq. (2.2) are verified, complex constraints, if present, are satisfied and the performance index (2.3), evaluated for $x^o(\cdot) := \varphi(\cdot, t_0^o, x^o(t_0^o), u^o(\cdot))$ and $u(\cdot) = u^o(\cdot)$ over the interval $[t_0^o, t_f^o]$, takes on the least possible value.   ∎

In a different, yet obvious, way the optimal control problem amounts to finding among all *feasible* controls that one (or those ones) to which the least possible value of the performance index is associated. A control is feasible if it is consistent with the following definition.

Definition 2.1 (*Feasible control*)
A vector of functions $u(\cdot) \in \overline{\Omega}$ defined over the interval $[t_0, t_f]$ is said to be a **feasible control** if the corresponding state motion verifies Eq. (2.2) and satisfies all complex constraints, if present.

The interval $[t_0, t_f]$ is referred to as the *control interval*.

In the forthcoming Sect. 2.2 we present the necessary conditions of the Maximum Principle for problems characterized by the presence of only *simple* constraints. More in detail, in Sects. 2.2.1 and 2.2.2 we deal with problems where the only constraints to be satisfied are $u(\cdot) \in \overline{\Omega}$ and those resulting from Eq. (2.2), the sets S_0 and S_f being *regular varieties*, that is sets complying with Definition 2.6 (see the beginning of Sect. 2.2). In Sect. 2.2.3 we also consider non-standard constraints on the final state, namely we allow the set S_f to have a bit more general nature. In Sect. 2.2.4 we discuss the so-called minimum time problems where the performance index is simply the length of the control interval. On the contrary, Sect. 2.3 deals with problems where constraints of a somehow more complex nature have to be satisfied. Subsequently, we present problems which are characterized by the presence of the so-called *singular arcs* (Sect. 2.4).

The results of the Maximum Principle make reference to the following items which are specified in Definitions 2.2–2.5.

Definition 2.2 (*Hamiltonian function*)
With reference to the system (2.1) and to the performance index (2.3) the function

$$H(x, u, t, \lambda_0, \lambda) := \lambda_0 l(x, u, t) + \lambda' f(x, u, t) \tag{2.4}$$

where λ_0 is a scalar and $\lambda \in R^n$, is said to be the **hamiltonian function**. ∎

Definition 2.3 (*Auxiliary system*)
With reference to a pair of functions $(x(\cdot), u(\cdot))$ satisfying Eq. (2.1) and the hamiltonian function (2.4), the system of linear differential equations

$$\dot{\lambda}(t) := -\left. \frac{\partial H(x, u(t), t, \lambda_0(t), \lambda(t))'}{\partial x} \right|_{x=x(t)} \tag{2.5a}$$

$$\dot{\lambda}_0(t) := 0 \tag{2.5b}$$

is the **auxiliary system** of Problem 2.1. ∎

Equation (2.5b) simply amounts to stating that $\lambda_0(\cdot)$ is a constant function: thus only Eq. (2.5a) is often referred to as the auxiliary system and Eq. (2.5b) is taken into account by simply requiring that $\lambda_0(\cdot)$ is constant. Also note that Eq. (2.1) can be replaced by

$$\dot{x}(t) = \frac{\partial H(x(t), u(t), t, \lambda_0, \lambda)'}{\partial \lambda} \tag{2.6}$$

This fact is expedient in stating the following definition.

Definition 2.4 (*Hamiltonian system*)
Equations (2.6) and (2.5a) where λ_0 is a constant, constitute the **hamiltonian system** for Problem 2.1.   ∎

Definition 2.5 (*H-minimizing control*)
The function $u_h(\cdot, \cdot, \cdot, \cdot)$ such that

$$H(x, u_h(x, t, \lambda_0, \lambda), t, \lambda_0, \lambda) \leq H(x, u, t, \lambda_0, \lambda), \ \forall u \in \overline{U}$$

is said to be the H-**minimizing control**.   ∎

For the sake of simplicity in notation, the arguments of the hamiltonian function are omitted, unless strictly necessary. Furthermore $H(t)$ is the value of the hamiltonian function corresponding to $x(t), u(t), t, \lambda_0$ and $\lambda(t)$.

2.2 Necessary Conditions: Simple Constraints

As mentioned in the previous section, the optimal control problems where only simple constraints are present amount to controlling system (2.1) in such a way that the performance index (2.3) is minimized while satisfying the constraints $u(\cdot) \in \overline{\Omega}$ and (2.2). Furthermore in Sects. 2.2.1 and 2.2.2 we assume that the sets S_0 and S_f are regular varieties to be denoted by $\overline{S}_0$ and $\overline{S}_f$, respectively. We remove this assumption in Sect. 2.2.3 with reference to the set S_f. Finally, in Sect. 2.2.4 we face the particular class of the so-called minimum time problems.

A regular variety is a set defined as follows.

Definition 2.6 (*Regular variety*)
Consider the set

$$S := \left\{ x \mid x \in R^n, \ \alpha_i(x) = 0, \ i = 1, 2, \ldots, q \leq n \right\}$$

where the functions $\alpha_i(\cdot)$ are continuous together with their first derivatives. Moreover, consider the $q \times n$ matrix $\Sigma(\cdot)$

$$\Sigma(x) := \begin{bmatrix} \dfrac{d\alpha_1(x)}{dx} \\ \vdots \\ \dfrac{d\alpha_q(x)}{dx} \end{bmatrix}$$

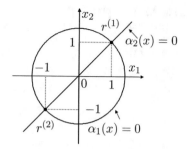

Fig. 2.1 Example 2.1. The set constituted by the two vectors $r^{(1)}$ and $r^{(2)}$ is a regular variety

If $\mathrm{rank}(\Sigma(x)) = q,\, \forall x \in S$, then the set S is said to be a **regular variety**. ∎

The nature of a regular variety is further clarified by the following example.

Example 2.1
Consider the set $S \subset R^2$ defined by the two functions

$$\alpha_1(x) = x_1^2 + x_2^2 - 2$$
$$\alpha_2(x) = x_1 - x_2$$

Figure 2.1 shows that it is constituted by the two vectors

$$r^{(1)} = \begin{bmatrix} 1 \\ 1 \end{bmatrix}, \quad r^{(2)} = \begin{bmatrix} -1 \\ -1 \end{bmatrix}$$

The set S is a regular variety. In fact the rank of the matrix

$$\Sigma(x) = \begin{bmatrix} 2x_1 & 2x_2 \\ 1 & -1 \end{bmatrix}$$

is 2 when either $x = r^{(1)}$ or $x = r^{(2)}$, that is for all $x \in S$.
   Now consider the set $S \subset R^2$ defined by the two functions

$$\alpha_1(x) = x_1^2 + x_2^2 - 2$$
$$\alpha_2(x) = (x_2 - 1)(x_2 - \sqrt{2})$$

Figure 2.2 shows that it is constituted by the three vectors

$$r^{(1)} = \begin{bmatrix} 0 \\ \sqrt{2} \end{bmatrix}, \quad r^{(2)} = \begin{bmatrix} -1 \\ 1 \end{bmatrix}, \quad r^{(3)} = \begin{bmatrix} 1 \\ 1 \end{bmatrix}$$

Fig. 2.2 Example 2.1. The
set constituted by the three
vectors $r^{(1)}$, $r^{(2)}$ e $r^{(3)}$ is not
a regular variety

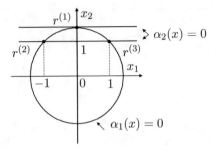

The set S is not a regular variety. In fact the rank of the matrix

$$\Sigma(x) = \begin{bmatrix} 2x_1 & 2x_2 \\ 0 & 2x_2 - 1 - \sqrt{2} \end{bmatrix}$$

is 1 when $x = r^{(1)}$.   ∎

Remark 2.1
Observe that any set S constituted by a single element of R^n and defined by

$$S = \{x \mid \alpha_i(x) = 0, \ i = 1, 2, \ldots, n\}, \ \alpha_i(x) = x_i - \bar{x}_i, \ i = 1, 2, \ldots, n$$

where $\bar{x}_i, \ i = 1, 2, \ldots, n$, are given, is a regular variety. In fact we have

$$\Sigma(x) = \begin{bmatrix} \dfrac{d\alpha_1(x)}{dx} \\ \dfrac{d\alpha_2(x)}{dx} \\ \vdots \\ \dfrac{d\alpha_n(x)}{dx} \end{bmatrix} = \begin{bmatrix} 1 & 0 & \cdots & 0 \\ 0 & 1 & \cdots & 0 \\ \vdots & \vdots & \ddots & \vdots \\ 0 & 0 & \cdots & 1 \end{bmatrix}$$

and, apparently, $\mathrm{rank}(\Sigma(x)) = n, \ \forall x$.   ∎

Consistently with Definition 2.6, we will describe the regular varieties in the following way

$$\bar{S}_0 := \left\{x \mid x \in R^n, \ \alpha_{0_i}(x) = 0, \ i = 1, 2, \ldots, q_0 \le n\right\} \tag{2.7a}$$

$$\bar{S}_f := \left\{x \mid x \in R^n, \ \alpha_{f_i}(x) = 0, \ i = 1, 2, \ldots, q_f \le n\right\} \tag{2.7b}$$

where all the functions $\alpha_{0_i}(\cdot)$ and $\alpha_{f_i}(\cdot)$ are continuous together with their first
derivatives. Furthermore $\Sigma_0(x)$ and $\Sigma_f(x)$ denote the matrices $\Sigma(x)$ corresponding
to the sets $\bar{S}_0$ and $\bar{S}_f$, respectively.

Optimal control problems with simple constraints are discussed in the forthcoming four Sects. 2.2.1–2.2.4. In the first one of them the performance index is purely integral (the term $m(x(t_f), t_f)$ is absent), whereas in the second one the performance index explicitly depends on the final event (the term $m(x(t_f), t_f)$ is present). In the third subsection we consider non-standard constraints on the final state, namely we allow the final state to belong to a set which is more general than a regular variety. Finally, in the last subsection we present problems where the objective is simply to minimize the length of the control interval.

2.2.1 Purely Integral Performance Index

We here consider problems where the functional J has the form

$$J = \int_{t_0}^{t_f} l(x(t), u(t), t)dt$$

With reference to this class of problems it is possible to state the set of NC collected in Theorem 2.1.

Theorem 2.1
Let $\overline{S}_0$ and $\overline{S}_f$ be two regular varieties given by Eq. (2.7). Let $u^o(\cdot) \in \overline{\Omega}$ be a control which is defined over the interval $\left[t_0^o, t_f^o\right]$, $t_0^o < t_f^o$, and transfers the state of the system (2.1) from a suitable point $x^o\left(t_0^o\right) \in \overline{S}_0$ to a suitable point $x^o\left(t_f^o\right) \in \overline{S}_f$ and $x^o(\cdot) := \varphi\left(\cdot; t_0^o, x^o\left(t_0^o\right), u^o(\cdot)\right)$. A necessary condition for the quadruple $\left(x^o(\cdot), u^o(\cdot), t_0^o, t_f^o\right)$ to be optimal is the existence of a solution $\lambda_0^o, \lambda^o(\cdot)$ of the auxiliary system (2.5) corresponding to $(x^o(\cdot), u^o(\cdot))$ such that

(i)

$$H\left(x^o(t), u^o(t), t, \lambda_0^o, \lambda^o(t)\right) \leq H\left(x^o(t), u, t, \lambda_0^o, \lambda^o(t)\right)$$
$$t \in \left[t_0^o, t_f^o\right], \text{a.e.}, \forall u \in \overline{U}$$

(ii)

$$H\left(x^o\left(t_f^o\right), u^o\left(t_f^o\right), t_f^o, \lambda_0^o, \lambda^o\left(t_f^o\right)\right) = 0$$

(iii)

$$H\left(x^o\left(t_0^o\right), u^o\left(t_0^o\right), t_0^o, \lambda_0^o, \lambda^o\left(t_0^o\right)\right) = 0$$

(iv)

$$\lambda_0^o \geq 0$$

(v)

$$\begin{bmatrix} \lambda_0^o \\ \lambda^o(\cdot) \end{bmatrix} \neq 0$$

(vi)

$$\lambda^o\left(t_f^o\right) = \sum_{i=1}^{q_f} \vartheta_{f_i} \left.\frac{d\alpha_{f_i}(x)}{dx}\right|'_{x=x^o\left(t_f^o\right)} \,, \quad \vartheta_{f_i} \in R, \ 1 \leq i \leq q_f$$

(vii)

$$\lambda^o\left(t_0^o\right) = \sum_{i=1}^{q_0} \vartheta_{0_i} \left.\frac{d\alpha_{0_i}(x)}{dx}\right|'_{x=x^o\left(t_0^o\right)} \,, \quad \vartheta_{0_i} \in R, \ 1 \leq i \leq q_0$$

■

Some comments are in order. They are collected in the forthcoming Remark 2.2 and their consequences are summarized in Remark 2.3.

Remark 2.2

(a) It is quite obvious that a control which differs from an optimal one over a set of zero measure is still an optimal control. This fact motivates the insertion of "a.e." which stands for *almost everywhere*.

(b) Condition *(i)* is usually referred to as the condition of the Maximum Principle and it amounts to saying that a control must minimize the hamiltonian function in order to be optimal. The seemingly contradiction resulting from the name (maximum) and the request (minimize) is due to the fact that the Maximum Principle was originally stated for problems where the performance index had to be maximized. In this event, the inequality sign in *(i)* must be reversed.

(c) In principle, condition *(i)* allows us to compute the H-minimizing control.

(d) Conditions *(ii)* and *(iii)* are referred to as **transversality conditions at the final and initial time**, respectively. They hold only if the final or initial time are not given. Thus, if the final (initial) time is specified in the optimal control problem at hand, any optimal solution might not comply with the relevant transversality condition.

(e) Conditions *(vi)* and *(vii)*, where the parameters $\vartheta_{f_i}, i = 1, 2, \ldots, q_f$ and ϑ_{0_i}, $i = 1, 2, \ldots, q_0$ are suitable constants, are mentioned as **orthogonality conditions at the final and initial time**, respectively. If the state at the final (initial) time is given, the corresponding orthogonality condition is anyhow satisfied.

(f) When the constant λ_0^o is zero the problem is said to be **pathological**. Whenever this is not the case, λ_0^o can be set equal to one and the problem is not pathological.

(g) If no constraint is imposed to the final state (i.e., it is "free"), the corresponding orthogonality condition *(vi)* simply becomes $\lambda^o(t_f^o) = 0$. Analogously, if no constraint is imposed to the initial state (i.e., it is "free"), the corresponding orthogonality condition *(vii)* simply becomes $\lambda^o(t_0^o) = 0$.

(h) If the optimal control problem is time-invariant, that is if both the functions $f(\cdot, \cdot, \cdot)$ and $l(\cdot, \cdot, \cdot)$ do not explicitly depend on time t, we can assume $t_0^o = 0$, without loss of generality. Furthermore the hamiltonian function is constant if evaluated along a trajectory which satisfies the NC. More precisely,

$$H\left(x(t), u(t), t, \lambda_0, \lambda(t)\right) = \text{const.}$$

provided that:

1. the pair $(x(\cdot), u(\cdot))$ satisfies Eq. (2.1);
2. the pair $(\lambda_0(\cdot), \lambda(\cdot))$ is a solution of the auxiliary system (2.5) corresponding to the pair $(x(\cdot), u(\cdot))$;
3. the quadruple $(x(\cdot), u(\cdot), \lambda_0, \lambda(\cdot))$ satisfies the condition *(i)* of the theorem.

Thus we can verify or impose (if it is the case) the transversality condition *(ii)* at any instant of the interval $\left[0, t_f\right]$. ∎

Remark 2.3
In principle, Theorem 2.1 allows us to find a control which satisfies the NC, thus constituting a *candidate* optimal control. In fact we assume that the problem is not pathological and set $\lambda_0 = 1$. Then we find the H-minimizing control as a function of x, λ and t. Subsequently, we substitute such a function into the $2n$ Eqs. (2.5a) and (2.6) of the hamiltonian system and solve them while satisfying the relevant $2n$ boundary conditions. They are constituted by the $n - q_f$ orthogonality conditions at the final time, the $n - q_0$ orthogonality conditions at the initial time, the q_0 relations resulting from the constraint $x(t_0) \in \overline{S}_0$ and the q_f relations resulting from the constraint $x(t_f) \in \overline{S}_f$. The obtained solution (obviously) depends on the values of t_0 and t_f which can be found by imposing the transversality conditions at t_0 and t_f. These considerations also apply to all the forthcoming theorems. ∎

2.2.2 Performance Index Function of the Final Event

We now consider problems where the performance index explicitly depends on the final event. In these problems the functional J takes on the form

$$J = \int_{t_0}^{t_f} l(x(t), u(t), t)dt + m\left(x\left(t_f\right), t_f\right), \quad m(\cdot, \cdot) \neq 0$$

For the sake of simplicity, we only consider the cases where the initial state x_0 and the initial time t_0 are given. Furthermore we make reference to the following assumptions on the set S_f of the admissible final states.

Assumption 2.1
The set S_f of the admissible final states is the whole state space, i.e., the final state x_f is free. ∎

Assumption 2.2

The set S_f of the admissible final states is a regular variety $\overline{S}_f$ of the kind (2.7b) and $l(\cdot, \cdot, \cdot) \neq 0$, i.e., the performance index also contains an integral type term.   ∎

For the class of problems considered in this subsection we have the following result.

Theorem 2.2

Let Assumption 2.1 or Assumption 2.2 hold. Furthermore let $u^o(\cdot) \in \overline{\Omega}$ be a control, defined over the interval $\left[t_0, t_f^o\right]$, $t_0 < t_f^o$, which transfers the state of the system (2.1) from x_0 to $x^o\left(t_f^o\right) \in S_f$ and $x^o(\cdot) := \varphi\left(\cdot; t_0, x_0, u^o(\cdot)\right)$.

A necessary condition for the triple $\left(x^o(\cdot), u^o(\cdot), t_f^o\right)$ to be optimal is the existence of a solution λ_0^o, $\lambda^o(\cdot)$ of the auxiliary system (2.5) corresponding to $(x^o(\cdot), u^o(\cdot))$ such that

(i)

$$H\left(x^o(t), u^o(t), t, \lambda_0^o, \lambda^o(t)\right) \leq H\left(x^o(t), u, t, \lambda_0^o, \lambda^o(t)\right)$$
$$t \in \left[t_0, t_f^o\right], \text{a.e.}, \forall u \in \overline{U}$$

(ii)

$$H\left(x^o\left(t_f^o\right), u^o\left(t_f^o\right), t_f^o, \lambda_0^o, \lambda^o\left(t_f^o\right)\right) + \lambda_0^o \left.\frac{\partial m\left(x^o\left(t_f^o\right), t\right)}{\partial t}\right|_{t=t_f^o} = 0$$

(iii)

$$\lambda_0^o \geq 0$$

(iv)

$$\begin{bmatrix} \lambda_0^o \\ \lambda^o(\cdot) \end{bmatrix} \neq 0$$

(v) If x_f is free

$$\lambda^o\left(t_f^o\right) - \lambda_0^o \left.\frac{\partial m\left(x, t_f^o\right)}{\partial x}\right|_{x=x^o\left(t_f^o\right)}^{\prime} = 0$$

If $x_f \in \overline{S}_f$

$$\lambda^o\left(t_f^o\right) - \lambda_0^o \left.\frac{\partial m\left(x, t_f^o\right)}{\partial x}\right|'_{x=x^o\left(t_f^o\right)} = \sum_{i=1}^{q_f} \vartheta_{f_i} \left.\frac{d\alpha_{f_i}(x)}{dx}\right|'_{x=x^o\left(t_f^o\right)}$$

$$\vartheta_{f_i} \in R, \ 1 \leq i \leq q_f$$

■

Remark 2.4
Observe that the presence of the term $m(\cdot, \cdot)$ simply implies a change in the transversality and orthogonality conditions at the final time. ■

Remark 2.5
The statement of Theorem 2.2 does not mention possible limitations to the value of the final time. Indeed the final time is implicitly assumed to be either given or free. However, more general situations can easily be tackled, even in the presence of complex constraints.

To this aim we introduce a new state variable x_{n+1} through the equations

$$\dot{x}_{n+1}(t) := 1, \quad x_{n+1}(t_0) = t_0$$

In so doing the system is "enlarged" because its state x_a is now

$$x_a := \begin{bmatrix} x_1 & x_2 & \cdots & x_n & x_{n+1} \end{bmatrix}$$

Obviously, we have

$$x_{n+1}(t) = x_{n+1}(t_0) + t - t_0 = t$$

and can easily express possible constraints on the final event in terms of suitable constraints on the final state $x_a(t_f)$ of the enlarged system. ■

2.2.3 Non-standard Constraints on the Final State

We can easily exploit the results of this section in dealing with problems where the final state is constrained in a non-standard way, namely when it must belong to a set which is not a regular variety. More in detail, we suppose that the final state x_f is not asked to comply with Eq. (2.2b), but rather to satisfy the condition $x_f \in \widehat{S}_f$, where

$$\widehat{S}_f := \left\{x \,|\, x \in S_f, \ a_i \leq x_i \leq b_i, \ i \in \mathcal{X}\right\}$$

In the above equation S_f is either a regular variety given by (2.7b) or the whole state space, a_i and b_i are given constants and $\mathcal{X}$ is a subset of $\{1, 2, \ldots, n\}$. For each value of the index i, either a_i or b_i, but not both, can be not finite. Thus requests of the kind $x_i \leq b_i$ or $a_i \leq x_i$ are allowed.

For this class of problems we first consider the case where only one state variable x_i must belong to a given interval $[a_i, b_i]$. Thus the set $\mathcal{X}$ is constituted by a single element. We handle this situation by *ignoring*, as a first attempt, the constraint $a_i \leq x_i(t_f) \leq b_i$ and then proceeding with the computation of a solution. If the resulting state motion is such that $x(t_f) \in \widehat{S}_f$, the solution at hand obviously satisfies the NC also for the original problem. If this does not happen, i.e., the constraint can not be ignored, we consider the appropriate alternative among those listed below.

(i) When both a_i and b_i are finite, we tackle the couple of problems where the final state must belong to the set $S_{fm} := \{x | x \in S_f, x_i = a_i\}$ or to the set $S_{fM} := \{x | x \in S_f, x_i = b_i\}$

(ii) When only a_i is finite, we tackle the problem where the final state must belong to the set $S_{fm} := \{x | x \in S_f, x_i = a_i\}$

(iii) When only b_i is finite, we tackle the problem where the final state must belong to the set $S_{fM} := \{x | x \in S_f, x_i = b_i\}$.

The sets S_{fm} and S_{fM}, if not empty, are regular varieties (possibly after some suitable and obvious rearrangements of the equations $\alpha_{f_j}(x) = 0$, $j = 1, 2, \ldots, q_f$, which define S_f). It is apparent that the solutions which are dominated by some others should be discarded.

The procedure outlined for the case where only one state variable must belong to a given interval can be generalized to the case where more than one component of the state vector must comply with requirements of this kind. Indeed it suffices to perform a combinatorial-type check of all possible scenarios.

2.2.4 Minimum Time Problems

We here discuss problems where the performance index is simply the time elapsed for the system state to be driven from the set S_0 of the admissible initial states to the set S_f of the admissible final states. Of course these two sets must be disjoint because, otherwise, the problem is trivial. Therefore we assume that the two sets are the regular varieties $\overline{S}_0$ and $\overline{S}_f$ and $\overline{S}_0 \cap \overline{S}_f = \emptyset$. Furthermore the problem is in general meaningful only if the control variable can not take on arbitrarily great values. Thus we always suppose that the set $\overline{U}$ of the admissible values of $u(t)$ is bounded.

A typical minimum time problem is defined by Eq. (2.1), the sets $\overline{U}, \overline{S}_0, \overline{S}_f$ and the performance index

$$J = \int_{t_0}^{t_f} l(x(t), u(t), t) \, dt, \quad l(x, u, t) = 1$$

Consequently, the hamiltonian function is

$$H(x, u, t, \lambda_0, \lambda) = \lambda_0 + \lambda' f(x, u, t)$$

and the relevant NC are those specified by Theorem 2.1 in Sect. 2.2.1.

Particularly significant results can be proved if we restrict the attention to the subclass of problems where

(i) The system is linear and time-invariant, i.e., it is described by

$$\dot{x}(t) := Ax(t) + Bu(t)$$

(ii) The set $\overline{U}$ is the closed, convex and bounded polyhedron defined by

$$\overline{U} := \{u \mid Su \leq a, \ a \in R^s\}$$

where S and a are a given $s \times m$ matrix and an s-dimensional vector, respectively.

The nature of a closed, convex and bounded polyhedron is clarified by the following example.

Example 2.2

Let the $s \times m$ matrix S and the s-dimensional vector a be

$$S = \begin{bmatrix} -1 & -1 \\ 1 & -2 \\ \dfrac{1}{2} & 0 \\ -\dfrac{1}{4} & 1 \end{bmatrix}, \quad a = \begin{bmatrix} -1 \\ 1 \\ 1 \\ 1 \end{bmatrix}$$

Then the corresponding closed, convex and bounded polyhedron is the four-sided polygon shown in Fig. 2.3. ∎

Fig. 2.3 The closed, convex and bounded polyhedron of Example 2.2

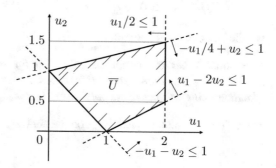

The results below require that the following assumption concerning *linear independence* is satisfied. We recall that a set of vectors $\{v_1, v_2, \ldots, v_k\}$ is linearly independent if the relation

$$\sum_{i=1}^{k} \beta_i v_i = 0$$

implies, besides being obviously implied, $\beta_1 = \beta_2 = \cdots = \beta_k = 0$.

Assumption 2.3
Given an arbitrary m-dimensional vector $v \neq 0$ aligned with any one of the edges of $\overline{U}$, the set of vectors $Bv, ABv, A^2 Bv, \cdots, A^{n-1} Bv$ is linearly independent.   ∎

The first important consequence of this assumption refers to the uniqueness of the H-minimizing control: here uniqueness has to be meant in the sense that it exists almost everywhere, so that two controls which only differ on a set of zero measure must be considered equal. This result is stated in the forthcoming Theorem 2.3 for the considered subclass of problems where the hamiltonian function is

$$H = \lambda_0 + \lambda'(Ax + Bu)$$

and the solutions of the auxiliary system are of the form

$$\lambda(t) = e^{-A't} \lambda(0), \quad \lambda_0(\cdot) = \text{cost}.$$

Observe that $\lambda(\cdot)$ can not be zero because, otherwise, also $\lambda_0 = 0$ (recall the transversality condition) and condition (v) of Theorem 2.1 is violated. Also note that the minimization of the hamiltonian function only concerns the term $\lambda'(t) Bu$ and introduce the following definition.

Definition 2.7 (*Extremal control*)
A control $u^*(\cdot)$ which minimizes the hamiltonian function for all t, corresponding to a given vector $\lambda^*(0) \neq 0$, i.e., such that

$$\lambda^{*\prime}(t) Bu^*(t) \leq \lambda^{*\prime}(t) Bu, \ \forall u \in \overline{U}, \ \lambda^{*\prime}(t) = e^{-A't} \lambda^*(0), \ 0 \leq t \leq t_f$$

is called an **extremal control**.   ∎

Theorem 2.3 *Let Assumption 2.3 hold. Then for any nonzero solution of Eq. (2.5a) a unique extremal control exists. It is piecewise constant and takes on values corresponding to the vertices of the polyhedron $\overline{U}$.*   ∎

The following example sheds light on the above result.

Fig. 2.4 The polyhedron $\overline{U}$ of Example 2.3

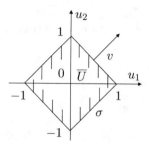

Example 2.3

Let

$$A = \begin{bmatrix} 1 & -1 & 0 \\ 0 & 0 & 1 \\ 0 & 0 & 0 \end{bmatrix}, \quad B = \begin{bmatrix} 0 & 1 \\ 0 & 0 \\ 1 & 0 \end{bmatrix}, \quad S = \begin{bmatrix} 1 & 1 \\ -1 & 1 \\ -1 & -1 \\ 1 & -1 \end{bmatrix}, \quad a = \begin{bmatrix} 1 \\ 1 \\ 1 \\ 1 \end{bmatrix},$$

Therefore the corresponding polyhedron $\overline{U}$ is the square shown in Fig. 2.4. If we select the vector $v = \begin{bmatrix} 1 & 1 \end{bmatrix}'$ which is aligned with the edge σ of $\overline{U}$ it is easy to check that Assumption 2.3 is not verified. In fact

$$Bv = \begin{bmatrix} 1 \\ 0 \\ 1 \end{bmatrix}, \quad ABv = \begin{bmatrix} 1 \\ 1 \\ 0 \end{bmatrix}, \quad A^2 Bv = \begin{bmatrix} 0 \\ 0 \\ 0 \end{bmatrix}$$

Obviously these three vectors constitute a set which is not linearly independent. Now observe that the general solution of Eq. (2.5a) is

$$\lambda_1(t) = \lambda_1(0)e^{-t}$$
$$\lambda_2(t) = \lambda_2(0) + \lambda_1(0)(1 - e^{-t})$$
$$\lambda_3(t) = \lambda_3(0) + \lambda_1(0)(1 - t - e^{-t}) - \lambda_2(0)t$$

so that it results

$$\lambda'(t)Bu = \left[\lambda_3(0) + \lambda_1(0)\left(1 - t - e^{-t}\right) - \lambda_2(0)t\right]u_1 + \lambda_1(0)e^{-t}u_2$$

The choice $\lambda_2(0) = \lambda_3(0) = -\lambda_1(0)$, $\lambda_1(0) > 0$ entails $\lambda'(t)Bu = \lambda_1(0)e^{-t}(u_2 - u_1)$ so that the hamiltonian function is minimized by all pairs (u_1, u_2) belonging to the edge σ.

Viceversa, let

$$A = \begin{bmatrix} 0 & 1 \\ 0 & 0 \end{bmatrix}, \quad B = \begin{bmatrix} 0 \\ 1 \end{bmatrix}, \quad S = \begin{bmatrix} 1 \\ -1 \end{bmatrix}, \quad a = \begin{bmatrix} 1 \\ 1 \end{bmatrix},$$

The polyhedron $\overline{U}$ is the line segment $-1 \le u \le 1$ and Assumption 2.3 is satisfied. In fact for $v = 1$ we have the two vectors

$$Bv = \begin{bmatrix} 0 \\ 1 \end{bmatrix}, \ ABv = \begin{bmatrix} 1 \\ 0 \end{bmatrix}$$

which constitute a linear independent set and the general solution of Eq. (2.5a) is

$$\lambda_1(t) = \lambda_1(0), \quad \lambda_2(t) = \lambda_2(0) - \lambda_1(0)t$$

so that

$$\lambda'(t)Bu = \lambda_2(0) - \lambda_1(0)t$$

If $\lambda_2(0)$ and $\lambda_1(0)$ are not both zero the H-minimizing control is uniquely determined and equal to 1 when $\lambda_2(0) - \lambda_1(0)t < 0$ and equal to -1 when $\lambda_2(0) - \lambda_1(0)t > 0$. Observe that the values taken on by the control correspond to the vertices of the polyhedron $\overline{U}$.   ∎

As a direct consequence of Theorem 2.3 a control which satisfies the NC is in general discontinuous. The times when $u(\cdot)$ commutes, i.e., switches from a vertex of the polyhedron to another one, are referred to as *switching times*.

If Assumption 2.3 holds, an optimal control is characterized by a *finite* number of switching times. In general this number depends on A, B, $\overline{S}_0$, $\overline{S}_f$ and $\overline{U}$ and, corresponding to given A, B, $\overline{U}$, such a number may turn out to be *unbounded* as $\overline{S}_0$ and/or $\overline{S}_f$ vary.

An upper limit to the number of switching times can be established for particular classes of problems. The relevant result is stated in the forthcoming Theorem 2.4, where we refer to an m-dimensional *parallelepiped* $\overline{P}$ which is a closed and bounded subset of R^m defined by

$$\overline{P} := \{u \,|\, a_i \le u_i \le b_i, \ i = 1, 2, \dots, m\}$$

Theorem 2.4
Let Assumption 2.3 hold, $\overline{U} = \overline{P}$ and all the eigenvalues of A be real. Then each component of an extremal control commutes at most $n - 1$ times.   ∎

Whenever the sets $\overline{S}_0$ and $\overline{S}_f$ shrink to a single point, namely when $\overline{S}_0 = \{x_0\}$ and $\overline{S}_f = \{x_f\}$, it is possible to state the following theorem.

Theorem 2.5
Let Assumption 2.3 hold, $\overline{S}_0 = \{x_0\}$ and $\overline{S}_f = \{x_f\}$. Then the optimal control, if it exists, is unique.   ∎

Note that this theorem is not an *existence* theorem but rather a *uniqueness* one.

It is obvious that an optimal control has to be sought within the set of extremal controls which are feasible as well, that is which transfer the system state from $\overline{S}_0$ to $\overline{S}_f$.

In some particular cases we have uniqueness of feasible extremal controls. One of such cases is specified in the following theorem.

Theorem 2.6
Let Assumption 2.3 hold, $\overline{S}_0 = \{x_0\}$ and $\overline{S}_f = \{0\}$. Furthermore let the origin of R^m be an interior point of $\overline{U}$. Then at most one extremal feasible control exists. ■

Whenever it is possible to resort to this theorem (which is again a *uniqueness* and not an *existence* theorem), the extremal feasible control which has been found is also an optimal control, obviously provided that an optimal control exists. We can claim that such a control exists in the following two cases.

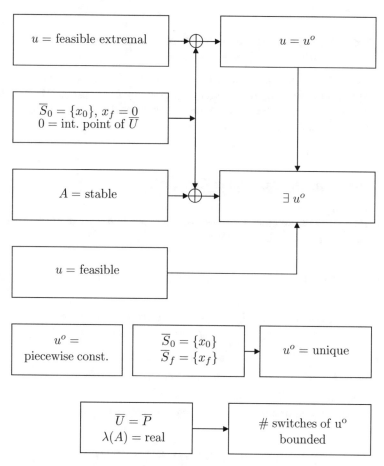

Fig. 2.5 Summary of the results for time optimal problems when Assumption 2.3 holds

Theorem 2.7
Let Assumption 2.3 hold. If a feasible control exists, then an optimal control exists
too.   ∎

Theorem 2.8
Let Assumption 2.3 hold. Furthermore let (i) $\overline{S}_f = \{0\}$, (ii) *the origin of* R^m *be an*
interior point of $\overline{U}$ *and* (iii) *the matrix A be stable. Then the optimal control exists*
for each initial state x_0.   ∎

We have used the term *stable* in the statement of this theorem to express the fact that
all the eigenvalues of the matrix A have negative real parts.

All the above results concerning minimum time problems are synthetically collected in Fig. 2.5.

2.3 Necessary Conditions: Complex Constraints

We discuss optimal control problems where at least one of the constraints described at
the beginning of the forthcoming subsection (points (a)–(c)) is present. Furthermore
we assume that also requirements on state and control variables of the kind previously
considered can be added to the problem statement.

2.3.1 Description of Complex Constraints

A large set of realistic situations may be taken into account if we allow the presence
of the requests defined in the following list.

(a) *Integral constraints*
   We consider both equality and inequality integral constraints. More specifically

   (a1) *Integral equality constraints*
       The state and/or control variables must satisfy the equation

$$\int_{t_0}^{t_f} w_e(x(t), u(t), t)dt = \overline{w}_e$$

   (a2) *Integral inequality constraints*
       The state and/or control variables must satisfy the equation

$$\int_{t_0}^{t_f} w_d(x(t), u(t), t)dt \le \overline{w}_d$$

(b) *Punctual and isolated constraints*

More precisely, these constraints are referred to as *Punctual and isolated equality constraints*. In fact the state variables must satisfy the equations

$$w^{(i)}(x(t_i), t_i) = 0, \ i = 1, 2, \ldots, s$$
$$t_0 < t_1 < \cdots < t_i < \cdots < t_s < t_f$$

where the $s \geq 1$ instants of time t_i may or may not be given.

(c) *Punctual and global constraints*

We consider both equality and inequality punctual and global constraints. More specifically,

(c1) *Punctual and global equality constraints*

The state and/or control variables must satisfy the equation

$$w(x(t), u(t), t) = 0, \ t_0 \leq t \leq t_f$$

(c2) *Punctual and global inequality constraints*

The state and/or control variables must satisfy the equation

$$w(x(t), u(t), t) \leq 0, \ t_0 \leq t \leq t_f$$

In the above equations the vectors $\overline{w}_e$ and $\overline{w}_d$ are given constants. Furthermore the vector functions $w_e(\cdot, \cdot, \cdot)$, $w_d(\cdot, \cdot, \cdot)$, $w^{(i)}(\cdot, \cdot)$ and $w(\cdot, \cdot, \cdot)$ are continuous together with their first partial derivatives with respect to x and t. With reference to cases (c1) and (c2), the functions $w(\cdot, \cdot, \cdot)$ are required to have also derivatives of higher order (to be specified from time to time) if they do not explicitly depend on u. Finally, the notation $\beta \leq \gamma$ where β and γ are r-dimensional vectors, stands for $\beta_i \leq \gamma_i$, $i = 1, 2, \ldots, r$.

2.3.2 Integral Constraints

We now discuss problems where the state and control variables must comply with integral-type constraints (see points (a1) and (a2) of Sect. 2.3.1).

Integral Equality Constraints

We consider two r_e-dimensional vectors. The first one is a vector $w_e(\cdot, \cdot, \cdot)$ of functions, whereas the second one is a vector $\overline{w}_e$ of constants. Then integral equality constraints are defined by

$$\int_{t_0}^{t_f} w_e(x(t), u(t), t) \, dt = \overline{w}_e \tag{2.8}$$

These constraints can easily be handled by adding to the original state variables an r_e-dimensional vector $\bar{x} := \begin{bmatrix} x_{n+1} & x_{n+2} & \cdots & x_{n+r_e} \end{bmatrix}'$ of new state variables. In fact if they are defined by the equations

$$\dot{\bar{x}}(t) := w_e(x(t), u(t), t), \quad \bar{x}(t_0) = 0$$

the constraints (2.8) are satisfied if $\bar{x}(t_f) = \overline{w}_e$. Thus the given problem has been restated in terms of a problem with simple constraints relevant to an *enlarged* system with state $x_a := \begin{bmatrix} x' & \bar{x}' \end{bmatrix}'$.

Integral Inequality Constraints

We can proceed in a similar manner when integral inequality constraints are present.

We consider two r_d-dimensional vectors. The first one is a vector $w_d(\cdot, \cdot, \cdot)$ of functions, whereas the second one is a vector $\overline{w}_d$ of constants. Then integral inequality constraints are defined by

$$\int_{t_0}^{t_f} w_d(x(t), u(t), t) \, dt \leq \overline{w}_d \tag{2.9}$$

These constraints can easily be handled by adding to the original state variables an r_d-dimensional vector $\bar{x} := \begin{bmatrix} x_{n+1} & x_{n+2} & \cdots & x_{n+r_d} \end{bmatrix}'$ of new state variables. In fact if they are defined by the equations

$$\dot{\bar{x}}(t) := w_d(x(t), u(t), t), \quad \bar{x}(t_0) = 0$$

the constraints (2.9) are satisfied if $\bar{x}(t_f) \leq \overline{w}_d$. Thus the given problem has been restated in terms of a problem with non-standard constraints on the final state. The new problem is relevant to an *enlarged* system with state $x_a := \begin{bmatrix} x' & \bar{x}' \end{bmatrix}'$ and can be handled as shown in Sect. 2.2.3.

2.3.3 Punctual and Isolated Constraints

We now deal with problems where the state variables must satisfy constraints of the form described at point (b) of Sect. 2.3.1. Such constraints are referred to as *punctual* and *isolated* because they concern the values taken on by the state variables at *a single time instant* inside the control interval $[t_0, t_f]$. Their formal statement is

$$w^{(i)}(x(t_i), t_i) = 0, \quad t_0 < t_1 < \cdots < t_i < \cdots < t_s < t_f$$

where the time instants t_i, $i = 1, 2, \ldots s$, may be given or not and $w^{(i)}(\cdot, \cdot)$ are vectors of functions with a number of components not greater than the system order.

Because of the presence of these constraints, both $\lambda(\cdot)$ and the hamiltonian function can be discontinuous at t_i's. More precisely, if $x^o(t)$ e $u^o(t)$ are optimal and $\lambda_0^o, \lambda^o(t)$ are the corresponding solutions of the auxiliary system, we have

$$\lim_{t \to t_i^-} \lambda^o(t) = \lim_{t \to t_i^+} \lambda^o(t) + \left. \frac{\partial w^{(i)}(x, t_i)}{\partial x} \right|_{x = x^o(t_i)}' \mu^{(i)} \qquad (2.10)$$

In the above equation $\mu^{(i)}$ is a suitable vector of the same size than $w^{(i)}$, whereas $\lim_{t \to t_i^-}$ and $\lim_{t \to t_i^+}$ stand for the limits when t approaches t_i from the left or the right, respectively. Furthermore if t_i is not given, we must also have

$$\lim_{t \to t_i^-} H(x^o(t), u^o(t), t, \lambda_0^o, \lambda^o(t)) =$$

$$= \lim_{t \to t_i^+} H(x^o(t), u^o(t), t, \lambda_0^o, \lambda^o(t)) - \left. \frac{\partial w^{(i)}(x^o(t_i), t)}{\partial t} \right|_{t = t_i}' \mu^{(i)} \qquad (2.11)$$

Equations (2.10) and (2.11) should hold for $i = 1, 2, \ldots, s$.

It has to stressed that, depending on the nature of the functions $w^{(i)}(\cdot, \cdot)$ and on the t_i's being given or not, the hamiltonian function and $\lambda(\cdot)$ may actually be discontinuous or not at those times.

2.3.4 Punctual and Global Constraints

We now face the class of constraints described at point (c) in Sect. 2.3.1. First, we consider punctual and global equality constraints (point (c1)), then we discuss punctual and global inequality constraints (point (c2)). All such constraints are referred to as *punctual* and *global* because they concern the values taken on by the state and control variables at *each time instant* inside the control interval $[t_0, t_f]$. The way we handle them makes reference also to the material in Sect. 2.3.3.

Punctual and Global Equality Constraints

The formal statement of these constraints is

$$w(x(t), u(t), t) = 0, \ t_0 \leq t \leq t_f \qquad (2.12)$$

where $w(\cdot, \cdot, \cdot)$ is an r-dimensional vector of functions. This statement does not distinguish between the following two situations.

In the first one, all the components of $w(\cdot, \cdot, \cdot)$ explicitly depend on the control variable u. Then Eq. (2.12) is a tool for limiting the choice of the values that the H-minimizing control is allowed to assume and we can proceed by defining a different hamiltonian function H_e which includes, by means of a suitable r-dimensional vector μ of multipliers, the constraint (2.12). We obtain

$$H_e(x, u, t, \lambda_0, \lambda, \mu) := H(x, u, t, \lambda_0, \lambda) + \mu'w(x, u, t) \qquad (2.13)$$

and we are led to the case of simple constraints, provided that reference is made to the function $H_e(\cdot, \cdot, \cdot, \cdot, \cdot, \cdot)$.

On the contrary, in the second situation at least one of the components of $w(\cdot, \cdot, \cdot)$ does not explicitly depend on u. If the ith component is such, then it is of the form $w_i(\cdot, \cdot)$. Observe that the request $w_i(x(t), t) = 0$, $t_0 \le t \le t_f$ is equivalent to asking that all the *total* time derivatives of $w_i(\cdot, \cdot)$ are identically zero. Assume that all the state variables are influenced by the control, either directly or indirectly (i.e., through other state variables). Then, for some $q \ge 1$, the qth order total time derivative of $w_i(\cdot, \cdot)$ must be an explicit function of u because $\dot{x}(\cdot)$ is such. Thus we can set

$$\widehat{w}_i(x(t), u(t), t) := \frac{d^q w_i(x(t), t)}{dt^q}$$

If ν_i denotes the smallest value of q corresponding to which the qth order derivative explicitly depends on u, the constraint $w_i(\cdot, \cdot) = 0$ is equivalent to the $\nu_i + 1$ constraints

$$\widehat{w}_i(x(t), u(t), t) = 0, \quad t_0 \le t \le t_f$$

$$\left. \frac{d^j w_i(x(t), t)}{dt^j} \right|_{t=\tau} = 0, \quad j = 0, 1, 2, \ldots, \nu_i - 1$$

where τ is any time belonging to the interval $[t_0, t_f]$.

In so doing, we now face a problem where punctual and isolated constraints are present (see Sect. 2.3.3) and, as in the first situation, the (appropriate) hamiltonian function includes the term $\mu_i \widehat{w}_i(x, u, t)$. Notice that the fulfilment of the last ν_i equations above may imply, as it was noticed in Sect. 2.3.3, the discontinuity of the hamiltonian function and of $\lambda(\cdot)$.

Finally, observe that the choice $\tau = t_0$ ($\tau = t_f$) puts into evidence that the set S_0 (S_f) might have to be shrunk or even the infeasibility of the problem.

Punctual and Global Inequality Constraints

These constraints were introduced at point (c2) of Sect. 2.3.1 and are handled in a way which is similar to that followed in dealing with punctual and global equality constraints.

The formal statement of these constraints is

$$w(x(t), u(t), t) \le 0, \, t_0 \le t \le t_f \qquad (2.14)$$

where $w(\cdot, \cdot, \cdot)$ is an r-dimensional vector of functions. Also for these constraints we consider two different situations.

In the first one all the components of the function $w(\cdot, \cdot, \cdot)$ explicitly depend on the control variable u. Then Eq. (2.14) can properly be interpreted as a tool for limiting the choice of the values which can be given to the H-minimizing control. Thus we can proceed by defining a *different* hamiltonian function H_d which includes, by means of a suitable r-dimensional vector μ of multipliers, the constraint (2.14). We obtain

$$H_d(x; u, t, \lambda_0, \lambda, \mu) := H(x, u, t, \lambda_0, \lambda) + \mu' w(x, u, t)$$

As known, the components μ_i, $i = 1, 2, \ldots, r$, of the vector μ must satisfy the relations

$$\mu_i \geq 0, \text{ if } w_i(x, u, t) = 0, \quad \mu_i = 0, \text{ if } w_i(x, u, t) < 0$$

Thus we can tackle the constraint (2.14) by referring to the *different* hamiltonian function $H_d(\cdot, \cdot, \cdot, \cdot, \cdot, \cdot)$ rather than to customary hamiltonian function $H(\cdot, \cdot, \cdot, \cdot, \cdot)$.

The second situation is fairly more complex. We have to deal with it whenever at least one of the components of the function $w(\cdot, \cdot, \cdot)$ does not explicitly depend on u. If the ith component is such, it takes on the form $w_i(\cdot, \cdot)$. Observe that the presence of the constraint is negligible if $w_i(x^o(t), t) < 0$, that is if the state motion is *far* from the boundary corresponding to $w_i(x, t) = 0$. If, on the contrary, $w_i(x(t), t) = 0$ holds over some (time) interval, i.e., the system motion *lies* on the constraint or, in other terms, the constraint is *binding*, then all time derivatives of $w_i(\cdot, \cdot) = 0$ must be identically zero over that interval. As it was done when discussing the case of punctual and global equality constraints, we assume that all the state variables are either directly or indirectly affected by the control. Thus it must result that for some $q \geq 1$ the qth order total time derivative of $w_i(\cdot, \cdot)$ is an explicit function of u because $\dot{x}(\cdot)$ is such. Then we can set

$$\widehat{w}_i(x(t), u(t), t) := \frac{d^q w_i(x(t), t)}{dt^q}$$

Let ν_i denote the smallest value q corresponding to which the qth derivative turns out be an explicit function of u. The fact that

$$w_i(x(t), t) = 0, \ t_0 \leq t_{i,j} \leq t \leq t_{i,j+1} \leq t_f$$

for some $t_{i,j}, t_{i,j+1}$ is equivalent to the $\nu_i + 1$ constraints

$$\widehat{w}_i(x(t), u(t), t) \leq 0, \ t_{i,j} \leq t \leq t_{i,j+1}$$

$$\left. \frac{d^h w_i(x(t), t)}{dt^h} \right|_{t=t_{i,j}} = 0, \ h = 0, 1, 2, \ldots, \nu_i - 1$$

In so doing, we now face a problem where punctual and isolated constraints are present (see Sect. 2.3.3) and, as in the first situation, the (appropriate) hamiltonian function includes the term $\mu_i \widehat{w}_i(x, u, t)$. We notice that the first one of the above equations allows the constraint $w_i \leq 0$ to hold in a non-binding way after the time $t_{i,j+1}$, whereas the fulfilment of the remaining ν_i equations may imply, as it was noticed in Sect. 2.3.3, the discontinuity of the hamiltonian function and of $\lambda(\cdot)$.

2.4 Necessary Conditions: Singular Arcs

In many optimal control problems the hamiltonian function is *linear* with respect to the control variable u, i.e., it takes the form

$$H(x, u, t, \lambda_0, \lambda) = a'(x, t, \lambda_0, \lambda)u + b(x, t, \lambda_0, \lambda) \tag{2.15}$$

In such cases it may happen that, corresponding to a given control $u^*(\cdot)$, we have a constant $\lambda_0^* \geq 0$ and a pair of functions $(x^*(\cdot), \lambda^*(\cdot))$ which are solutions of Eqs. (2.6) and (2.5a), i.e., of

$$\dot{x}^*(t) = \left. \frac{\partial H(x^*(t), u^*(t), t, \lambda_0^*, \lambda)}{\partial \lambda} \right|'_{\lambda = \lambda^*(t)}$$

$$\dot{\lambda}^*(t) = -\left. \frac{\partial H(x, u^*(t), t, \lambda_0^*, \lambda^*)}{\partial x} \right|'_{x = x^*(t)}$$

and such that one or more of the components of the vector $a(\cdot, \cdot, \cdot, \cdot)$ are zero when $t \in [t_1, t_2], t_1 < t_2$. During this time interval the state trajectory is said to be a *singular arc* and the control components with index equal to that of the zero components of $a(\cdot, \cdot, \cdot, \cdot)$ are referred to as *singular components of the control*. When the whole vector $a(\cdot, \cdot, \cdot, \cdot)$ is zero, the control is said to be *singular*. Finally, a pair of functions $(x^*(\cdot), u^*(\cdot))$ corresponding to which a singular arc exists over the interval $[t_1, t_2]$ is referred to as *singular solution* over such interval.

The forthcoming theorem presents a condition that a singular solution must satisfy in order to be part of an optimal solution. There we implicitly assume that the functions $a(\cdot, \cdot, \cdot, \cdot)$ and $b(\cdot, \cdot, \cdot, \cdot)$ possess suitable differentiability properties. Furthermore we let $a^{(i)}(x(t), t, \lambda_0, \lambda(t), u(t))$, $i = 0, 1, 2, \ldots$, be the total ith time derivative of the function $a(\cdot, \cdot, \cdot, \cdot)$, after the expressions of $\dot{x}(\cdot)$ and $\dot{\lambda}(\cdot)$ have been taken into account. As an example we have

$$a^{(1)}(x(t), t, \lambda_0, \lambda(t), u(t)) := \left.\frac{\partial a(x, t, \lambda_0, \lambda)}{\partial t}\right|_{x=x(t), \lambda=\lambda(t)} +$$

$$+ \left.\frac{\partial a(x, t, \lambda_0, \lambda(t))}{\partial x}\right|_{x=x(t)} \left.\frac{\partial H(x(t), u(t), t, \lambda_0, \lambda)}{\partial \lambda}\right|'_{\lambda=\lambda(t)} +$$

$$- \left.\frac{\partial a(x(t), t, \lambda_0, \lambda)}{\partial \lambda}\right|_{\lambda=\lambda(t)} \left.\frac{\partial H(x, u(t), t, \lambda_0, \lambda(t))}{\partial x}\right|'_{x=x(t)}$$

Theorem 2.9

Let the hamiltonian function be of the form (2.15) and $u^o(\cdot)$ an optimal control which originates the state motion $x^o(\cdot)$. Furthermore let $(\lambda_0^o, \lambda^o(\cdot))$ be a solution of the auxiliary system (2.5) corresponding to the pair $(x^o(\cdot), u^o(\cdot))$. In order that the pair $(x^o(\cdot), u^o(\cdot))$ includes a singular solution over the interval $[t_1, t_2]$ it is necessary that

(i)

$$\frac{d^r a(x^o(t), t, \lambda_0^o, \lambda^o(t))}{dt^r} = 0, \ r = 0, 1, 2, \dots, \ t \in [t_1, t_2]$$

(ii)

$$\frac{\partial a^{(p)}(x^o(t), t, \lambda_0^o, \lambda^o(t), u)}{\partial u} = 0, \ p = 1, 3, 5, \dots, \ t \in [t_1, t_2]$$

(iii)

$$(-1)^q \frac{\partial a^{(2q)}(x^o(t), t, \lambda_0^o, \lambda^o(t), u)}{\partial u} \geq 0, \ q = 1, 2, 3, \dots, \ t \in [t_1, t_2]$$

Finally, if $\bar{q}$ is the smallest value of q such that condition (iii) holds with the strictly inequality sign, the check of the conditions above must be carried on up to the values $\bar{r} := 2\bar{q}, \ \bar{p} := 2\bar{q} - 1, \ \bar{q}$. ∎

2.5 The Considered Problems

As a rule we assume that the considered problems are not pathological, i.e., that the constant λ_0 is not zero, so that it can be set to one. However, for the sake of completeness, we present pathological problems at the end of Sects. 3.3 (Problem 3.15) and 3.4 (Problems 3.19 and 3.20).

Table 2.1 shows the kind of problems with simple constraints which are discussed in the forthcoming Chaps. 3–6.

We deal with time optimal problems in Chap. 7 (Problems 7.1–7.7). However, we present problems where the performance index is precisely the length of the control interval also before Chap. 7. There is no doubt that such problems can be viewed as time optimal control problems. To be specific, they are inserted in Sect. 3.2 (Problem

Table 2.1 Problems with simple constraints: the presence (absence) of x means that the statement on the top of the column holds (does not hold)

Problems	x_0 = given	x_f = given	t_0 = given	t_f = given	$m = 0$	$x_f \in \widehat{S}_f$
3.1–3.4	x	x	x	x	x	
3.5–3.9	x	x	x		x	
3.10–3.15	x		x	x	x	
3.16–3.20	x		x		x	
4.1–4.3		x	x	x	x	
4.4–4.6			x	x	x	
4.7–4.10					x	
5.1–5.4	x		x	x		
5.5–5.8	x		x			
6.1, 6.5	x		x	x	x	x
6.2, 6.4, 6.6	x		x		x	x
6.3	x		x	x		x

3.9-4), in Sect. 3.4 (Problem 3.20) and in Chap. 6 (Problems 6.2, 6.4). We have not collected them in Chap. 7 because they either include constraints which are not dealt with in the discussion of Sect. 2.2.4 or are considered more suitable to illustrate other aspects of optimal control.

Table 2.2 shows the kind of problems with complex constraints which are discussed in the forthcoming Chaps. 8–10. In this table we adopt the following abbreviations:

Table 2.2 Problems with complex constraints: the presence (absence) of x means that the constraint mentioned on the top of the column must be satisfied

Problems	V. $\int$ =	V. $\int \leq$	V.= P.I.	V.= P.G.	V.$\leq$ P.G.
8.1–8.6	x				
8.7–8.10		x			
9.1–9.8			x		
10.1–10.4				x	
10.5, 10.6					x
10.7, 10.8			x		x
10.9					x
10.10		x			x

$$\text{V.} \int = \quad Integral\ equality\ constraints$$

$$\text{V.} \int \le \quad Integral\ inequality\ constraint$$

V. = P.I. *Punctual and isolated equality constraint*

V. = P.G. *Punctual and global equality constraint*

V. ≤ P.G. *Punctual and global inequality constraint*

We take under consideration problems with singular arcs in Chap. 11 (Problems 11.1–11.4).

We end this short presentation of the material aimed at illustrating the NC in the forthcoming nine chapters by strengthening that the conditions of the Maximum Principle presented in Sect. 2.2 must always be taken into account, no matter what the considered problems are and not just in the presence of simple constraints only.

Problems 12.1–12.4 in the last chapter of the book (Chap. 12) constitute significant applications of the local sufficient conditions of optimality.

Finally, in many figures we have plotted the *state trajectories* of the controlled system. Here we recall that a state trajectory of a second-order system is a function which describes the way the second state variable (velocity for a double integrator) x_2 depends on the first state variable (position for a double integrator) x_1. Corresponding to a given control, it is well known that they can be deduced from the state motion of the system, namely from the functions $x_1(\cdot)$ and $x_2(\cdot)$.

Chapter 3
Simple Constraints: $J = \int$, $x(t_0) =$ Given

Abstract The necessary conditions of the Maximum Principle are successfully applied to the problems of this chapter. Their main features are: (1) the initial time is zero; (2) the initial state is given; (3) the performance index is of a purely integral type; (4) only simple constraints act on the system, namely we only require that (a) the final state is free or belongs to a regular variety, (b) the final time is either free or given, (c) the control variable is free or takes on values in a closed subset of the real numbers. Accordingly, we consider problems where (i) both the final time and state are given; (ii) the final time is free and the final state is given; (iii) the final time is given and the final state is not given; (iv) the final time is free and the final state is not given.

The NC of the Maximum Principle are successfully applied to the problems of this chapter. Their main features are: (1) the initial time is zero; (2) the initial state is given; (3) the performance index is of a purely integral type; (4) only simple constraints act on the system, namely we only require that (a) the final state is free or belongs to a regular variety, (b) the final time is either free or given, (c) the control variable is free or takes on values in a closed subset of the real numbers. Accordingly, we consider problems where: (i) both the final time and state are given (Sect. 3.1); (ii) the final time is free and the final state is given (Sect. 3.2); (iii) the final time is given and the final state is not given (Sect. 3.3); (iv) the final time is free and the final state is not given (Sect. 3.4).

From a technical point of view we stress that the material presented in Sect. 2.2.1 is expedient in dealing with the forthcoming problems and recall that, thanks to Remark 2.1, we need checking whether the set of the admissible final states is a regular variety only when it does not shrink to a single point. Moreover, according to what has been anticipated in Chap. 1, the system to be controlled is always a double integrator, that is the system

$$\dot{x}_1(t) := x_2(t)$$
$$\dot{x}_2(t) := u(t)$$

Thus the above equations will not be explicitly recalled in the sequel.

© Springer International Publishing Switzerland 2017

A. Locatelli, *Optimal Control of a Double Integrator*, Studies in Systems,
Decision and Control 68, DOI 10.1007/978-3-319-42126-1_3

3.1 $(x(t_f), t_0, t_f) =$ **Given**

This section hosts four problems. The control variable is unconstrained (the set $\overline{U}$ is the set of real numbers) in the first two of them, whereas it is constrained to have absolute value not greater than 1 in the last two problems.

Problem 3.1
Let

$$x(0) = \begin{bmatrix} \beta \\ \gamma \end{bmatrix}, \quad x(t_f) = \begin{bmatrix} 0 \\ 0 \end{bmatrix}$$

$$J = \int_0^{t_f} l(x(t), u(t), t)\, dt, \ l(x, u, t) = \frac{u^2}{2}, \ u(t) \in R, \ \forall t$$

where the parameters β, γ and the final time t_f are given.

The hamiltonian function and the H-minimizing control are

$$H = \frac{1}{2}u^2 + \lambda_1 x_2 + \lambda_2 u, \quad u_h = -\lambda_2$$

whereas Eq. (2.5a) and its general solution over an interval beginning at time 0 are

$$\dot{\lambda}_1(t) = 0 \qquad\qquad\qquad \lambda_1(t) = \lambda_{10}$$
$$\dot{\lambda}_2(t) = -\lambda_1(t) \qquad\qquad \lambda_2(t) = \lambda_{20} - \lambda_{10}t$$

Thus

$$x_1(t) = \beta + \gamma t - \frac{1}{2}\lambda_{20}t^2 + \frac{1}{6}\lambda_{10}t^3, \qquad x_2(t) = \gamma - \lambda_{20}t + \frac{1}{2}\lambda_{10}t^2$$

By enforcing feasibility, that is $x_1\left(t_f\right) = x_2\left(t_f\right) = 0$, we obtain λ_{10} and λ_{20} as functions of β, γ, t_f.

Table 3.1 reports the values given to these three parameters and the corresponding state trajectories are plotted in Figs. 3.1 and 3.2. Corresponding to the same initial state, Fig. 3.1 puts into evidence that the absolute value of the velocity (the second state variable) increases as the final time t_f decreases. This is particularly apparent as far as the maximum absolute value is concerned. In a similar way, when the final time is the same, Fig. 3.2 shows that $x_2(\cdot)$ is strongly dependent on the initial state. ∎

Table 3.1 Problem 3.1

	β	γ	t_f
Tr.1	1	1	1
Tr.2	1	1	1/2
Tr.3	1	1	5
Tr.4	$1/\sqrt{2}$	0	1
Tr.5	0	$1/\sqrt{2}$	1

The values given to the parameters β, γ, t_f

Fig. 3.1 Problem 3.1. State trajectories corresponding to the choices of the parameters shown in Table 3.1. Tr.1 (*solid line*), Tr.2 (*dashed line*), Tr.3 (*dash-single-dotted line*)

Fig. 3.2 Problem 3.1. State trajectories corresponding to the choices of the parameters shown in Table 3.1. Tr.1 (*solid line*), Tr.4 (*dashed line*), Tr.5 (*dotted line*)

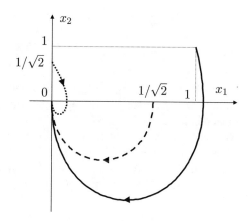

Problem 3.2

Let

$$x(0) = \begin{bmatrix} 1 \\ 0 \end{bmatrix}, \quad x(t_f) = \begin{bmatrix} 2 \\ 0 \end{bmatrix}, \quad t_f = 1$$

$$J = \int_0^{t_f} l(x(t), u(t), t)\, dt, \; l(x, u, t) = \frac{u^2}{2(1 + \beta t)} + x_1, \; u(t) \in R, \; \forall t$$

where $\beta \geq 0$ is a given integer parameter. The hamiltonian function and the H-minimizing control are

Table 3.2 Problem 3.2

β	λ_{10}	λ_{20}	J
0	−11.50	−5.92	7.50
5	−3.57	−2.44	3.51
30	−0.47	−0.55	1.93

The values of the performance index and of the two parameters which characterize the solution

$$H = \frac{u^2}{2(1 + \beta t)} + x_1 + \lambda_1 x_2 + \lambda_2 u, \quad u_h = -(1 + \beta t)\lambda_2$$

so that Eq. (2.5a) and its general solution over an interval beginning at time 0 are

$$\dot{\lambda}_1(t) = -1 \qquad\qquad\qquad \lambda_1(t) = \lambda_{10} - t$$

$$\dot{\lambda}_2(t) = -\lambda_1(t) \qquad\qquad \lambda_2(t) = \lambda_{20} - \lambda_{10}t + \frac{t^2}{2}$$

It follows

$$x_1(t) = 1 - \frac{\lambda_{20}}{2}t^2 + \frac{\lambda_{10} - \beta\lambda_{20}}{6}t^3 + \frac{2\beta\lambda_{10} - 1}{24}t^4 - \frac{\beta}{40}t^5$$

$$x_2(t) = -\lambda_{20}t + \frac{\lambda_{10} - \beta\lambda_{20}}{2}t^2 + \frac{2\beta\lambda_{10} - 1}{6}t^3 - \frac{\beta}{8}t^4$$

By enforcing feasibility, i.e., $x_1(t_f) = 2$ and $x_2(t_f) = 0$, we find λ_{10} and λ_{20}. We choose three different β's and report the relevant values of λ_{10}, λ_{20} and the performance index in Table 3.2. When $\beta = 0$ or $\beta = 30$ we obtain the state trajectories plotted in Fig. 3.3. In accordance to the nature of the performance index, when β increases the control variable is expected to take on greater values while approaching the final time: this fact clearly results by looking again at Fig. 3.3. Indeed the absolute value of the derivative (dx_2/dx_1) of the trajectory close to the final state is

Fig. 3.3 Problem 3.2. State trajectories when $\beta = 0$ (*solid line*) and $\beta = 30$ (*dashed line*)

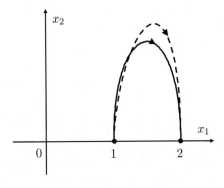

greater when $\beta = 30$. This entails that the absolute value of the control variable is greater, too.   ∎

Problem 3.3
Let

$$x(0) = \begin{bmatrix} 0 \\ 0 \end{bmatrix}, \quad x(t_f) = \begin{bmatrix} 1 \\ 0 \end{bmatrix}$$

$$J = \int_0^{t_f} l(x(t), u(t), t)\, dt, \quad l(x, u, t) = |u|, \quad u(t) \in \overline{U}, \ \forall t, \ \overline{U} = \{u | -1 \le u \le 1\}$$

where the final time t_f is given.

The hamiltonian function and the H-minimizing control are

$$H = |u| + \lambda_1 x_2 + \lambda_2 u, \quad u_h = \begin{cases} 1, & \lambda_2 < -1 \\ 0, & |\lambda_2| < 1 \\ -1, & \lambda_2 > 1 \end{cases}$$

Thus Eq. (2.5a) and its general solution over an interval beginning at time 0 are

$$\dot{\lambda}_1(t) = 0 \qquad\qquad \lambda_1(t) = \lambda_{10}$$
$$\dot{\lambda}_2(t) = -\lambda_1(t) \qquad\qquad \lambda_2(t) = \lambda_{20} - \lambda_{10}t$$

The function $\lambda_2(\cdot)$ is linear with respect to t so that the control $u(\cdot)$ can switch at most twice. However, as a consequence of the shapes of the trajectories corresponding to the allowed values of the control $(1, 0, -1)$ we notice that no feasible control exists when the number of switchings is less than two (see also Fig. 3.4). Thus let us denote the two switching times with τ_1 and τ_2. It is apparent that $u(0) = 1$, because $x_1(t_f) > x_1(0)$ and $x_2(0) = 0$, so that

Fig. 3.4 Problem 3.3. State trajectories corresponding to $u(\cdot) = 1$ (*dash-double-dotted line*), $u(\cdot) = 0$ (*solid line*), $u(\cdot) = -1$ (*dash-single-dotted line*)

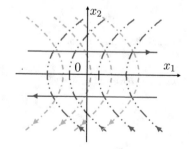

$$
\begin{array}{lll}
0 \le t < \tau_1 & \tau_1 < t < \tau_2 & \tau_2 < t \le t_f \\
u(t) = 1 & u(t) = 0 & u(t) = -1 \\
0 \le t \le \tau_1 & \tau_1 \le t \le \tau_2 & \tau_2 \le t \le t_f \\
x_1(t) = \dfrac{1}{2}t^2 & x_1(t) = \tau_1 t - \dfrac{1}{2}\tau_1^2 & x_1(t) = -\dfrac{1}{2}(\tau_1^2 + \tau_2^2) + \\
 & & \quad +(\tau_1 + \tau_2)t - \dfrac{1}{2}t^2 \\
x_2(t) = t & x_2(t) = \tau_1 & x_2(t) = \tau_1 - (t - \tau_2)
\end{array}
$$

By enforcing control feasibility $(x_1\left(t_f\right) = 1, x_2\left(t_f\right) = 0)$ and recalling that $0 \le \tau_1 \le \tau_2 \le t_f$, we conclude that the NC can be verified only if $t_f \ge 2$ and

$$
\tau_1 = \frac{t_f - \sqrt{t_f^2 - 4}}{2}, \quad \tau_2 = \frac{t_f + \sqrt{t_f^2 - 4}}{2}
$$

The lower bound for the final time can easily be determined by resorting to the material of Chap. 7. In fact $t_f = 2$ is the minimum time required to drive the system from $x(0)$ to $x(t_f)$ if the control is constrained as in the present problem. The state trajectories are plotted in Fig. 3.5 when $t_f = 2$, $t_f = 3$, $t_f = 6$. According to an obvious expectation, we notice that the maximum required absolute value of the velocity (second state variable) decreases as the length of the control interval increases.   ∎

Problem 3.4
This problem is similar to Problem 3.3. In fact we modify only the performance index which is now

$$
J = \int_0^{t_f} l(x(t), u(t), t)\, dt, \quad l(x, u, t) = t x_2
$$

The final time t_f is given and, as already noticed for Problem 3.3, $t_f \ge 2$.

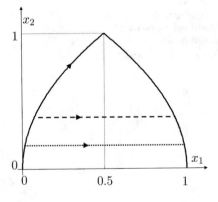

Fig. 3.5 Problem 3.3. State trajectories when $t_f = 2$ (*solid line*), $t_f = 3$ (*dashed line*), $t_f = 6$ (*dotted line*)

The hamiltonian function and the H-minimizing control are

$$H = tx_2 + \lambda_1 x_2 + \lambda_2 u, \quad u_h = -\text{sign}(\lambda_2)$$

Thus Eq. (2.5a) and its general solution over an interval beginning at time 0 are

$$\dot{\lambda}_1(t) = 0 \qquad\qquad \lambda_1(t) = \lambda_{10}$$
$$\dot{\lambda}_2(t) = -\lambda_1(t) - t \qquad\qquad \lambda_2(t) = \lambda_{20} - \lambda_{10}t - \frac{t^2}{2}$$

The function $\lambda_2(\cdot)$ is a parabola. Thus, inside the interval $(0, t_f)$, it can either have always the same sign or commute once or twice. The first occurrence leads to a non-feasible control because the state trajectories which pass through the origin and correspond either to $u(\cdot) = 1$ or to $u(\cdot) = -1$ do not pass through the point with coordinates $x(t_f)$. Also the control consistent with the switching of $\lambda_2(\cdot)$ from a positive to a negative value is not feasible. This conclusion can be drawn by looking at the shape of the state trajectories when $u(t) = -1, 0 \le t < \tau, u(t) = 1, \tau < t \le t_f$ (see Fig. 7.1a). Thus we have to consider only the following two occurrences: $\lambda_2(\cdot)$ switches once from a negative to a positive value or switches twice from a negative to a positive value and subsequently from a positive to a negative value. The first occurrence leads to the control $u(t) = 1, 0 \le t < \tau, u(t) = -1, \tau < t \le t_f$ which is feasible only if $\tau = 1$ and $t_f = 2$. On the contrary, the second occurrence leads to $u(t) = 1, 0 \le t < \tau_1, u(t) = -1, \tau_1 < t < \tau_2, u(t) = 1, \tau_2 < t \le t_f$ which is a feasible control whenever $t_f \ge 2$. From the equations

$$0 \le t < \tau_1, \quad u(t) = 1$$
$$0 \le t \le \tau_1$$
$$x_1(t) = \frac{t^2}{2}, \quad x_2(t) = t$$

$$\tau_1 < t < \tau_2, \quad u(t) = -1$$
$$\tau_1 \le t \le \tau_2$$
$$x_1(t) = x_1(\tau_1) + x_2(\tau_1)(t - \tau_1) - \frac{(t - \tau_1)^2}{2}, \quad x_2(t) = x_2(\tau_1) - (t - \tau_1)$$

$$\tau_2 < t \le t_f, \quad u(t) = 1$$
$$\tau_2 \le t \le t_f$$
$$x_1(t) = x_1(\tau_2) + x_2(\tau_2)(t - \tau_2) + \frac{(t - \tau_2)^2}{2}, \quad x_2(t) = x_2(\tau_2) + (t - \tau_2)$$

we obtain, by enforcing feasibility $(x_1(t_f) = 1, x_2(t_f) = 0)$,

Fig. 3.6 Problem 3.4. State
trajectories when $t_f = 2$
(*solid line*), $t_f = 3$ (*dashed
line*), $t_f = 6$ (*dotted line*)

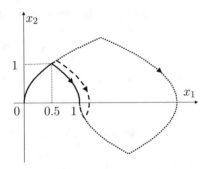

$$\tau_1 = \frac{4 + t_f^2}{4t_f}, \quad \tau_2 = \tau_1 + \frac{t_f}{2}$$

When $t_f = 2$ we note that only one switching takes place since $\tau_1 = 1$, $\tau_2 = t_f = 2$.
These are the values which have been previously computed when we have assumed
a single switching. The state trajectories are plotted in Fig. 3.6 when the final time
takes on the same values considered in Problem 3.3. When $t_f \neq 2$, it is interesting
to note that the change of the performance index substantially modifies the shape of
the trajectories.   ∎

3.2 $(x(t_f), \ t_0) =$ **Given, $t_f =$ Free**

Here we discuss five problems. The control variable is unconstrained in the first three
of them, whereas its absolute value must be not greater than 1 in the forth one. The
performance index of the second problem (which is a generalization of the preceding
one) is defined by a function explicitly dependent on time, whereas it depends on
time and state in the third one. Finally, the last problem takes into consideration four
rather similar scenarios which differ because of the form of the performance index
and the nature of the control variable constraints.

Problem 3.5
Let

$$x(0) = \begin{bmatrix} \alpha \\ \beta \end{bmatrix}, \quad x(t_f) = \begin{bmatrix} 0 \\ 0 \end{bmatrix}$$

$$J = \int_0^{t_f} l(x(t), u(t), t)\, dt, \quad l(x, u, t) = \gamma + \frac{u^2}{2}, \quad u(t) \in R, \ \forall t$$

where the parameters $\alpha, \beta, \gamma > 0$ are given and the final time t_f is free.

The hamiltonian function and the H-minimizing control are

$$H = \gamma + \frac{u^2}{2} + \lambda_1 x_2 + \lambda_2 u, \quad u_h = -\lambda_2$$

Thus Eq. (2.5a) and its general solution over an interval beginning at time 0 are

$$\dot{\lambda}_1(t) = 0 \qquad\qquad \lambda_1(t) = \lambda_{10}$$
$$\dot{\lambda}_2(t) = -\lambda_1(t) \qquad\qquad \lambda_2(t) = \lambda_{20} - \lambda_{10}t$$

The transversality condition at the final time

$$H(t_f) = \gamma + \frac{u^2(t_f)}{2} + \lambda_1(t_f)x_2(t_f) + \lambda_2(t_f)u(t_f) = \gamma + \frac{-\lambda_2^2(t_f)}{2} = 0$$

together with control feasibility $(x_1(t_f) = x_2(t_f) = 0)$ lead to the equations

$$x_1\left(t_f\right) = \alpha + \beta t_f - \frac{\lambda_{20}}{2}t_f^2 + \frac{\lambda_{10}}{6}t_f^3 = 0$$
$$x_2\left(t_f\right) = \beta - \lambda_{20}t_f + \frac{\lambda_{10}}{2}t_f^2 = 0$$
$$H\left(t_f\right) = \gamma - \frac{\left(\lambda_{20} - \lambda_{10}t_f\right)^2}{2} = 0$$

where we have exploited the expressions of $\lambda_1(\cdot)$, $\lambda_2(\cdot)$ and $u_h(\cdot)$. Corresponding to the initial state $\alpha = \beta = 1$ and some values of the parameter γ, we obtain the results collected in Table 3.3.

The relevant state motions and control functions are plotted in Fig. 3.7. According to an obvious expectation, we note that the length of the control interval decreases and the required control effort increases as γ becomes larger. ■

Table 3.3 Problem 3.5

γ	λ_{10}	λ_{20}	t_f
0.5	0.70	1.55	3.65
1	1.22	2.11	2.88
5	4.32	4.32	1.73

The parameters λ_{10}, λ_{20}, t_f corresponding to some values of γ when $\alpha = \beta = 1$

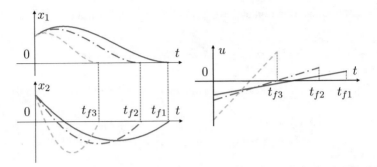

Fig. 3.7 Problem 3.5. State motions $x_1(\cdot)$, $x_2(\cdot)$ and control $u(\cdot)$ when $\alpha = \beta = 1$. They correspond to $\gamma = 0.5$ (*solid line*), $\gamma = 1$ (*dash-single-dotted line*), $\gamma = 5$ (*dashed line*)

Problem 3.6
Let

$$x(0) = \begin{bmatrix} \alpha \\ \beta \end{bmatrix}, \quad x(t_f) = \begin{bmatrix} 0 \\ 0 \end{bmatrix}$$

$$J = \int_0^{t_f} l(x(t), u(t), t) \, dt, \quad l(x, u, t) = \gamma t^\delta + \frac{u^2}{2}, \quad u(t) \in R, \ \forall t$$

where the parameters α, β, $\gamma > 0$ and $\delta \geq 0$ are given, whereas the final time t_f is free.

We proceed in the same way as we did when dealing with Problem 3.5. Therefore we first consider the hamiltonian function and the H-minimizing control which are

$$H = \gamma t^\delta + \frac{u^2}{2} + \lambda_1 x_2 + \lambda_2 u, \quad u_h = -\lambda_2$$

Thus Eq. (2.5a) and its general solution over an interval beginning at time 0 are

$$\begin{aligned} \dot{\lambda}_1(t) &= 0 & \lambda_1(t) &= \lambda_{10} \\ \dot{\lambda}_2(t) &= -\lambda_1(t) & \lambda_2(t) &= \lambda_{20} - \lambda_{10}t \end{aligned}$$

We require feasibility ($x_1(t_f) = x_2(t_f) = 0$) and enforce the transversality condition at the final time. Note that such a condition must necessarily be written for $t = t_f$ because the problem is not time-invariant. We obtain the equations

$$x_1(t_f) = \alpha + \beta t_f - \frac{\lambda_{20}}{2} t_f^2 + \frac{\lambda_{10}}{6} t_f^3 = 0$$

$$x_2(t_f) = \beta - \lambda_{20} t_f + \frac{\lambda_{10}}{2} t_f^2 = 0$$

$$H(t_f) = \gamma t_f^\delta - \frac{(\lambda_{20} - \lambda_{10} t_f)^2}{2} = 0$$

Table 3.4 Problem 3.6

	$\delta = 0$	$\delta = 2$	$\delta = 5$
$\gamma = 0.5$	3.65	2.18	1.64
$\gamma = 1.0$	2.88	1.91	1.51
$\gamma = 5.0$	1.73	1.41	1.25

The values of t_f as a function of γ and δ when $\alpha = \beta = 1$

where we have exploited the expressions of $\lambda_1(\cdot)$, $\lambda_2(\cdot)$ and $u_h(\cdot)$ and of the constraint on $x(t_f)$. From these equations we compute the three unknowns λ_{10}, λ_{20} and t_f. As we did for the preceding problem, we choose the initial state $\alpha = \beta = 1$. The value of the final time is reported in Table 3.4 corresponding to various pairs of the parameters γ and δ. Note that we have selected γ in the same way as in Problem 3.5. Obviously, we obtain the same results when $\delta = 0$. According to the role of the two parameters γ and δ, the final time decreases when either γ or δ increase. ∎

Problem 3.7

Let

$$x(0) = \begin{bmatrix} 1 \\ 0 \end{bmatrix}, \quad x(t_f) = \begin{bmatrix} 2 \\ 0 \end{bmatrix}$$

$$J = \int_0^{t_f} l(x(t), u(t), t)\, dt, \quad l(x, u, t) = t^\gamma x_1 + \frac{u^2}{2}, \quad u(t) \in R, \ \forall t$$

The final time t_f is free, whereas the integer parameter $\gamma \geq 0$ is given.

The hamiltonian function and the H-minimizing control are

$$H = t^\gamma x_1 + \frac{u^2}{2} + \lambda_1 x_2 + \lambda_2 u, \quad u_h = -\lambda_2$$

Thus Eq. (2.5a) and its general solution over an interval beginning at time 0 are

$$\dot{\lambda}_1(t) = -t^\gamma \qquad\qquad \lambda_1(t) = \lambda_{10} - \frac{t^{\gamma+1}}{\gamma + 1}$$

$$\dot{\lambda}_2(t) = -\lambda_1(t) \qquad\qquad \lambda_2(t) = \lambda_{20} - \lambda_{10}t + \frac{t^{\gamma+2}}{(\gamma + 1)(\gamma + 2)}$$

The state motion can easily be obtained, namely

$$x_1(t) = 1 - \frac{\lambda_{20}}{2}t^2 + \frac{\lambda_{10}}{6}t^3 - \frac{t^{\gamma+4}}{(\gamma + 1)(\gamma + 2)(\gamma + 3)(\gamma + 4)}$$

$$x_2(t) = -\lambda_{20}t + \frac{\lambda_{10}}{2}t^2 - \frac{t^{\gamma+3}}{(\gamma + 1)(\gamma + 2)(\gamma + 3)}$$

Table 3.5 Problem 3.7

γ	λ_{10}	λ_{20}	t_f
0	−6.00	−3.74	1.25
2	−8.70	−4.85	1.10
50	−11.72	−5.91	1.01

The values of the parameters which specify the solution

The transversality condition at the final time, written at this time,

$$H(t_f) = t_f^\gamma x_1(t_f) + \frac{u^2(t_f)}{2} + \lambda_1(t_f)x_2(t_f) + \lambda_2(t_f)u(t_f) =$$

$$= 2t_f^\gamma - \frac{\lambda_2^2(t_f)}{2} = 2t_f^\gamma - \frac{1}{2}\left(\lambda_{20} - \lambda_{10}t_f + \frac{t_f^{\gamma+2}}{(\gamma+1)(\gamma+2)}\right)^2 = 0$$

together with the equations which express the satisfaction of the constraint on the final state, constitute a set of three equations for the three unknown parameters λ_{10}, λ_{20} and t_f. Their values are reported in Table 3.5 as functions of γ. The corresponding state trajectories are plotted in Fig. 3.8.

Notice that the present problem with $\gamma = 0$ and Problem 3.2 with $\beta = 0$ only differ because in the latter the final time is given and equal to 1. Therefore it should be expected that the performance index (J_a) of Problem 3.2 is greater than the one (J_b) of the problem under consideration: in fact we have $J_a = 9.50$ and $J_b = 7.38$. ∎

Problem 3.8
As in Problem 3.3, let

$$x(0) = \begin{bmatrix} 0 \\ 0 \end{bmatrix}, \quad x(t_f) = \begin{bmatrix} 1 \\ 0 \end{bmatrix}, \quad u(t) \in \overline{U}, \ \forall t, \ \overline{U} = \{u| -1 \le u \le 1\}$$

Fig. 3.8 Problem 3.7. State trajectories when $\gamma = 0$ (*solid line*), $\gamma = 2$ (*dashed line*), $\gamma = 50$ (*dash-single-dotted line*)

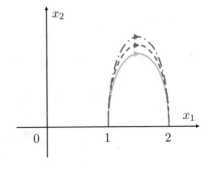

whereas the performance index is

$$J = \int_0^{t_f} l(x(t), u(t), t)\, dt, \quad l(x, u, t) = 1 + |u|$$

where the final time t_f is free.

The hamiltonian function and the H-minimizing control are

$$H = 1 + |u| + \lambda_1 x_2 + \lambda_2 u, \quad u_h = \begin{cases} 1, & \lambda_2 < -1 \\ 0, & |\lambda_2| < 1 \\ -1, & \lambda_2 > 1 \end{cases}$$

Thus Eq. (2.5a) and its general solution over an interval beginning at time 0 are

$$\dot\lambda_1(t) = 0 \qquad\qquad \lambda_1(t) = \lambda_{10}$$
$$\dot\lambda_2(t) = -\lambda_1(t) \qquad\qquad \lambda_2(t) = \lambda_{20} - \lambda_{10} t$$

Considerations similar to those reported in Problem 3.3 lead us to conclude that $u(\cdot)$ can switch at most twice and the number of switchings times is precisely two. They occur at

$$\tau_1 := \frac{t_f - \sqrt{t_f^2 - 4}}{2}, \quad \tau_2 := \frac{t_f + \sqrt{t_f^2 - 4}}{2}$$

and it must be

$$\lambda_2(\tau_1) = \lambda_{20} - \lambda_{10}\tau_1 = -1, \quad \lambda_2(\tau_2) = \lambda_{20} - \lambda_{10}\tau_2 = 1$$

because a switching occurs when λ_2 is equal to -1 or 1. It is still obvious that $u(0) = 1$. Thus the transversality condition at the final time (written for $t = 0$ because the problem is time-invariant) is

$$H(0) = 1 + |u(0)| + \lambda_1(0)x_2(0) + \lambda_2(0)u(0) = 2 + \lambda_{20} = 0$$

and supplies $\lambda_{20} = -2$. The equations for the switching times τ_1 and τ_2 together with those which specify the value of $\lambda_2(\cdot)$ at those times, completely characterize the solution. We have

$$\lambda_{10} = -\sqrt{3}, \quad \tau_1 = \frac{1}{\sqrt{3}}, \quad \tau_2 = \frac{3}{\sqrt{3}}, \quad t_f = \frac{4}{\sqrt{3}}$$

■

Problem 3.9

We consider four fairly similar problems. Their performance indexes are different and the absolute value of the control variable is bounded in all of them, but in the third one where $u(t)$ can assume any value. The initial as well as the final states are identical and given by

$$x(0) = \begin{bmatrix} 10 \\ 0 \end{bmatrix}, \quad x(t_f) = \begin{bmatrix} 0 \\ 0 \end{bmatrix}$$

The final time t_f is free in all cases. The performance indexes

$$J = \int_0^{t_f} l(x(t), u(t), t)\, dt$$

and the constraints on the control variable are defined as follows.

Problem 3.9-1

$$l(x, u, t) = 1 + \frac{u^2}{2}, \quad u(t) \in \overline{U}, \ \forall t, \ \overline{U} = \{u| -1 \le u \le 1\}$$

Problem 3.9-2

$$l(x, u, t) = 1 + |u|, \quad u(t) \in \overline{U}, \ \forall t, \ \overline{U} = \{u| -1 \le u \le 1\}$$

Problem 3.9-3

$$l(x, u, t) = 1 + \frac{u^2}{2}, \quad u(t) \in R, \ \forall t$$

Problem 3.9-4

$$l(x, u, t) = 1, \quad u(t) \in \overline{U}, \ \forall t, \ \overline{U} = \{u| -1 \le u \le 1\}$$

Obviously, the four hamiltonian functions relevant to these problems are different, whereas Eq. (2.5a) and its general solution over an interval beginning at time 0 do not change, so that, in all cases,

$$\dot{\lambda}_1(t) = 0 \qquad\qquad\qquad \lambda_1(t) = \lambda_{10}$$
$$\dot{\lambda}_2(t) = -\lambda_1(t) \qquad\qquad \lambda_1(t) = \lambda_{20} - \lambda_{10}t$$

Thus the function $\lambda_2(\cdot)$ is anyway linear with respect to t.

We now discuss the various problems.

Problem 3.9-1 $l(x, u, t) = 1 + \dfrac{u^2}{2}, \quad u(t) \in \overline{U}, \ \forall t, \ \overline{U} = \{u| -1 \le u \le 1\}$

The hamiltonian function and the H-minimizing control are

$$H = 1 + \frac{u^2}{2} + \lambda_1 x_2 + \lambda_2 u, \qquad u_h = \begin{cases} 1, & \lambda_2 \le 1 \\ -\lambda_2, & |\lambda_2| \le 1 \\ -1, & \lambda_2 \ge 1 \end{cases}$$

It is apparent that $u(0) < 0$, so that the transversality condition at the final time (written for $t = 0$ because the problem is time-invariant)

$$H(0) = 1 + \frac{u^2(0)}{2} + \lambda_2(0)u(0) = 0 \Rightarrow \lambda_2(0)u(0) = \lambda_{20}u(0) < -1 \Rightarrow \begin{cases} u(0) = -1 \\ \lambda_{20} = \dfrac{3}{2} \end{cases}$$

It is also apparent that $u(t_f) > 0$, so that, considering again the transversality condition at the final time (written for such a time), we have

$$H(t_f) = 1 + \frac{u^2(t_f)}{2} + \lambda_2(t_f)u(t_f) = 0 \Rightarrow \lambda_2(t_f)u(t_f) < -1 \Rightarrow \begin{cases} u(t_f) = 1 \\ \lambda_2(t_f) = -\dfrac{3}{2} \end{cases}$$

We conclude that the control has the form

$$u(t) = \begin{cases} -1, & 0 \le t \le \tau_1 \\ -\lambda_2(t), & \tau_1 \le t \le \tau_2 \\ 1, & \tau_2 \le t \le t_f \end{cases}$$

where the times τ_1 and τ_2 with $\tau_1 < \tau_2 < t_f$ are free. Now we can find $x_1(\cdot)$ and $x_2(\cdot)$ which depend on the four parameters $\lambda_{10}, \tau_1, \tau_2, t_f$. By enforcing feasibility $(x(t_f) = 0)$ and the satisfaction of the constraints on $\lambda_2(\cdot)$ which result from the presence of the switchings $(\lambda_2(\tau_1) = 1, \lambda_2(\tau_2) = -1)$, we obtain the values for such parameters. They are collected in Table 3.6.

Problem 3.9-2 $l(x, u, t) = 1 + |u|, \quad u(t) \in \overline{U}, \; \forall t, \; \overline{U} = \{u | -1 \le u \le 1\}$

The hamiltonian function and the H-minimizing control are

$$H = 1 + |u| + \lambda_1 x_2 + \lambda_2 u, \qquad u_h = \begin{cases} -1, & \lambda_2 > 1 \\ 0, & |\lambda_2| < 1 \\ 1, & \lambda_2 < -1 \end{cases}$$

Table 3.6 Problem 3.9

Problem	τ_1	τ_2	t_f	λ_{10}	J_u
3.9-1	6.85	1.14	5.71	0.44	4.57
3.9-2	8.49	1.41	7.07	0.35	2.83
3.9-3	–	–	6.51	0.43	4.61
3.9-4	3.16	–	6.32	0.32	6.32

The values of the parameters which specify the solution

Likewise in Problem 3.9-1, it is apparent that $u(0) < 0$ and $u(t_f) > 0$. Thus the transversality condition at the final time first written for $t = 0$ (the problem is time-invariant) and then for $t = t_f$ implies again

$$H(0) = 1 + |u(0)| + \lambda_{20}u(0) = 0 \Rightarrow \lambda_{20}u(0) < -1 \Rightarrow \begin{cases} u(0) = -1 \\ \lambda_{20} = \dfrac{3}{2} \end{cases}$$

and

$$H(t_f) = 1 + |u(t_f)| + \lambda_2(t_f)u(t_f) = 0 \Rightarrow \lambda_2(t_f)u(t_f) < -1 \Rightarrow \begin{cases} u(t_f) = 1 \\ \lambda_2(t_f) = -\dfrac{3}{2} \end{cases}$$

We conclude that the control $u(\cdot)$ is specified by

$$u(t) = \begin{cases} -1, & 0 \leq t < \tau_1 \\ 0, & \tau_1 < t < \tau_2 \\ 1, & \tau_2 < t \leq t_f \end{cases}$$

where the times τ_1 and τ_2, $\tau_1 < \tau_2 < t_f$ are not given. Now we can find $x_1(\cdot)$ and $x_2(\cdot)$ which depend on the four parameters $\lambda_{10}, \tau_1, \tau_2, t_f$. By enforcing feasibility $(x(t_f) = 0)$ and the satisfaction of the constraints on $\lambda_2(\cdot)$ which result from the presence of the switchings $(\lambda_2(\tau_1) = 1, \lambda_2(\tau_2) = -1)$, we obtain the values of such parameters. They are collected in Table 3.6.

Problem 3.9-3 $l(x, u, t) = 1 + \dfrac{u^2}{2}$, $u(t) \in R$, $\forall t$

The hamiltonian function and the H-minimizing control are

$$H = 1 + \frac{u^2}{2} + \lambda_1 x_2 + \lambda_2 u, \quad u_h = -\lambda_2$$

Likewise in Problems 3.9-1 and 3.9-2, it is apparent that $u(0) < 0$. Therefore the transversality condition at the final time, written for $t = 0$ (the problem is time-invariant), is

$$H(0) = 1 + \frac{u^2(0)}{2} + \lambda_1(0)x_2(0) + \lambda_2(0)u(0) = 1 - \frac{\lambda_{20}^2}{2} = 0$$

and supplies $\lambda_{20} = \sqrt{2}$. We can conclude that $u(\cdot)$ is specified by

$$u(t) = -\lambda_2(t) = \sqrt{2} - \lambda_{10}t, \quad 0 \leq t \leq t_f$$

Now we can find $x_1(\cdot)$ and $x_2(\cdot)$ which depend on the two parameters λ_{10}, t_f. By enforcing feasibility $(x(t_f) = 0)$, we obtain the values of these parameters. They are collected in Table 3.6.

Problem 3.9-4 $l(x, u, t) = 1$, $u(t) \in \overline{U}$, $\forall t$, $\overline{U} = \{u| -1 \le u \le 1\}$

The hamiltonian function and the H-minimizing control are

$$H = 1 + \lambda_1 x_2 + \lambda_2 u, \quad u_h = \begin{cases} -1, & \lambda_2 > 0 \\ 1, & \lambda_2 < 0 \end{cases}$$

Likewise in Problem 3.9-1, it is apparent that $u(0) < 0$. Therefore the transversality condition at the final time, written for $t = 0$ (the problem is time-invariant), is

$$H(0) = 1 + \lambda_1(0)x_2(0) + \lambda_2(0)u(0) = 1 + \lambda_{20}u(0) = 0$$

and supplies $\lambda_{20} = 1$, so that $u(0) = -1$. We can conclude that $u(\cdot)$ is defined either by

$$u(t) = -1, \ 0 \le t \le t_f$$

or by

$$u(t) = -1, \ 0 \le t < \tau_1; \quad u(t) = 1, \ \tau_1 < t \le t_f$$

where the time τ_1, $\tau_1 < t_f$ is not given. It is straightforward to check that the first alternative is not feasible. On the contrary, the second alternative allows us to find $x_1(\cdot)$ and $x_2(\cdot)$ which depend on the three parameters λ_{10}, τ_1, t_f. By enforcing feasibility $(x(t_f) = 0)$ and the satisfaction of the constraint on $\lambda_2(\cdot)$ which results from the presence of the switchings $(\lambda_2(\tau_1) = 0)$, we obtain the values of these parameters. They are collected in Table 3.6.

We observe that it is legitimate to interpret the performance indexes and constraints on $u(\cdot)$ of the four problems above as tools aimed at limiting the control effort without excessively increasing the length of the control interval. Different compromises are thus achieved and we can easily recognize their characteristics. In particular, we note that the maximum increment of the final time t_f, approximately 24%, takes place when passing from Problems 3.9-1 and 3.9-2. As far as the control effort is concerned, a meaningful comparison can be performed by looking at Fig. 3.9 where the resulting functions $u(\cdot)$ are plotted. Even more significant is considering the (reasonable) measure of the control effort supplied by

$$J_u := \int_0^{t_f} |u(t)| \, dt$$

Its values are shown in the last column of Table 3.6. Observe that the maximum increment of J_u, approximately 166%, takes place when passing from Problems 3.9-1–3.9-4. According to the importance assigned to the two parameters t_f and J_u,

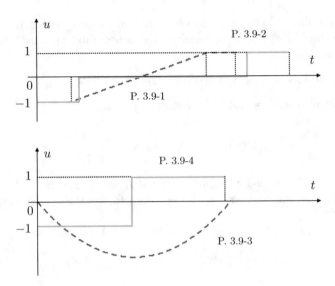

Fig. 3.9 Problem 3.9. Plots of the control variable

the choice of the most convenient problem statement clearly results by looking at Table 3.6.   ∎

3.3 $x(t_f) = $ Not Given, $(t_0, t_f) = $ Given

We present six problems. The first four of them are characterized by the following facts: (a) the control variable is unconstrained; (b) the function $l(\cdot, \cdot, \cdot)$ depends on the control only in the first and second problem, on the control and state in the third problem, on the control, state and time in the fourth problem; (c) two different sets $\overline{S}_f$ are considered in the second problem: in both cases we have more than one solution which satisfy the NC. Two scenarios where the control variable is either constrained or unconstrained are compared in the fifth problem.

Finally, the sixth problem is worth mentioning because it is pathological (the NC are satisfied only if $\lambda_0^o = 0$).

Problem 3.10
Let

$$x(0) = \begin{bmatrix} \beta \\ \gamma \end{bmatrix}, \quad x(t_f) \in \overline{S}_f = \{x \mid \alpha_f(x) = 0\}, \quad \alpha_f(x) = x_1 + x_2$$

$$J = \int_0^{t_f} l(x(t), u(t), t)\, dt, \quad l(x, u, t) = \frac{u^2}{2}, \quad u(t) \in R, \quad \forall t$$

where the parameters β, γ and the final time t_f are given.

The set $\overline{S}_f$ of the admissible final states is a regular variety because

$$\Sigma_f(x) = \frac{d\alpha_f(x)}{dx} = \begin{bmatrix} 1 & 1 \end{bmatrix}$$

so that rank $\left(\Sigma_f(x)\right) = 1, \forall x \in \overline{S}_f$.

The hamiltonian function and the H-minimizing control are

$$H = \frac{u^2}{2} + \lambda_1 x_2 + \lambda_2 u, \quad u_h = -\lambda_2$$

Thus Eq. (2.5a) and its general solution over an interval beginning at time 0 are

$$\dot{\lambda}_1(t) = 0 \qquad\qquad\qquad \lambda_1(t) = \lambda_{10}$$
$$\dot{\lambda}_2(t) = -\lambda_1(t) \qquad\qquad \lambda_2(t) = \lambda_{20} - \lambda_{10} t$$

The orthogonality condition at the final time

$$\begin{bmatrix} \lambda_1(t_f) \\ \lambda_2(t_f) \end{bmatrix} = \begin{bmatrix} \lambda_{10} \\ \lambda_{20} - \lambda_{10} t_f \end{bmatrix} = \vartheta_f \left. \frac{d\alpha_f(x)}{dx} \right|'_{x = x(t_f)} = \vartheta_f \begin{bmatrix} 1 \\ 1 \end{bmatrix}$$

supplies

$$\lambda_1\left(t_f\right) = \lambda_2\left(t_f\right) \Rightarrow \lambda_{20} = \lambda_{10}\left(1 + t_f\right)$$

Therefore

$$x_2(t) = \gamma - \lambda_{10} \left(\left(1 + t_f\right) t - \frac{t^2}{2} \right)$$

$$x_1(t) = \beta + \gamma t - \lambda_{10} \left(\left(1 + t_f\right) \frac{t^2}{2} - \frac{t^3}{6} \right)$$

If we enforce control feasibility, i.e., $x_1\left(t_f\right) + x_2\left(t_f\right) = 0$, we obtain

$$\lambda_{10} = 3 \frac{\beta + \gamma\left(1 + t_f\right)}{3 t_f + 3 t_f^2 + t_f^3}$$

Some state trajectories are plotted in Fig. 3.10 when $\beta = \gamma = 1$. They refer to three different values of t_f and show that the variations of the state variable x_2 become more and more rapid as the final time becomes smaller. Consistently, we should expect that the control variable takes on more and more large absolute values, though for shorter time intervals. Furthermore the performance index is likely to increase because it strongly penalizes $u(\cdot)$ being different from zero. All these facts are fairly obvious and are confirmed by Table 3.7 where we have reported the value

Fig. 3.10 Problem 3.10.
State trajectories when
$\beta = \gamma = 1$. They refer to
$t_f = 0.1$ (*solid line*), $t_f = 1$
(*dashed line*), $t_f = 5$
(*dash-single-dotted line*)

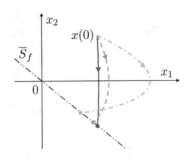

Table 3.7 Problem 3.10

t_f	J	$u(0)$	$u(t_f)$
0.1	19.98	−20.94	−19.03
1	1.93	−2.57	−1.29
5	0.34	−0.59	−0.10

The values of the performance index and of the control variable at the initial and final times when
$\beta = \gamma = 1$

J of the performance index, $u(0)$ and $u(t_f)$. The last two values are expedient in
evaluating the involvement of the control variable which is a linear function with
respect to time.   ■

Problem 3.11
Let

$$x(0) = \begin{bmatrix} 0 \\ 0 \end{bmatrix}, \quad J = \int_0^{t_f} l(x(t), u(t), t)\, dt, \quad l(x, u, t) = \frac{u^2}{2}, \quad u(t) \in R, \; \forall t, \quad t_f = 1$$

We assume that the set $\overline{S}_f$ of the admissible final states is specified by one of the
following two equations

$$\overline{S}_f = \left\{ x \,|\, \alpha_f(x) = 0 \right\}, \quad \alpha_f(x) = 5(x_1^2 - 1) - x_2$$
$$\overline{S}_f = \left\{ x \,|\, \alpha_f(x) = 0 \right\}, \quad \alpha_f(x) = x_1 - \frac{1}{3}x_2^3 - 1$$

The resulting scenarios are dealt with in Problems 3.11-1 and 3.11-2, respectively.
In both cases the set $\overline{S}_f$ is a regular variety. In fact

$$\Sigma_f(x) = \frac{d\alpha_f(x)}{dx} = \begin{bmatrix} 10x_1 & -1 \end{bmatrix}$$

so that rank $\left(\Sigma_f(x) \right) = 1$, $\forall x \in \overline{S}_f$ in the first case, whereas in the second one
we have

$$\Sigma_f(x) = \frac{d\alpha_f(x)}{dx} = \left[\, 1 \;\; -x_2^2 \,\right]$$

with rank $\left(\Sigma_f(x)\right) = 1, \forall x \in \overline{S}_f$.

Obviously, the hamiltonian function and the H-minimizing control are equal and given by

$$H = \frac{u^2}{2} + \lambda_1 x_2 + \lambda_2 u, \quad u_h = -\lambda_2$$

Thus Eq. (2.5a) and its general solution over an interval beginning at time 0 are

$$\dot{\lambda}_1(t) = 0 \qquad\qquad \lambda_1(t) = \lambda_{10}$$
$$\dot{\lambda}_2(t) = -\lambda_1(t) \qquad\qquad \lambda_2(t) = \lambda_{20} - \lambda_{10}t$$

for both scenarios.

Problem 3.11-1 $\alpha_f(x) = 5(x_1^2 - 1) - x_2$.

In view of the form of the functions $\lambda_1(\cdot)$ and $\lambda_2(\cdot)$, the orthogonality condition at the final time is

$$\begin{bmatrix} \lambda_1(t_f) \\ \lambda_2(t_f) \end{bmatrix} = \begin{bmatrix} \lambda_{10} \\ \lambda_{20} - \lambda_{10}t_f \end{bmatrix} = \vartheta_f \left.\frac{d\alpha_f(x)}{dx}\right|'_{x=x(t_f)} = \vartheta_f \begin{bmatrix} 10x_1(t_f) \\ -1 \end{bmatrix}$$

We can easily compute the state motion (see the expressions of the H-minimizing control and $\lambda_2(\cdot)$) and we obtain

$$x_1(t) = -\frac{\lambda_{20}}{2}t^2 + \frac{\lambda_{10}}{6}t^3$$
$$x_2(t) = -\lambda_{20}t + \frac{\lambda_{10}}{2}t^2$$

These results and the orthogonality condition lead to

$$\frac{\lambda_{10}}{\lambda_{20} - \lambda_{10}t_f} = -10\left(\frac{\lambda_{10}}{6}t_f^3 - \frac{\lambda_{20}}{2}t_f^2\right)$$

which, together with feasibility, namely

$$x_2(t_f) = -\lambda_{20}t_f + \frac{\lambda_{10}}{2}t_f^2 = 5\left(\left(-\frac{\lambda_{20}}{2}t_f^2 + \frac{\lambda_{10}}{6}t_f^3\right)^2 - 1\right) = 5\left(x_1^2(t_f) - 1\right)$$

constitute a system of two equations for the two unknowns λ_{10} and λ_{20}. We get the three pairs of values reported in Table 3.8. There we also give $x_1(t_f)$ and the values of the performance index. It is clear that the first solution is to be preferred.

Table 3.8 Problem 3.11-1

Solution	λ_{10}	λ_{20}	J	$x_1(t_f)$
#1	2.18	2.44	1.10	-0.85
#2	-3.98	-3.63	2.01	1.15
#3	-31.13	-10.68	52.31	0.15

The values of the parameters which characterize the three solutions

Fig. 3.11 Problem 3.11-1.
State trajectories
corresponding to the solution
#1 (*solid line*), #2 (*dashed
line*), #3 (*dash-single-dotted
line*)

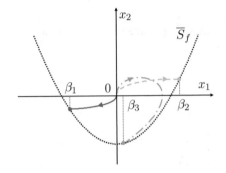

The regular variety $\overline{S}_f$ and the state trajectories corresponding to the three solutions are shown in Fig. 3.11. In this figure β_i, $i = 1, 2, 3$ denotes the final states $x_1(t_f)$ relevant to the computed solutions.

The significance of these results is made more clear if we tackle the similar problem where the final state is given and belongs to $\overline{S}_f$.

Let $\beta := x_1(t_f)$. Then we can easily find λ_{10}, λ_{20} and J as functions of β. We get

$$\lambda_{10} = 6(5\beta^2 - 2\beta - 5), \quad \lambda_{20} = 4\left(\frac{\lambda_{10}}{6} - \beta\right), \quad J = \frac{(\lambda_{10} - \lambda_{20})^3 + \lambda_{20}^3}{6\lambda_{10}}$$

The function $J(\cdot)$ is plotted in Fig. 3.12. The values of β where this function attains a relative maximum or minimum are precisely those which satisfy the NC.

Fig. 3.12 Problem 3.11-1.
The plot of the performance
index as a function of β. The
points β_i, $i = 1, 2, 3$
correspond to the values of
$x_1(t_f)$ relevant to the three
solutions which satisfy the
NC

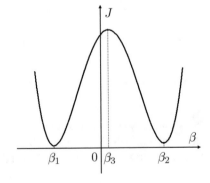

Problem 3.11-2 $\alpha_f(x) = x_1 - \dfrac{x_2^3}{3} - 1.$

In view of the expressions of $\lambda_1(\cdot)$ and $\lambda_2(\cdot)$, the orthogonality condition at the final time is

$$\begin{bmatrix} \lambda_1(t_f) \\ \lambda_2(t_f) \end{bmatrix} = \begin{bmatrix} \lambda_{10} \\ \lambda_{20} - \lambda_{10}t_f \end{bmatrix} = \vartheta_f \left.\frac{d\alpha_f(x)}{dx}\right|'_{x=x(t_f)} = \vartheta_f \begin{bmatrix} 1 \\ -x_2^2(t_f) \end{bmatrix}$$

By recalling the form of the H-minimizing control and $\lambda_2(\cdot)$, the state motion is

$$x_1(t) = -\frac{\lambda_{20}}{2}t^2 + \frac{\lambda_{10}}{6}t^3$$

$$x_2(t) = -\lambda_{20}t + \frac{\lambda_{10}}{2}t^2$$

If these results are inserted into the orthogonality condition, we get

$$\frac{\lambda_{10}}{\lambda_{20} - \lambda_{10}t_f} = -\frac{1}{\left(\dfrac{\lambda_{10}}{2}t_f^2 - \lambda_{20}\right)^2}$$

whereas, enforcing feasibility, we obtain

$$x_1(t_f) = -\frac{\lambda_{20}}{2}t_f^2 + \frac{\lambda_{10}}{6}t_f^3 = \frac{1}{3}\left(-\lambda_{20}t_f + \frac{\lambda_{10}}{2}t_f^2\right)^3 + 1 = \frac{1}{3}x_2^3(t_f) + 1$$

These two equations constitute a system for the two unknowns λ_{10} and λ_{20}. Three couples of values are found. They are reported in Table 3.9 where also $x_2(t_f)$ and the value of the performance index are shown. It is obvious that the third solution has to be preferred. The regular variety $\overline{S}_f$ and the state trajectories corresponding to the three solutions are plotted in Fig. 3.13. In this figure β_i, $i = 1, 2, 3$ denote the final states $x_2(t_f)$ relevant to the computed solutions. As for Problem 3.11-1, the significance of these results is made more clear if we tackle the similar problem where the final state is given and belongs to $\overline{S}_f$. Let $\beta := x_2(t_f)$. Then we can easily find λ_{10}, λ_{20} and J as functions of β. We get

Table 3.9 Problem 3.11-2

Solution	λ_{10}	λ_{20}	J	$x_2(t_f)$
#1	−9.19	−5.25	3.74	0.65
#2	−14.82	−6.67	9.42	−0.74
#3	−0.61	1.43	1.57	−1.76

The parameters which identify the three solutions

Fig. 3.13 Problem 3.11-2.
State trajectories
corresponding to the solution
#1 (*solid line*), #2 (*dashed
line*), #3 (*dash-single-dotted
line*)

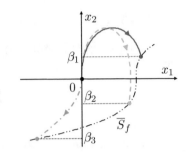

Fig. 3.14 Problem 3.11-2.
The plot of the performance
index as a function of β. The
points β_i, $i = 1, 2, 3$
correspond tos the values of
$x_2(t_f)$ relevant to the three
solutions which satisfy the
NC

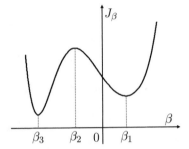

$$\lambda_{10} = 12\left(\frac{\beta}{2} - \frac{\beta^3}{3}\right), \quad \lambda_{20} = \left(\frac{\lambda_{10\beta}}{2} - \beta\right), \quad J = \frac{(\lambda_{10} - \lambda_{20})^3 + \lambda_{20}^3}{6\lambda_{10}}$$

The function $J(\cdot)$ is plotted in Fig. 3.14. The values of β where this function attains
a relative maximum or minimum are precisely those which satisfy the NC.   ■

Problem 3.12
Let

$$x(0) = \begin{bmatrix} \eta \\ \vartheta \end{bmatrix}, \quad x(t_f) = \text{free}$$

$$J = \int_0^{t_f} l(x(t), u(t), t)\, dt, \quad l(x, u, t) = \frac{u^2}{2} + x_1, \quad u(t) \in R, \ \forall t$$

where the parameters η, ϑ and the final time t_f are given.
    The hamiltonian function and the H-minimizing control are

$$H = \frac{u^2}{2} + x_1 + \lambda_1 x_2 + \lambda_2 u, \quad u_h = -\lambda_2$$

Thus Eq. (2.5a) and its general solution over an interval beginning at time 0 are

$$\dot{\lambda}_1(t) = -1 \qquad\qquad \lambda_1(t) = \lambda_{10} - t$$
$$\dot{\lambda}_2(t) = -\lambda_1(t) \qquad\qquad \lambda_2(t) = \lambda_{20} - \lambda_{10}t + \frac{1}{2}t^2$$

We have $\lambda_1\left(t_f\right) = \lambda_2\left(t_f\right) = 0$ because the final state is free, so that

$$\lambda_{10} = t_f, \quad \lambda_{20} = \frac{t_f^2}{2}, \quad u = -\frac{t_f^2}{2} + t_f t - \frac{1}{2}t^2$$

and

$$x_2(t) = \vartheta - \frac{t_f^2}{2}t + \frac{t_f}{2}t^2 - \frac{1}{6}t^3$$
$$x_1(t) = \eta + \vartheta t - \frac{t_f^2}{4}t^2 + \frac{t_f}{6}t^3 - \frac{1}{24}t^4$$

The state motions and the corresponding controls $u(\cdot)$ are plotted in Fig. 3.15 when $\eta = \vartheta = 1$. They refer to three different values of t_f. Whenever the final time is sufficiently large it is clear that it is convenient to make the first state variable $x_1(\cdot)$ to assume small values even though large control actions have to be used. This fact can easily be verified by looking at Fig. 3.15. The maximum absolute values of the control variable occur at the initial time and are equal to $0.5t_f^2$. Notice the significant differences between them. ∎

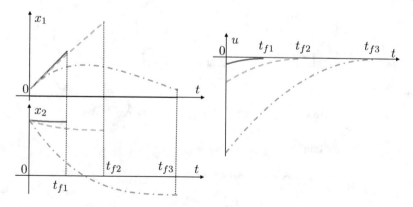

Fig. 3.15 Problem 3.12. Plots of $u(\cdot)$, $x_1(\cdot)$ and $x_2(\cdot)$ when $\eta = \vartheta = 1$. They correspond to $t_f = t_{f1} = 0.5$ (*solid line*), $t_f = t_{f2} = 1$ (*dashed line*), $t_f = t_{f3} = 2$ (*dash-single-dotted line*)

Problem 3.13

Let

$$x(0) = \begin{bmatrix} 1 \\ 0 \end{bmatrix}, \quad x(t_f) \in \overline{S}_f = \{x | \alpha_f(x) = 0\}, \quad \alpha_f(x) = x_1 - 2, \quad t_f = 1$$

$$J = \int_0^{t_f} l(x(t), u(t), t) \, dt, \quad l(x, u, t) = \frac{u^2}{2(1 + \varphi t)} + x_1, \quad u(t) \in R, \quad \forall t$$

where $\varphi \geq 0$ is a given integer parameter.

The set $\overline{S}_f$ of the admissible final states is a regular variety because

$$\Sigma_f(x) = \frac{d\alpha_f(x)}{dx} = \begin{bmatrix} 1 & 0 \end{bmatrix}$$

so that rank $(\Sigma_f(x)) = 1$, $\forall x \in \overline{S}_f$.

The hamiltonian function and the H-minimizing control are

$$H = \frac{u^2}{2(1 + \varphi t)} + x_1 + \lambda_1 x_2 + \lambda_2 u, \quad u_h = -(1 + \varphi t)\lambda_2$$

Thus Eq. (2.5a) and its general solution over an interval beginning at time 0 are

$$\dot{\lambda}_1(t) = -1 \qquad\qquad \lambda_1(t) = \lambda_{10} - t$$

$$\dot{\lambda}_2(t) = -\lambda_1(t) \qquad\qquad \lambda_2(t) = \lambda_{20} - \lambda_{10}t + \frac{t^2}{2}$$

The orthogonality condition at the final time

$$\begin{bmatrix} \lambda_1(t_f) \\ \lambda_2(t_f) \end{bmatrix} = \begin{bmatrix} \lambda_{10} - t_f \\ \lambda_{20} - \lambda_{10}t_f + \frac{t_f^2}{2} \end{bmatrix} = \vartheta_f \left. \frac{d\alpha_f(x)}{dx} \right|_{x=x(t_f)}^{'} = \vartheta_f \begin{bmatrix} 1 \\ 0 \end{bmatrix}$$

leads to

$$\lambda_{20} - \lambda_{10}t_f + \frac{t_f^2}{2} = \lambda_{20} - \lambda_{10} + \frac{1}{2} = 0 \Rightarrow \lambda_{20} = \lambda_{10} - \frac{1}{2}$$

Then the state motion is

$$x_1(t) = 1 - \frac{\lambda_{20}}{2}t^2 + \frac{\lambda_{10} - \varphi\lambda_{20}}{6}t^3 + \frac{2\varphi\lambda_{10} - 1}{24}t^4 - \frac{\varphi}{40}t^5$$

$$x_2(t) = -\lambda_{20}t + \frac{\lambda_{10} - \varphi\lambda_{20}}{2}t^2 + \frac{2\varphi\lambda_{10} - 1}{6}t^3 - \frac{\varphi}{8}t^4$$

By enforcing feasibility, that is $x_1(t_f) = 2$, and recalling how λ_{10} is tied to λ_{20}, we compute these two parameters.

Thus Eq. (2.5a) and its general solution over an interval beginning at time 0 are

$$\dot{\lambda}_1(t) = -1 \qquad\qquad \lambda_1(t) = \lambda_{10} - t$$
$$\dot{\lambda}_2(t) = -\lambda_1(t) \qquad\qquad \lambda_2(t) = \lambda_{20} - \lambda_{10}t + \frac{1}{2}t^2$$

We have $\lambda_1\left(t_f\right) = \lambda_2\left(t_f\right) = 0$ because the final state is free, so that

$$\lambda_{10} = t_f, \quad \lambda_{20} = \frac{t_f^2}{2}, \quad u = -\frac{t_f^2}{2} + t_f t - \frac{1}{2}t^2$$

and

$$x_2(t) = \vartheta - \frac{t_f^2}{2}t + \frac{t_f}{2}t^2 - \frac{1}{6}t^3$$
$$x_1(t) = \eta + \vartheta t - \frac{t_f^2}{4}t^2 + \frac{t_f}{6}t^3 - \frac{1}{24}t^4$$

The state motions and the corresponding controls $u(\cdot)$ are plotted in Fig. 3.15 when $\eta = \vartheta = 1$. They refer to three different values of t_f. Whenever the final time is sufficiently large it is clear that it is convenient to make the first state variable $x_1(\cdot)$ to assume small values even though large control actions have to be used. This fact can easily be verified by looking at Fig. 3.15. The maximum absolute values of the control variable occur at the initial time and are equal to $0.5t_f^2$. Notice the significant differences between them. ∎

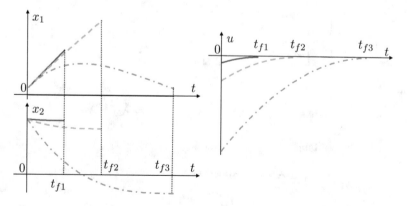

Fig. 3.15 Problem 3.12. Plots of $u(\cdot)$, $x_1(\cdot)$ and $x_2(\cdot)$ when $\eta = \vartheta = 1$. They correspond to $t_f = t_{f1} = 0.5$ (*solid line*), $t_f = t_{f2} = 1$ (*dashed line*), $t_f = t_{f3} = 2$ (*dash-single-dotted line*)

Problem 3.13

Let

$$x(0) = \begin{bmatrix} 1 \\ 0 \end{bmatrix}, \quad x(t_f) \in \overline{S}_f = \{x \mid \alpha_f(x) = 0\}, \quad \alpha_f(x) = x_1 - 2, \quad t_f = 1$$

$$J = \int_0^{t_f} l(x(t), u(t), t)\, dt, \quad l(x, u, t) = \frac{u^2}{2(1 + \varphi t)} + x_1, \quad u(t) \in R, \quad \forall t$$

where $\varphi \geq 0$ is a given integer parameter.

The set $\overline{S}_f$ of the admissible final states is a regular variety because

$$\Sigma_f(x) = \frac{d\alpha_f(x)}{dx} = \begin{bmatrix} 1 & 0 \end{bmatrix}$$

so that rank $\left(\Sigma_f(x)\right) = 1$, $\forall x \in \overline{S}_f$.

The hamiltonian function and the H-minimizing control are

$$H = \frac{u^2}{2(1 + \varphi t)} + x_1 + \lambda_1 x_2 + \lambda_2 u, \quad u_h = -(1 + \varphi t)\lambda_2$$

Thus Eq. (2.5a) and its general solution over an interval beginning at time 0 are

$$\dot{\lambda}_1(t) = -1 \qquad\qquad \lambda_1(t) = \lambda_{10} - t$$

$$\dot{\lambda}_2(t) = -\lambda_1(t) \qquad\qquad \lambda_2(t) = \lambda_{20} - \lambda_{10}t + \frac{t^2}{2}$$

The orthogonality condition at the final time

$$\begin{bmatrix} \lambda_1(t_f) \\ \lambda_2(t_f) \end{bmatrix} = \begin{bmatrix} \lambda_{10} - t_f \\ \lambda_{20} - \lambda_{10}t_f + \dfrac{t_f^2}{2} \end{bmatrix} = \vartheta_f \left.\frac{d\alpha_f(x)}{dx}\right|'_{x=x(t_f)} = \vartheta_f \begin{bmatrix} 1 \\ 0 \end{bmatrix}$$

leads to

$$\lambda_{20} - \lambda_{10}t_f + \frac{t_f^2}{2} = \lambda_{20} - \lambda_{10} + \frac{1}{2} = 0 \Rightarrow \lambda_{20} = \lambda_{10} - \frac{1}{2}$$

Then the state motion is

$$x_1(t) = 1 - \frac{\lambda_{20}}{2}t^2 + \frac{\lambda_{10} - \varphi\lambda_{20}}{6}t^3 + \frac{2\varphi\lambda_{10} - 1}{24}t^4 - \frac{\varphi}{40}t^5$$

$$x_2(t) = -\lambda_{20}t + \frac{\lambda_{10} - \varphi\lambda_{20}}{2}t^2 + \frac{2\varphi\lambda_{10} - 1}{6}t^3 - \frac{\varphi}{8}t^4$$

By enforcing feasibility, that is $x_1(t_f) = 2$, and recalling how λ_{10} is tied to λ_{20}, we compute these two parameters.

Table 3.10 Problem 3.13

φ	λ_{10}	λ_{20}	J	$x_2(t_f)$
0	−2.37	−2.87	2.87	1.52
5	−0.67	−1.17	2.00	1.85
30	0.34	−0.16	1.47	2.22

The values of the performance index and of some parameters which characterize the solution

Fig. 3.16 Problem 3.13. State trajectories corresponding to $\varphi = 0$ (*solid line*), $\varphi = 5$ (*dashed line*) and $\varphi = 30$ (*dash-single-dotted line*)

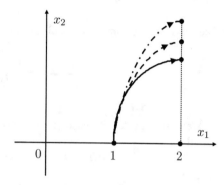

Table 3.10 reports their values together with those of the performance index and $x_2(t_f)$. All of them refer to various choices of φ.

The state trajectories are plotted in Fig. 3.16.

Note that the magnitude of x_2 increases with φ. In fact an increase of φ causes the use of control to be less penalized when approaching the final time. Consistently, it is convenient that the variable x_1 (which should be small during the transient, as expressed by the performance index) tends to $x_1(t_f) = 2$ as late as possible. But, then, we need large values of $x_2(t)$ at the end of the control interval.

It is interesting to compare the problem at hand, denoted for convenience with the letter P, with Problems 3.12 and 3.2.

First, we observe that Problem 3.2 only differs from P because the second state variable is given at the final time $(x_2(t_f) = 0)$, rather than being free. Therefore the performance index of P should not be greater than that of Problem 3.2 whenever $\beta = \varphi$. Tables 3.10 and 3.2 confirm this statement.

When $\varphi = 0$ we note, as a second point, that if we choose $\eta = 1$, $\vartheta = 0$, $t_f = 1$, then P differs from Problem 3.12 only because the state of the latter is free at the final time rather then being partially specified $(x_1(t_f) = 2)$. Thus it is not surprising that the value of the performance index substantially increases from 0.97 (Problem 3.12) to 2.87 (P). ∎

Problem 3.14

Let

$$x(0) = \begin{bmatrix} 0 \\ 0 \end{bmatrix}, \quad x(t_f) \in \overline{S}_f = \{x | \alpha_f(x) = 0\}, \quad \alpha_f(x) = x_1 - 1$$

$$J = \int_0^{t_f} l(x(t), u(t), t)\, dt, \quad l(x, u, t) = \frac{u^2}{2}, \quad t_f = 1.5$$

The set $\overline{S}_f$ of the admissible final states is a regular variety because

$$\Sigma_f(x) = \frac{d\alpha_f(x)}{dx} = \begin{bmatrix} 1 & 0 \end{bmatrix}$$

so that rank $(\Sigma_f(x)) = 1$, $\forall x \in \overline{S}_f$.

We first consider the case (Problem 3.14-1) where the control is constrained, namely

$$u(t) \in \overline{U}, \quad \forall t, \quad \overline{U} = \{u | -1 \le u \le 1\}$$

and then the case (Problem 3.14-2) where the control is unconstrained, namely

$$u(t) \in R, \quad \forall t$$

The hamiltonian function is

$$H = \frac{u^2}{2} + \lambda_1 x_2 + \lambda_2 u$$

in both cases. Thus Eq. (2.5a) and its general solution over an interval beginning at time 0 are, all the same,

$$\dot{\lambda}_1(t) = 0 \qquad\qquad \lambda_1(t) = \lambda_{10}$$
$$\dot{\lambda}_2(t) = -\lambda_1(t) \qquad\qquad \lambda_2(t) = \lambda_{20} - \lambda_{10}t$$

Similarly, the orthogonality condition at the final time

$$\begin{bmatrix} \lambda_1(t_f) \\ \lambda_2(t_f) \end{bmatrix} = \begin{bmatrix} \lambda_{10} \\ \lambda_{20} - \lambda_{10}t_f \end{bmatrix} = \vartheta_f \left. \frac{d\alpha_f(x)}{dx} \right|'_{x=x(t_f)} = \vartheta_f \begin{bmatrix} 1 \\ 0 \end{bmatrix}$$

is equal and requires $\lambda_2(t_f) = 0$, from which it follows

$$\lambda_2(t) = \lambda_{10}(t_f - t)$$

so that the function $\lambda_2(\cdot)$ is always either non-negative or non-positive.

On the contrary, the H-minimizing control is substantially different in the two considered cases.

Problem 3.14-1 $u(t) \in \overline{U}$, $\forall t$, $\overline{U} = \{u| -1 \le u \le 1\}$.

The H-minimizing control is

$$u_h = \begin{cases} 1, & \lambda_2 \le -1 \\ -\lambda_2, & |\lambda_2| \le 1 \\ -1, & \lambda_2 \ge 1 \end{cases}$$

This outcome implies that $\lambda_2(\cdot)$ can not be non-negative. In fact the control would be $u(\cdot) \le 0$ which is not feasible because $x_1(\cdot) \le 0$. Therefore we consider the second possibility ($\lambda_2(\cdot) \le 0$). It is easy to conclude that $\lambda_2(\cdot) \le -1$ must be discarded because we have already found that $\lambda_2(t_f) = 0$. Analogously, $|\lambda_2(\cdot)| \le 1$ must be discarded as we have

$$u(t) = -\lambda_2(t) = \lambda_{10}(t - t_f) \Rightarrow \begin{cases} x_2(t) = \lambda_{10}\left(\dfrac{t^2}{2} - t_f t\right) \\ x_1(t) = \lambda_{10}\left(\dfrac{t^3}{6} - t_f\dfrac{t^2}{2}\right) \end{cases}$$

and, enforcing feasibility ($x_1(t_f) = 1$),

$$\lambda_{10} = -\frac{3}{t_f^3} \Rightarrow \lambda_{20} = -\frac{3}{t_f^2} = -\frac{3}{1.5^2} < -1$$

which is a contradiction. Therefore the only left possibility is $\lambda_2(\cdot)$ less than -1 up to a time $\tau < t_f$ and included between -1 and 0 for times greater than τ. Consistently, when $0 \le t \le \tau$, the state motion is

$$x_1(t) = \frac{1}{2}t^2, \quad x_2(t) = t$$

because $u(t) = 1$, whereas, when $\tau \le t \le t_f$,

$$x_1(t) = \frac{1}{2}\tau^2 + \tau(t - \tau) + \frac{t_f}{2(t_f - \tau)}(t - \tau)^2 + \frac{\tau^2}{2(t_f - \tau)}(t - \tau) +$$
$$-\frac{1}{6(t_f - \tau)}(t^3 - \tau^3)$$

$$x_2(t) = \tau + \frac{t_f}{t_f - \tau}(t - \tau) - \frac{1}{2(t_f - \tau)}(t^2 - \tau^2)$$

because

$$u(t) = -\lambda_2(t) = \frac{t_f - t}{t_f - \tau}$$

Fig. 3.17 Problem 3.14. State motion and the function $u(\cdot)$ of Problem 3.14-1 (*solid line*) and of Problem 3.14-2 (*dashed line*)

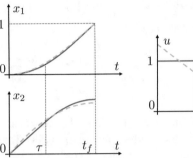

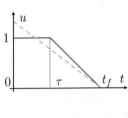

These expressions have been derived by taking into account that $\lambda_2(\tau) = -1$, so that

$$\lambda_{10} = -\frac{1}{t_f - \tau}$$

By enforcing feasibility ($x_1(t_f) = 1$), we get $\tau = 0.63$ and $\lambda_{10} = -1.15$.

Solid lines in Fig. 3.17 correspond to the state motion and the control.

It is interesting to observe that the final time can not be less than $\sqrt{2}$ which is the length of the control interval necessary to have $x_1(t_f) = 1$ when $u(\cdot) = 1$. Therefore if $t_f < \sqrt{2}$ the problem does not admit a solution. Also observe that the constraint on the control variable is not binding when $t_f \geq \sqrt{3}$. In fact the equation $x_1(t_f) = 1$ in the unknown τ gives

$$\tau = t_f \pm \sqrt{3t_f^2 - 6}$$

so that $0 \leq \tau \leq t_f$ only if $\sqrt{2} \leq t_f \leq \sqrt{3}$.

Problem 3.14-2 $u(t) \in R$, $\forall t$.

The H-minimizing control is

$$u_h = -\lambda_2$$

Thus, by recalling that $\lambda_2(t) = \lambda_{10}(t - t_f)$, the state motion is

$$x_1(t) = \frac{\lambda_{10}}{6}t^3 - \frac{\lambda_{10}t_f}{2}t^2, \quad x_2(t) = \frac{\lambda_{10}}{2}t^2 - \lambda_{10}t_f t$$

In order that the control is feasible ($x_1(t_f) = 1$) it must be

$$\lambda_{10} = -\frac{3}{t_f^3}$$

Dashed lines in Fig. 3.17 correspond to the state motion and the control. Note that the state motions are fairly similar for both problems. ∎

As mentioned at the beginning of this section, the forthcoming problem is patholog-ical.

Problem 3.15
Let

$$x(0) = \begin{bmatrix} 0 \\ 0 \end{bmatrix}, \quad x(t_f) \in \overline{S}_f = \{x \mid \alpha_f(x) = 0\}, \quad \alpha_f(x) = x_1 - 2x_2 + 2, \quad t_f = 2$$

$$J = \int_0^{t_f} l(x(t), u(t), t)\, dt, \quad l(x, u, t) = u, \quad u(t) \in \overline{U}, \; \forall t, \; \overline{U} = \{u \mid 0 \le u \le 1\}$$

The set $\overline{S}_f$ of the admissible final states is a regular variety because

$$\Sigma_f(x) = \frac{d\alpha_f(x)}{dx} = \begin{bmatrix} 1 & -2 \end{bmatrix}$$

so that rank $(\Sigma_f(x)) = 1, \forall x \in \overline{S}_f$.
It is easy to ascertain that the only feasible (and thus optimal) control is $u(t) = 1$, $0 \le t \le t_f$. This implies

$$x_2(t) = t, \quad x_1(t) = \frac{t^2}{2}$$

The corresponding state trajectory c is plotted in Fig. 3.18 together with the set $\overline{S}_f$. Different control functions within the interval $0 \le t \le t_f$ (of course complying with the constraint on $u(\cdot)$) lead to $x_2(t_f) < 2$, so that it is not possible to satisfy the request $x(t_f) \in \overline{S}_f$. In fact suppose that at time $\bar{t}$, $0 \le \bar{t} \le t_f$ the state is P, a point of the trajectory c. The derivative of the trajectory when the state is P is proportional to $u(\bar{t})$ $(dx_2/dx_1 = u/x_2)$, so that we conclude that, if $u(\bar{t}) < 1$, the derivative of the resulting trajectory c^* is smaller than that of c $(dx_2/dx_1 = 1/x_2)$. Therefore c^* diverges from c and points towards the interior of the region included between c and the x_1 axis.

Fig. 3.18 Problem 3.15. The state trajectory c corresponding to $u(\cdot) = 1$ and the set $\overline{S}_f$ of the admissible final states

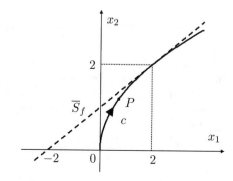

In order to prove that the problem is pathological, we first consider the hamiltonian function

$$H = \lambda_0 u + \lambda_1 x_2 + \lambda_2 u$$

and the resulting H-minimizing control

$$u_h = \begin{cases} 0, & \lambda_0 + \lambda_2 > 0 \\ 1, & \lambda_0 + \lambda_2 < 0 \end{cases}$$

The Eq. (2.5a) and its general solution over an interval beginning at time 0 are

$$\dot{\lambda}_1(t) = 0 \qquad\qquad\qquad \lambda_1(t) = \lambda_{10}$$
$$\dot{\lambda}_2(t) = -\lambda_1(t) \qquad\qquad \lambda_2(t) = \lambda_{20} - \lambda_{10}t$$

Therefore the orthogonality condition at the final time is

$$\begin{bmatrix} \lambda_1(t_f) \\ \lambda_2(t_f) \end{bmatrix} = \begin{bmatrix} \lambda_{10} \\ \lambda_{20} - \lambda_{10}t_f \end{bmatrix} = \vartheta_f \left.\frac{d\alpha_f(x)}{dx}\right|'_{x=x(t_f)} = \vartheta_f \begin{bmatrix} 1 \\ -2 \end{bmatrix}$$

and

$$\frac{\lambda_{10}}{\lambda_{20} - \lambda_{10}t_f} = \frac{\lambda_{10}}{\lambda_{20} - 2\lambda_{10}} = -\frac{1}{2}$$

This implies $\lambda_{20} = 0$.

In summary, we have $u(\cdot) = 1$ only if

$$\lambda_0 + \lambda_2(t) = \lambda_0 - \lambda_{10}t < 0, \ \forall t$$

If $\lambda_0 = 1$ (recall that we can always set $\lambda_0 = 1$ whenever $\lambda_0 \neq 0$) the above equation can not be verified.

Therefore the NC are satisfied only if $\lambda_0 = 0$ and $\lambda_{10} > 0$.   ■

3.4 $x(t_f) =$ Not Given, $t_0 =$ Given, $t_f =$ Free

In this section we consider five problems. The first and second of them (Problems 3.16 and 3.17) have the same performance indexes, whereas the sets of the admissible final states are different. Furthermore two solutions which satisfy the NC can be found for Problem 3.16. The third problem deals with two different specifications of the set of admissible final states. Finally, the last two problems are pathological. Thus they require that $\lambda_0 = 0$ in order that the NC are satisfied.

Problem 3.16

Let

$$x(0) = \begin{bmatrix} 0 \\ 0 \end{bmatrix}, \quad x(t_f) \in \overline{S}_f = \{x \mid \alpha_f(x) = 0\}, \quad \alpha_f(x) = x_1^2 - 1$$

$$J = \int_0^{t_f} l(x(t), u(t), t) \, dt, \quad l(x, u, t) = 1 + \frac{u^2}{2}, \quad u(t) \in R, \; \forall t$$

where the final time t_f is free.

The set $\overline{S}_f$ of the admissible final states is a regular variety because

$$\Sigma_f(x) = \frac{d\alpha_f(x)}{dx} = \begin{bmatrix} 2x_1 & 0 \end{bmatrix}$$

and rank $\left(\Sigma_f(x) \right) = 1, \; \forall x \in \overline{S}_f$.

The hamiltonian function and the H-minimizing control are

$$H = 1 + \frac{u^2}{2} + \lambda_1 x_2 + \lambda_2 u, \quad u_h = -\lambda_2$$

Thus Eq. (2.5a) and its general solution over an interval beginning at time 0 are

$$\begin{aligned} \dot{\lambda}_1(t) &= 0 & \lambda_1(t) &= \lambda_{10} \\ \dot{\lambda}_2(t) &= -\lambda_1(t) & \lambda_2(t) &= \lambda_{20} - \lambda_{10}t \end{aligned}$$

It follows that the state motion is

$$x_1(t) = -\frac{\lambda_{20}}{2}t^2 + \frac{\lambda_{10}}{6}t^3$$

$$x_2(t) = -\lambda_{20}t + \frac{\lambda_{10}}{2}t^2$$

and feasibility requires

$$x_1^2(t_f) - 1 = \left(-\frac{\lambda_{20}}{2}t_f^2 + \frac{\lambda_{10}}{6}t_f^3 \right)^2 - 1 = 0$$

The orthogonality condition at the final time is

$$\begin{bmatrix} \lambda_1(t_f) \\ \lambda_2(t_f) \end{bmatrix} = \begin{bmatrix} \lambda_{10} \\ \lambda_{20} - \lambda_{10}t_f \end{bmatrix} = \vartheta_f \left. \frac{d\alpha_f(x)}{dx} \right|'_{x=x(t_f)} = \begin{bmatrix} 2x_1(t_f) \\ 0 \end{bmatrix}$$

where the expression of $\lambda(\cdot)$ has been taken into account. Thus

$$\lambda_{20} = \lambda_{10}t_f$$

Fig. 3.19 Problem 3.16. The
state trajectories T_a and T_b
corresponding to the two
solutions and the set $\overline{S}_f$ of
the admissible final states

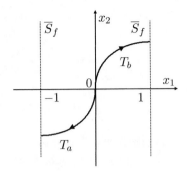

The transversality condition at the final time, which can be written for $t = 0$ because
the problem is time-invariant, is

$$H(0) = 1 + \frac{u^2(0)}{2} + \lambda_1(0)x_2(0) + \lambda_2(0)u(0) = 1 - \frac{\lambda_{20}^2}{2} = 0$$

In view of these two conditions and feasibility we compute the three unknowns λ_{10},
λ_{20}, t_f. Two solutions exist

$$t_f = t_{fa} := \sqrt{\frac{3}{\sqrt{2}}}, \quad \lambda_{10} = \lambda_{10a} := \frac{\sqrt{2}}{t_f}, \quad \lambda_{20} = \lambda_{20a} := \sqrt{2}$$

$$t_f = t_{fb} := t_{fa}, \quad \lambda_{10} = \lambda_{10b} := -\lambda_{10a}, \quad \lambda_{20} = \lambda_{20b} := -\lambda_{20a}$$

We ascertain that the two resulting controls $u_a(\cdot)$ and $u_b(\cdot)$ as well as the values
J_a and J_b of the performance index are such that $u_b(\cdot) = -u_a(\cdot)$, $J_b = J_a$. This
outcome had to be expected because of the symmetry of $\overline{S}_f$ with respect to the origin
and the absence of constraints on the control variable.

Therefore the state trajectories T_a and T_b corresponding to the two solutions are
symmetric with respect to the origin, as it appears in Fig. 3.19.   ∎

Problem 3.17
Let

$$x(0) = \begin{bmatrix} 1 \\ 1 \end{bmatrix}, \quad x(t_f) \in \overline{S}_f = \{x \mid \alpha_f(x) = 0\}, \quad \alpha_f(x) = x_1 + x_2$$

$$J = \int_0^{t_f} l(x(t), u(t), t)\, dt, \quad l(x, u, t) = 1 + \frac{u^2}{2}, \quad u(t) \in R, \ \forall t$$

where the final time t_f is free.

The set $\overline{S}_f$ of the admissible final states is a regular variety because

$$\Sigma_f(x) = \frac{d\alpha_f(x)}{dx} = \begin{bmatrix} 1 & 1 \end{bmatrix}$$

and rank $\left(\Sigma_f(x)\right) = 1, \forall x \in \overline{S}_f$.

The hamiltonian function and the H-minimizing control are

$$H = 1 + \frac{u^2}{2} + \lambda_1 x_2 + \lambda_2 u, \quad u_h = -\lambda_2$$

Thus Eq. (2.5a) and its general solution over an interval beginning at time 0 are

$$\dot{\lambda}_1(t) = 0 \qquad\qquad \lambda_1(t) = \lambda_{10}$$
$$\dot{\lambda}_2(t) = -\lambda_1(t) \qquad\qquad \lambda_2(t) = \lambda_{20} - \lambda_{10}t$$

It follows that the state motion is

$$x_1(t) = 1 + t - \frac{\lambda_{20}}{2}t^2 + \frac{\lambda_{10}}{6}t^3$$
$$x_2(t) = 1 - \lambda_{20}t + \frac{1}{2}\lambda_{10}t^2$$

and the control is feasible if

$$x_1(t_f) + x_2(t_f) = 2 - \lambda_{20}t_f\left(1 + \frac{1}{2}t_f\right) + \frac{\lambda_{10}}{2}t_f^2\left(1 + \frac{1}{3}t_f\right) + t_f = 0$$

The orthogonality condition at the final time is

$$\begin{bmatrix} \lambda_1(t_f) \\ \lambda_2(t_f) \end{bmatrix} = \begin{bmatrix} \lambda_{10} \\ \lambda_{20} - \lambda_{10}t_f \end{bmatrix} = \vartheta_f \left.\frac{d\alpha_f(x)}{dx}\right|_{x=x(t_f)}' = \begin{bmatrix} 1 \\ 1 \end{bmatrix}$$

and implies

$$\lambda_{20} = \lambda_{10}(1 + t_f)$$

Finally, the transversality condition at the final time, which can be written for $t = 0$ because the problem is time-invariant, is

$$H(0) = 1 + \frac{u^2(0)}{2} + \lambda_1(0)x_2(0) + \lambda_2(0)u(0) = 1 + \lambda_{10} - \frac{\lambda_{20}^2}{2} = 0$$

From the above three conditions (feasibility, orthogonality and transversality) we get

$$t_f = 1.42, \quad \lambda_{10} = 0.78, \quad \lambda_{20} = 1.89$$

∎

Problem 3.18

We now consider Problems 3.18-1 and 3.18-2 which only differ because of the shapes of the sets of the admissible final states. We have for both problems

$$x(0) = \begin{bmatrix} 0 \\ 0 \end{bmatrix}, \quad J = \int_0^{t_f} l(x(t), u(t), t)\, dt, \quad l(x, u, t) = 1 + |u|$$

$$u(t) \in \overline{U}, \; \forall t, \; \overline{U} = \{u | -1 \le u \le 1\}$$

where the final time t_f is free.

As for Problem 3.18-1, the final state must belong to the set

$$\overline{S}_{f_1} = \left\{ x | \alpha_{f_1}(x) = 0 \right\}, \; \alpha_{f_1}(x) = x_1 - x_2 - 2$$

whereas, for Problem 3.18-2, it must be an element of the set

$$\overline{S}_{f_2} = \left\{ x | \alpha_{f_2}(x) = 0 \right\}, \; \alpha_{f_2}(x) = \frac{(x_1 - 2)^2}{2} + x_2$$

Both $\overline{S}_{f_1}$ and $\overline{S}_{f_2}$ are regular varieties. In fact we have

$$\Sigma_{f_1}(x) = \frac{d\alpha_{f_1}(x)}{dx} = \begin{bmatrix} 1 & -1 \end{bmatrix}$$

$$\Sigma_{f_2}(x) = \frac{d\alpha_{f_2}(x)}{dx} = \begin{bmatrix} x_1 - 2 & 1 \end{bmatrix}$$

so that rank $\left(\Sigma_{f_1}(x) \right) = 1, \forall x \in \overline{S}_{f_1}$ and rank $\left(\Sigma_{f_2}(x) \right) = 1, \forall x \in \overline{S}_{f_2}$. These regular varieties are shown in Figs. 3.20 and 3.21, respectively.

The hamiltonian function and the H-minimizing control are the same for both problems, namely

$$H = 1 + |u| + \lambda_1 x_2 + \lambda_2 u, \quad u_h = \begin{cases} 1, & \lambda_2 < -1 \\ 0, & |\lambda_2| < 1 \\ -1, & \lambda_2 > 1 \end{cases}$$

Thus also Eq. (2.5a) and its general solution over an interval beginning at time 0 are equal for both problems

$$\dot{\lambda}_1(t) = 0 \qquad\qquad \lambda_1(t) = \lambda_{10}$$
$$\dot{\lambda}_2(t) = -\lambda_1(t) \qquad\qquad \lambda_2(t) = \lambda_{20} - \lambda_{10} t$$

and $\lambda_2(\cdot)$ is linear with respect to t and the control is either constant or switches once or twice. More precisely, the six following cases can occur

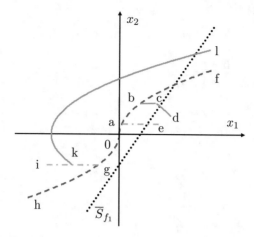

Fig. 3.20 Problem 3.18-1. The set $\overline{S}_{f_1}$ of the admissible final states. When the control does not switch the relevant trajectories are (0–a–b–f), if $u = 1$, and (0–g–h), if $u = -1$. When the control switches once the relevant trajectories are (0–a–e), if $u(t) = 1, t < \tau, u(t) = 0, \tau < t$, and (0–g–i), if $u(t) = -1, t < \tau, u(t) = 0, \tau < t$. When the control switches twice the relevant trajectories are (0–b–c–d), if $u(t) = 1, t < \tau_1, u(t) = 0, \tau_1 < t < \tau_2, u(t) = -1, \tau_2 < t$, and (0–g–k–l,) if $u(t) = -1, t < \tau_1, u(t) = 0, \tau_1 < t < \tau_2, u(t) = 1, \tau_2 < t$

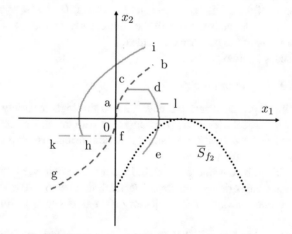

Fig. 3.21 Problem 3.18-2. The set $\overline{S}_{f_2}$ of the admissible final states. When the control does not switch the relevant trajectories are (0–a–c–b), if $u = 1$, and (0–f–g), if $u = -1$. When the control switches once the relevant trajectories are (0–a–l), if $u(t) = 1, t < \tau, u(t) = 0, \tau < t$, and (0–f–h–k), if $u(t) = -1, t < \tau, u(t) = 0, \tau < t$. When the control switches twice the relevant trajectories are (0–c–d–e), if $u(t) = 1, t < \tau_1, u(t) = 0, \tau_1 < t < \tau_2, u(t) = -1, \tau_2 < t$, and (0–f–h–i), if $u(t) = -1, t < \tau_1, u(t) = 0, \tau_1 < t < \tau_2, u(t) = 1, \tau_2 < t$

Case 1

$0 \le t \le t_f$	$\lambda_2(t) < -1$	$u(t) = 1$

Case 2

$0 \le t \le \tau$	$\lambda_2(t) < -1$	$u(t) = 1$
$\tau < t \le t_f$	$|\lambda_2(t)| < 1$	$u(t) = 0$

Case 3

$0 \le t < \tau_1$	$\lambda_2(t) < -1$	$u(t) = 1$
$\tau_1 < t < \tau_2$	$|\lambda_2(t)| < 1$	$u(t) = 0$
$\tau_2 < t \le t_f$	$\lambda_2(t) > 1$	$u(t) = -1$

Case 4

$0 \le t \le t_f$	$\lambda_2(t) > 1$	$u(t) = -1$

Case 5

$0 \le t \le \tau$	$\lambda_2(t) > 1$	$u(t) = -1$
$\tau < t \le t_f$	$|\lambda_2(t)| < 1$	$u(t) = 0$

Case 6

$0 \le t < \tau_1$	$\lambda_2(t) > 1$	$u(t) = -1$
$\tau_1 < t < \tau_2$	$|\lambda_2(t)| < 1$	$u(t) = 0$
$\tau_2 < t \le t_f$	$\lambda_2(t) < -1$	$u(t) = 1$

In the above list τ, τ_1 and τ_2 are suitable time instants to be found. Furthermore we have not considered the cases where $-1 < \lambda_2(t) < 1$, $0 < t < \tau$ because they entail that $u(\cdot)$ is zero at the beginning of the control interval, thus causing a useless increase of the value of the performance index.

We now discuss the two problems.

Problem 3.18-1 $\alpha_{f_1}(x) = x_1 - x_2 - 2$.

We first examine Fig. 3.20 where we have plotted the regular variety $\overline{S}_{f_1}$ and the shapes of the state trajectories which are consistent with the above presented cases.

It is apparent that *Case 4* and *Case 5* should not be considered as the corresponding trajectories (0–g–i) and (0–g–h) do not intersect $\overline{S}_{f_1}$.

Case 6 should not be considered as well. In fact it is easy to verify that the value of the performance index relevant to the corresponding trajectory (0–g–k–l) is surely greater than the one relevant to the trajectory (0–a–b–f) which is consistent with *Case 1*.

We have $u(0) = 1$ in *Cases 1, 2, 3*, so that the transversality condition at the final time, written for $t = 0$ because the problem is time-invariant,

$$H(0) = 1 + |u(0)| + \lambda_1(0)x_2(0) + \lambda_2(0)u(0) = 2 + \lambda_{20} = 0$$

implies $\lambda_{20} = -2$, whereas the orthogonality condition at the final time is

$$\begin{bmatrix} \lambda_1(t_f) \\ \lambda_2(t_f) \end{bmatrix} = \begin{bmatrix} \lambda_{10} \\ \lambda_{20} - \lambda_{10}t_f \end{bmatrix} = \vartheta_{f_1} \frac{d\alpha_{f_1}(x)}{dx}\bigg|'_{x=x(t_f)} = \vartheta_{f_1} \begin{bmatrix} 1 \\ -1 \end{bmatrix}$$

From the last equation we get

$$\frac{\lambda_{10}}{\lambda_{20} - \lambda_{10} t_f} = -1$$

and

$$\lambda_{10} = \frac{2}{1 - t_f}$$

if we take into account the value of λ_{20}.

The control in *Case 1* $(u(\cdot) = 1)$ does not satisfy the NC. In fact we have

$$x_2(t) = t, \quad x_1(t) = \frac{1}{2}t^2, \quad t_f = 1 + \sqrt{5}, \quad \lambda_{10} = -\frac{2}{\sqrt{5}}$$

where the value of t_f results from enforcing feasibility $(x_2(t_f) = x_1(t_f) - 2)$. However it follows that

$$\lambda_{10} = -\frac{2}{\sqrt{5}}, \quad \lambda_2(t_f) = \frac{2}{\sqrt{5}} > 0$$

which is inconsistent with $u(t_f) = 1$.

The control in *Case 2* (one switching from 1 to 0 at time τ) satisfies the NC. In fact it is

$$
\begin{aligned}
&0 \le t < \tau & &\tau < t \le t_f \\
&u(t) = 1 & &u(t) = 0 \\
&0 \le t \le \tau & &\tau \le t \le t_f \\
&x_2(t) = t & &x_2(t) = x_2(\tau) \\
&x_1(t) = \frac{1}{2}t^2 & &x_1(t) = x_1(\tau) + x_2(\tau)(t - \tau)
\end{aligned}
$$

If we enforce feasibility, the switching condition $(\lambda_2(\tau) = -1)$ and recall the link between λ_{10} and t_f, we get

$$t_f = 1 + \sqrt{5}, \quad \tau = \frac{\sqrt{5}}{2}, \quad \lambda_{10} = -\frac{2}{\sqrt{5}}, \quad \lambda_2(t_f) = \frac{2}{\sqrt{5}}$$

which is consistent with $u(t_f) = 0$. The corresponding state trajectory has the shape (0–a–e) in Fig. 3.20. It obviously ends when $\overline{S}_{f_1}$ is reached.

The control in *Case 3* (one switching from 1 to 0 at time τ_1 and a switching from 0 to -1 at time τ_2) does not satisfy the NC. In fact it is

$0 \le t < \tau_1$	$\tau_1 < t \le \tau_2$	$\tau_2 < t \le t_f$
$u(t) = 1$	$u(t) = 0$	$u(t) = -1$
$0 \le t \le \tau_1$	$\tau_1 \le t \le \tau_2$	$\tau_2 \le t \le t_f$
$x_2(t) = t$	$x_2(t) = x_2(\tau_1)$	$x_2(t) = x_2(\tau_2) +$
		$-(t - \tau_2)$
$x_1(t) = \dfrac{1}{2}t^2$	$x_1(t) = x_1(\tau_1) +$	$x_1(t) = x_1(\tau_2) +$
	$+ x_2(\tau_1)(t - \tau_1)$	$+ x_2(\tau_2)(t - \tau_2) - \dfrac{1}{2}(t - \tau_2)^2$

If we enforce feasibility, the switching conditions ($\lambda_2(\tau_1) = -1$ and $\lambda_2(\tau_2) = 1$) and the link between λ_{10} and t_f, we get

$$t_f = 1 + \sqrt{6}, \quad \tau_1 = \frac{\sqrt{6}}{2}, \quad \tau_2 = \frac{3\sqrt{6}}{2}$$

This result should be disregarded as it implies $\tau_2 > t_f$. In summary, the only control which satisfies the NC is that in *Case* 2.

Problem 3.18-2 $\alpha_{f_2}(x) = \dfrac{(x_1 - 2)^2}{2} + x_2 - 2$.

We first examine Fig. 3.21 where the regular variety $\overline{S}_{f_2}$ and the shapes of the state trajectories which are consistent with the above presented cases are plotted.

It is apparent that *Cases 1, 2, 4, 5, 6* should not be taken into consideration because the corresponding state trajectories (0–a–c–b), (0–a–l), (0–f–g), (0–f–h–k) and (0–f–h–i) do not intersect $\overline{S}_{f_2}$.

Therefore only the state trajectories of *Case 3* are feasible. They are characterized by two control switchings from 1 to 0 and from 0 to -1.

Thus we have $u(0) = 1$ as in Problem 3.18-1. The transversality condition at the final time, written for $t = 0$ because the problem is time-invariant, is

$$H(0) = 1 + |u(0)| + \lambda_1(0)x_2(0) + \lambda_2(0)u(0) = 2 + \lambda_{20} = 0$$

and again implies $\lambda_{20} = -2$, whereas the orthogonality condition at the final time is

$$\begin{bmatrix} \lambda_1(t_f) \\ \lambda_2(t_f) \end{bmatrix} = \begin{bmatrix} \lambda_{10} \\ \lambda_{20} - \lambda_{10}t_f \end{bmatrix} = \vartheta_{f_2} \frac{d\alpha_{f_2}(x)}{dx} \Bigg|_{x = x(t_f)} = \vartheta_{f_2} \begin{bmatrix} x_1(t_f) - 2 \\ 1 \end{bmatrix}$$

From the last equation we get

$$\frac{\lambda_{10}}{2 + \lambda_{10}t_f} = 2 - x_1(t_f)$$

if we take into account the value of λ_{20}. When we adopt the control in *Case 3* (one switching from 1 to 0 at time τ_1 and one switching form 0 to -1 at time τ_2) it results

$0 \leq t < \tau_1$	$\tau_1 < t \leq \tau_2$	$\tau_2 < t \leq t_f$
$u(t) = 1$	$u(t) = 0$	$u(t) = -1$
$0 \leq t \leq \tau_1$	$\tau_1 \leq t \leq \tau_2$	$\tau_2 \leq t \leq t_f$
$x_2(t) = t$	$x_2(t) = x_2(\tau_1)$	$x_2(t) = x_2(\tau_2) +$ $-(t - \tau_2)$

$$x_1(t) = \frac{1}{2}t^2 \qquad x_1(t) = x_1(\tau_1) + \qquad x_1(t) = x_1(\tau_2) +$$
$$+x_2(\tau_1)(t - \tau_1) \qquad +x_2(\tau_2)(t - \tau_2) - \frac{1}{2}(t - \tau_2)^2$$

The switching conditions ($\lambda_2(\tau_1) = -1$ and $\lambda_2(\tau_2) = 1$) give

$$\tau_1 = -\frac{1}{\lambda_{10}}, \quad \tau_2 = -\frac{3}{\lambda_{10}}$$

These two equations together with the feasibility condition $((x_1(t_f)-2)^2+2x_2(t_f)-4 = 0)$, which easily follows from the above reported expressions of the state motion, and the orthogonality condition at the final time constitute a system of four equations which allow the computation of the four unknowns $\lambda_{10}, \tau_1, \tau_2, t_f$ and, subsequently, $x_1(t_f), x_2(t_f)$. We get

$$\lambda_{10} = -1.47, \quad \tau_1 = 0.68, \quad \tau_2 = 2.04, \quad t_f = 2.92$$
$$x_1(t_f) = 1.36, \quad x_2(t_f) = -0.20$$

∎

The two forthcoming problems are pathological.

Problem 3.19

Let

$$x(0) = \begin{bmatrix} 0 \\ 2 \end{bmatrix}, \quad x(t_f) \in \overline{S}_f = \{x \mid \alpha_f(x) = 0\}, \quad \alpha_f = (x_1 - 2)^2 - x_2 + 2$$

$$J = \int_0^{t_f} l(x(t), u(t), t)\, dt, \quad l(x, u, t) = 1 + \frac{u^2}{2}$$

$$u(t) \in \overline{U}, \forall t, \quad \overline{U} = \{u \mid -1 \leq u \leq 0\}$$

where the final time t_f is free. The set $\overline{S}_f$ of the admissible final states is shown in Fig. 3.22. It is a regular variety because

$$\Sigma_f(x) = \frac{d\alpha_f(x)}{dx} = \begin{bmatrix} 2(x_1 - 2) & -1 \end{bmatrix}$$

and rank $\left(\Sigma_f(x)\right) = 1, \forall x \in \overline{S}_f$.

Fig. 3.22 Problem 3.19. The set $\overline{S}_f$ of the admissible final states and the state trajectory c corresponding to $u(\cdot) = 0$

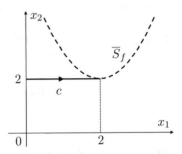

Only the control $u(t) = 0$, $\forall t$, is feasible and thus also optimal. It follows that

$$x_2(t) = 2, \quad x_1(t) = 2t, \quad t_f = 1$$

The corresponding state trajectory c is plotted in Fig. 3.22. Different controls over the interval $0 \le t < \tau$ would entail, for each $\tau \ne 0$, $x_2(\tau) < 2$ and the request $x(t_f) \in \overline{S}_f$ can not satisfied. This statement can be proved as done in dealing with Problem 3.15.

The hamiltonian function is

$$H = \lambda_0 \left(1 + \frac{u^2}{2} \right) + \lambda_1 x_2 + \lambda_2 u$$

and, if $\lambda_0 \ne 0$, that is if $\lambda_0 = 1$, the H-minimizing control is

$$u_h = \begin{cases} 0, & \lambda_2 \le 0 \\ -\lambda_2, & 0 \le \lambda_2 \le 1 \\ -1, & 1 \le \lambda_2 \end{cases}$$

Thus Eq. (2.5a) and its general solution over an interval beginning at time 0 are

$$\dot{\lambda}_1(t) = 0 \qquad\qquad\qquad \lambda_1(t) = \lambda_{10}$$
$$\dot{\lambda}_2(t) = -\lambda_1(t) \qquad\qquad \lambda_2(t) = \lambda_{20} - \lambda_{10}t$$

so that the orthogonality condition at the final time is

$$\begin{bmatrix} \lambda_1(t_f) \\ \lambda_2(t_f) \end{bmatrix} = \begin{bmatrix} \lambda_{10} \\ \lambda_{20} - \lambda_{10}t_f \end{bmatrix} = \vartheta_f \left. \frac{d\alpha_f(x)}{dx} \right|_{x=x(t_f)}' =$$

$$= \vartheta_f \begin{bmatrix} 2(x_1(t_f) - 2) \\ -1 \end{bmatrix} = \vartheta_f \begin{bmatrix} 0 \\ -1 \end{bmatrix}$$

where we have taken into account that $x_1(t_f) = 2$. We get

$$\lambda_{10} = 0, \quad \lambda_2(t) = \lambda_{20}$$

Therefore it must be $\lambda_{20} \leq 0$ because the control is $u(\cdot) = 0$. The transversality condition at the final time, written for $t = 0$ because the problem is time-invariant, is

$$H(0) = \lambda_0 \left(1 + \frac{u^2(0)}{2}\right) + \lambda_1(0)x_2(0) + \lambda_2(0)u(0) = \lambda_0 + 2\lambda_{10} = \lambda_0 = 0$$

and the NC are satisfied only if

$$\lambda_0 = \lambda_{10} = 0, \quad \lambda_{20} < 0$$

which is consistent with $u_h = 0$.   ∎

Problem 3.20
Let

$$x(0) = \begin{bmatrix} 0 \\ 0 \end{bmatrix}, \quad x(t_f) \in \overline{S}_f = \{x \mid \alpha_f(x) = 0\}, \quad \alpha_f(x) = x_1 - 2x_2 + 2$$

$$J = \int_0^{t_f} l(x(t), u(t), t)\, dt, \quad l(x, u, t) = 1, \quad u(t) \in \overline{U}, \ \forall t, \ \overline{U} = \{u \mid -1 \leq u \leq 1\}$$

where the final time t_f is free. The set $\overline{S}_f$ of the admissible final states is a regular variety because

$$\Sigma_f(x) = \frac{d\alpha_f(x)}{dx} = \begin{bmatrix} 0.5 & -1 \end{bmatrix}$$

and rank $(\Sigma_f(x)) = 1$, $\forall x \in \overline{S}_f$. The hamiltonian function and the H-minimizing control are

$$H = \lambda_0 + \lambda_1 x_2 + \lambda_2 u, \quad u_h = -\text{sign}(\lambda_2)$$

Thus Eq. (2.5a) and its general solution over an interval beginning at time 0 are

$$\begin{array}{ll} \dot{\lambda}_1(t) = 0 & \lambda_1(t) = \lambda_{10} \\ \dot{\lambda}_2(t) = -\lambda_1(t) & \lambda_2(t) = \lambda_{20} - \lambda_{10}t \end{array}$$

It follows that the transversality condition at the final time, written for $t = 0$ because the problem is time-invariant, is

$$H(0) = \lambda_0 + \lambda_1(0)x_2(0) + \lambda_2(0)u(0) = \lambda_0 + \lambda_{20}u(0) = 0$$

Therefore the following three pairs $(\lambda_0, \lambda_{20})$ are allowed.

Case1
$$\lambda_0 = 1 \qquad\qquad\qquad \lambda_{20} = 1$$

Case2
$$\lambda_0 = 1 \qquad\qquad\qquad \lambda_{20} = -1$$

Case3
$$\lambda_0 = 0 \qquad\qquad\qquad \lambda_{20} = 0$$

In view of the solutions of Eq. (2.5a), the orthogonality condition at the final time is

$$\begin{bmatrix} \lambda_1(t_f) \\ \lambda_2(t_f) \end{bmatrix} = \begin{bmatrix} \lambda_{10} \\ \lambda_{20} - \lambda_{10} t_f \end{bmatrix} = \vartheta_f \left. \frac{d\alpha_f(x)}{dx} \right|'_{x=x(t_f)} = \vartheta_f \begin{bmatrix} 1 \\ -2 \end{bmatrix}$$

Note that $\lambda_2(t_f) = \lambda_{20} - \lambda_{10} t_f \neq 0$ because, otherwise, we have $\vartheta_f = 0$ and, consequently, $\lambda_{10} = 0$ and $\lambda_{20} = 0$. This causes the transversality condition to be satisfied by $\lambda_0 = 0$, against condition (v) of Theorem 2.1. Thanks to the orthogonality condition, we can conclude that

$$\lambda_{20} = \lambda_{10}(t_f - 2) \Rightarrow \lambda_2(t) = \lambda_{10}(t_f - t - 2)$$

Then *Case 1* (*Case 2*) implies $\lambda_{10} \neq 0$, so that the function $\lambda_2(\cdot)$ can either be always non-negative (non-positive) or change sign at time τ. On the contrary, *Case 3* entails $\lambda_{10} = 0$ and/or $t_f = 2$. However, $\lambda_{10} = 0$ is not allowed as it implies $\lambda_2(\cdot) = 0$ and also $\lambda_0 = 0$ from the transversality condition. This outcome violates condition (v) of Theorem 2.1. Therefore *Case 3* is consistent with $t_f = 2$ and $\lambda_2(\cdot)$ positive (negative) only, so that $u(\cdot) = -1$ ($u(\cdot) = 1$). But

$$u(\cdot) = -1 \Rightarrow \begin{cases} x_1(t_f) = -2 \\ x_2(t_f) = -2 \end{cases} \Rightarrow x(t_f) \notin \overline{S}_f$$

whereas

$$u(\cdot) = 1 \Rightarrow \begin{cases} x_1(t_f) = 2 \\ x_2(t_f) = 2 \end{cases} \Rightarrow x(t_f) \in \overline{S}_f$$

Thus the necessary conditions are satisfied if $\lambda_0 = \lambda_{20} = 0$, $\lambda_{10} > 0$. Consequently, $u(\cdot) = 1$, $t_f = 2$.

However, in order to state that the problem is pathological it is necessary to show that *Case 1* and *Case 2* are not feasible. This is now done in detail.

Case 1 with $\lambda_2(\cdot) \geq 0$ and therefore $u(t) = -1$, $0 \leq t < t_f$.

By enforcing feasibility ($x(t_f) \in \overline{S}_f$) and recalling the orthogonality condition, we get

$$t_f = 2 + \sqrt{8}, \quad \lambda_{10} = \frac{1}{\sqrt{8}}, \quad \lambda_2(t_f) = -\frac{2}{\sqrt{8}}$$

which is a contradiction as $\lambda_2(t_f) < 0$, whereas it has been assumed that $\lambda_2(\cdot) \geq 0$.

Fig. 3.23 Problem 3.20. The set $\overline{S}_f$ of the admissible final states and the state trajectory (0–a) which satisfies the NC

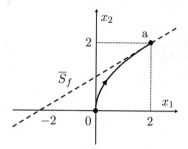

Case 1 with $\lambda_2(\cdot)$ changing sign at time τ, so that $u(t) = -1, 0 \le t < \tau, u(t) = 1,$ $\tau < t \le t_f$.

Recall the expression of $\lambda_2(\cdot)$. Then, if $\lambda_2(\tau) = \lambda_{20} - \lambda_{10}\tau = 0$, it must be $\lambda_{10} \ne 0$. This fact, together with the orthogonality and switching conditions, implies $\tau = t_f - 2$. By enforcing feasibility, we get $\tau = 0$. Therefore no switching exists, against the assumption.

Case 2 with $\lambda_2(\cdot) \le 0$, so that $u(t) = 1, 0 \le t < t_f$.

By enforcing feasibility, we get $t_f = 2$. Therefore the orthogonality condition implies $\lambda_{20} = 0$, a contradiction because, by assumption, $\lambda_{20} = -1$.

Case 2 with $\lambda_2(\cdot)$ changing sign at time τ so that $u(t) = 1, 0 \le t < \tau, u(t) = -1,$ $\tau < t \le t_f$.

The existence of a switching requires that $\lambda_2(\tau) = \lambda_{20} - \lambda_{10}\tau = 0$. Consequently, it implies (recall the expression of $\lambda_2(\cdot)$) that $\lambda_{10} \ne 0$. This fact, together with the orthogonality and switching conditions, leads to $\tau = t_f - 2$. It is easy to verify that

$$x_2(t_f) = \tau - 2, \quad x_1(t_f) = \frac{\tau^2}{2} + 2\tau - 2$$

Thus, by enforcing feasibility, we obtain that τ should satisfy the equation $\tau^2 + 8 = 0$ which does not admit a real solution.

In summary, the problem is pathological and the state trajectory (0–a) of Fig. 3.23 is the one which satisfies the NC. ∎

Chapter 4
Simple Constraints:
$J = \int$, $x(t_0) = $ Not Given

Abstract The main features of the optimal control problems considered here are:
(1) the initial state is not given; (2) the performance index is of a purely integral type;
(3) only simple constraints act on the system, namely we only require that (a) the
initial and final states are free or belong to a regular variety, (b) the final time is either
free or given, (c) the control variable is free or takes on values in a closed subset of
the real numbers. Accordingly, we consider problems where (i) the final state and
both the initial and final times are given; (ii) the final state is not given and both the
initial and final times are given; (iii) the final state is not given and both the initial
and final times are free; (iv) the initial time is always zero but in the last section.

As in Chap. 3, we consider optimal control problems defined on a double integrator
which can be handled by making reference to the material in Sect. 2.2.1. Their main
features are: (1) the initial state is not given; (2) the performance index is of a purely
integral type; (3) only simple constraints act on the system, namely we only require
that (a) the initial and final states are free or belong to a regular variety, (b) the final
time is either free or given, (c) the control variable is free or takes on values in a
closed subset of the real numbers. Accordingly, we consider problems where: (i) the
final state and both the initial and final times are given (Sect. 4.1); (ii) the final state
is not given and both the initial and final times are given (Sect. 4.2); (iii) the final
state is not given and both the initial and final times are free (Sect. 4.3); (iv) the initial
time is always zero but in the last section.

4.1 $(x(t_f), t_0, t_f) = $ Given

The final state is the same for the three problems of this section. The substantially
different nature of the regular variety of the admissible initial states implies that
only one solution which satisfies the NC can be found for the first and third of them
(Problems 4.1 and 4.3). On the contrary, four solutions can be found for the second
one (Problem 4.2): two of them are optimal. In the performance index the function

© Springer International Publishing Switzerland 2017

A. Locatelli, *Optimal Control of a Double Integrator*, Studies in Systems,
Decision and Control 68, DOI 10.1007/978-3-319-42126-1_4

$l(\cdot, \cdot, \cdot)$ depends on the state and control in the first problem, on the control only in the second one, on the state, control and time in the third one.

Problem 4.1 Let

$$x(0) \in \overline{S}_0 = \{x \mid \alpha_0(x) = 0\}, \quad \alpha_0(x) = x_2 - 1, \quad x(t_f) = \begin{bmatrix} 0 \\ 0 \end{bmatrix}$$

$$J = \int_0^{t_f} l(x(t), u(t), t)\, dt, \quad l(x, u, t) = \frac{u^2}{2} + \beta x_1 + \gamma x_2, \quad u(t) \in R, \ \forall t$$

where the parameters $\beta \geq 0$, $\gamma \geq 0$ and the final time t_f are given.

The set $\overline{S}_0$ of the admissible initial states is shown in Fig. 4.1 and is a regular variety because

$$\Sigma_0(x) = \frac{d\alpha_0(x)}{dx} = \begin{bmatrix} 0 & 1 \end{bmatrix}$$

and rank $(\Sigma_0(x)) = 1, \forall x \in \overline{S}_0$.

The hamiltonian function and the H−minimizing control are

$$H = \frac{u^2}{2} + \beta x_1 + \gamma x_2 + \lambda_1 x_2 + \lambda_2 u, \quad u_h = -\lambda_2$$

Therefore Eq. (2.5a) and its general solution over a time interval beginning at time 0 are

$$\dot{\lambda}_1(t) = -\beta \qquad\qquad \lambda_1(t) = \lambda_{10} - \beta t$$
$$\dot{\lambda}_2(t) = -\gamma - \lambda_1(t) \qquad\qquad \lambda_2(t) = \lambda_{20} - \gamma t - \lambda_{10} t + \frac{\beta}{2} t^2$$

and the orthogonality condition at the initial time is

$$\begin{bmatrix} \lambda_1(0) \\ \lambda_2(0) \end{bmatrix} = \begin{bmatrix} \lambda_{10} \\ \lambda_{20} \end{bmatrix} = \vartheta_0 \left. \frac{d\alpha_0(x)}{dx} \right|'_{x=x(0)} = \vartheta_0 \begin{bmatrix} 0 \\ 1 \end{bmatrix}$$

Thus

$$\lambda_{10} = 0, \quad u(t) = -\lambda_2(t) = -\lambda_{20} + \gamma t - \frac{\beta}{2} t^2$$

and also

$$x_1(t) = x_1(0) + t - \frac{\lambda_{20}}{2} t^2 + \frac{\gamma}{6} t^3 - \frac{\beta}{24} t^4$$

$$x_2(t) = 1 - \lambda_{20} t + \frac{\gamma}{2} t^2 - \frac{\beta}{6} t^3$$

By enforcing feasibility $(x(t_f) = 0)$, a system of two equations for the two unknowns λ_{20} e $x_1(0)$ is obtained, specifically,

Fig. 4.1 Problem 4.1. The set $\overline{S}_0$ of the admissible initial states and the state trajectories when $t_f = 0.2$ (trajectory (a–0)), when $t_f = 1$ (trajectory (b–0)), when $t_f = 2$ (trajectory (c–0))

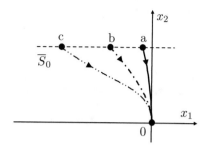

$$x_1(t_f) = x_1(0) + t_f - \frac{\lambda_{20}}{2}t_f^2 + \frac{\gamma}{6}t_f^3 - \frac{\beta}{24}t_f^4 = 0$$

$$x_2(t_f) = 1 - \lambda_{20}t_f + \frac{\gamma}{2}t_f^2 - \frac{\beta}{6}t_f^3 = 0$$

We get

$$x_1(0) = -\frac{12t_f - 2\gamma t_f^3 + \beta t_f^4}{24}$$

$$\lambda_{20} = \frac{6 + 3\gamma t_f^2 - \beta t_f^3}{6t_f}$$

Note that as t_f tends to zero λ_{20} tends to infinity and the control action becomes more intense ($|u(0)|$ tends to infinity). On the contrary, $x_1(0)$ tends to zero and the system initial state tends to lie on the x_2 axis. State trajectories corresponding to $\beta = \gamma = 1$ and various choices of the final time are shown in Fig. 4.1. ∎

Problem 4.2

Let

$$x(0) = \in \overline{S}_0 = \{x \mid \alpha_0(x) = 0\}, \quad \alpha_0(x) = x_1^2 + x_2^2 - 1, \quad x(t_f) = \begin{bmatrix} 0 \\ 0 \end{bmatrix}$$

$$J = \int_0^{t_f} l(x(t), u(t), t) \, dt, \quad l(x, u, t) = \frac{u^2}{2}, \quad u(t) \in R, \; \forall t$$

where the final time t_f is given.

The set $\overline{S}_0$ of the admissible initial states is shown in Fig. 4.3. It is a regular variety because

$$\Sigma_0(x) = \frac{d\alpha_0(x)}{dx} = \begin{bmatrix} 2x_1 & 2x_2 \end{bmatrix}$$

and rank $(\Sigma_0(x)) = 1$, $\forall x \in \overline{S}_0$. As a matter of fact, it is not possible that x_1 and x_2 are contemporarily zero, as it appears in the quoted figure.

The hamiltonian function and the H−minimizing control are

$$H = \frac{u^2}{2} + \lambda_1 x_2 + \lambda_2 u, \quad u_h = -\lambda_2$$

Therefore Eq. (2.5a) and its general solution over a time interval beginning at time 0 are

$$\dot{\lambda}_1(t) = 0 \qquad\qquad \lambda_1(t) = \lambda_{10}$$
$$\dot{\lambda}_2(t) = -\lambda_1(t) \qquad\qquad \lambda_2(t) = \lambda_{20} - \lambda_{10}t$$

We observe that the nature of $\overline{S}_0$ and of the performance index is such that the number of solutions which satisfy the NC must be even. Indeed the symmetry of the set $\overline{S}_0$ with respect to the origin entails that the pair $(-\bar{x}(\cdot), -\bar{u}(\cdot))$ satisfies Theorem 2.1 whenever the pair $(\bar{x}(\cdot), \bar{u}(\cdot))$ satisfies it. For this reason, we assume that $x_2(0) \geq 0$.

The orthogonality condition at the initial time is

$$\begin{bmatrix} \lambda_1(0) \\ \lambda_2(0) \end{bmatrix} = \begin{bmatrix} \lambda_{10} \\ \lambda_{20} \end{bmatrix} = \vartheta_0 \left. \frac{d\alpha_0(x)}{dx} \right|'_{x=x(0)} = \vartheta_0 \begin{bmatrix} 2x_1(0) \\ 2x_2(0) \end{bmatrix}$$

and, if $x_1(0) \neq 0$, it implies

$$\lambda_{20} = \frac{x_2(0)}{x_1(0)} \lambda_{10}$$

while, if $x_1(0) = 0$, we have $x_2(0) = 1$ (in order that $x(0) \in \overline{S}_0$), $\lambda_{10} = 0$, $u(\cdot) = -\lambda_{20}$. However, this eventuality must be discarded because, otherwise,

$$x_2(t_f) = 1 - \lambda_{20}t_f, \quad x_1(t_f) = t_f - \frac{\lambda_{20}}{2}t_f^2$$

and the constraint $x(t_f) = 0$ can not be satisfied. Thus

$$u(t) = -\lambda_2(t) = \lambda_{10}\left(t - \frac{x_2(0)}{x_1(0)}\right)$$

$$x_1(t) = x_1(0) + x_2(0)t - \frac{\lambda_{10}}{2}\left(\frac{1}{3}t^3 - \frac{x_2(0)}{x_1(0)}t^2\right)$$

$$x_2(t) = x_2(0) + \lambda_{10}\left(\frac{1}{2}t^2 - \frac{x_2(0)}{x_1(0)}t\right)$$

$$x_2(0) = \sqrt{1 - x_1(0)^2}$$

By enforcing feasibility, that is $x(t_f) = 0$, the unknown parameters $x_1(0)$, $x_2(0)$, λ_{10}, λ_{20} take on the set of values Q_1 and Q_2 which are reported in Table 4.1, if $t_f = 5$. According to what has been observed above, the sets Q_3 and Q_4 specify solutions

Table 4.1 Problem 4.2

	$x_1(0)$	$x_2(0)$	λ_{10}	λ_{20}	J
Q_1	-0.96	0.29	-0.02	0.01	0.01
Q_2	0.29	0.96	0.26	0.84	0.44
Q_3	0.96	-0.29	0.02	-0.01	0.01
Q_4	-0.29	-0.96	-0.26	-0.84	0.44

The values of the parameters $x_1(0)$, $x_2(0)$, λ_{10}, λ_{20} and of the performance index J when $t_f = 5$

Fig. 4.2 Problem 4.2. The performance index as a function of $x_1(0)$ when $x_2(0) = \sqrt{1 - x_1^2(0)}$ and the control satisfies the NC

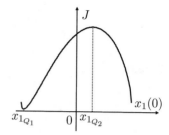

of the problem which satisfy the NC as well. The table gives also the values of the performance index. We conclude that the set Q_1 (and Q_3 as well) must be preferred.

A deeper insight into the above conclusions ca be gained by first recalling that $x_1(0) \in [-1, 1]$ and we have assumed $x_2(0) \geq 0$. Then the optimal value J^o of the performance index as a function of $x_1(0)$ can easily be computed through a straightforward application of the Hamilton–Jacoby theory (see [15]) and we find

$$J^o(x_1(0)) = \beta_1(x_1(0))x_1(0) + \beta_2(x_1(0))x_2(0) + \beta_3(x_1(0))$$

$$\beta_1(x_1(0)) = \frac{12(2x_1(0) + 5x_2(0))}{250}, \quad \beta_2(x_1(0)) = \frac{25\beta_1(x_1(0)) + 2x_2(0)}{10}$$

$$\beta_3(x_1(0)) = -\frac{(5\beta_1(x_1(0)) - \beta_2(x_1(0)))^3 + \beta_2^3(x_1(0))}{6\beta_1(x(0))}$$

The function $J^o(\cdot)$ is plotted in Fig. 4.2. Its minimum value is attained when $x_1(0) = x_{1_{Q_1}} := -0.96$, which is precisely the value corresponding to the set Q_1. The derivative of $J^o(\cdot)$ vanishes when $x_1(0) = x_{1_{Q_2}} = 0.29$ (the value corresponding to the set Q_2), consistently with the NC calling for the first variation of the performance index to vanish when the solution is optimal. As a conclusion the sets Q_1 and Q_3 yield an optimal solution. The state trajectories corresponding to the sets $Q_i, i = 1, 2, 3, 4$ are shown in Fig. 4.3. ∎

Fig. 4.3 Problem 4.2. The
state trajectories
corresponding to Q_i,
$i = 1, 2, 3, 4$

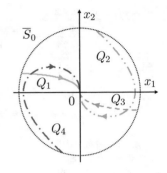

Problem 4.3
Let

$$x(0) \in \overline{S}_0 = \{x \mid \alpha_0(x) = 0\}, \ \alpha_0(x) = x_1 - x_2 + 1, \ x(t_f) = \begin{bmatrix} 0 \\ 0 \end{bmatrix}$$

$$J = \int_0^{t_f} l(x(t), u(t), t) \, dt, \ l(x, u, t) = \frac{u^2}{2} + tx_2, \ u(t) \in R, \ \forall t$$

where the final time t_f is given.

The set $\overline{S}_0$ of the admissible initial states is shown in Fig. 4.4. It is a regular variety because

$$\Sigma_0(x) = \frac{d\alpha_0(x)}{dx} = \begin{bmatrix} 1 & -1 \end{bmatrix}$$

and rank $(\Sigma_0(x)) = 1$, $\forall x \in \overline{S}_0$.

The hamiltonian function and the $H-$minimizing control are

$$H = \frac{u^2}{2} + tx_2 + \lambda_1 x_2 + \lambda_2 u, \quad u_h = -\lambda_2$$

Fig. 4.4 Problem 4.3. The
set $\overline{S}_0$ of the admissible
initial states and state
trajectories corresponding to
$t_f = 0.5$ (*solid line*),
$t_f = 1.5$ (*dash-single-dotted
line*), $t_f = 2.5$
(*dash-double-dotted line*)

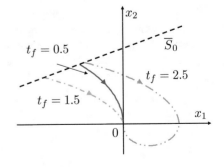

Thus Eq. (2.5a) and its general solution over a time interval beginning at time 0 are

$$\dot{\lambda}_1(t) = 0 \qquad\qquad \lambda_1(t) = \lambda_{10}$$
$$\dot{\lambda}_2(t) = -\lambda_1(t) - t \qquad\qquad \lambda_2(t) = \lambda_{20} - \lambda_{10}t - \frac{t^2}{2}$$

The orthogonality condition at the initial time

$$\begin{bmatrix} \lambda_1(0) \\ \lambda_2(0) \end{bmatrix} = \begin{bmatrix} \lambda_{10} \\ \lambda_{20} \end{bmatrix} = \vartheta_0 \left.\frac{d\alpha_0(x)}{dx}\right|'_{x=x(0)} = \vartheta_0 \begin{bmatrix} 1 \\ -1 \end{bmatrix}$$

implies $\lambda_{20} = -\lambda_{10}$, so that

$$u(t) = \lambda_{10} + \lambda_{10}t + \frac{t^2}{2}$$
$$x_1(t) = x_1(0) + x_2(0)t + \lambda_{10}\frac{t^2}{2} + \frac{t^3}{6} + \frac{t^4}{24}$$
$$x_2(t) = x_2(0) + \lambda_{10}t + \lambda_{10}\frac{t^2}{2} + \frac{t^3}{6}$$

By enforcing feasibility ($x_1(0) = x_2(0) - 1$, $x_1(t_f) = x_2(t_f) = 0$), it is easy to find $x_2(0)$ and λ_{10} as functions of t_f.

Table 4.2 collects the values of these two parameters and of the performance index J corresponding to five different choices of the final time t_f. Note that the performance index becomes negative as the final time increases. This outcome is a consequence of the particular nature of the adopted function $l(\cdot, \cdot, \cdot)$. When $t_f = 1.5$, $x_2(0)$ attains a minimum, while λ_{10} attains a maximum. The state trajectories corresponding to three different values of t_f are shown in Fig. 4.4. Observe that when t_f is sufficiently large, then it is convenient to generate a trajectory where the second state variable takes on negative values. This fact is consistent with the form of the function $l(\cdot, \cdot, \cdot)$ and is further put into evidence in Fig. 4.5 which shows the plots of

$$J_u := \int_0^{t_f} \frac{u^2(t)}{2}\, dt, \quad J_{tx} := \int_0^{t_f} tx_2(t)\, dt, \quad J = J_u + J_{tx}$$

Table 4.2 Problem 4.3

t_f	λ_{10}	$x_2(0)$	J
0.1	−9.06	0.95	4.53
0.5	−1.30	0.79	0.67
1.5	−0.45	0.62	0.30
5.0	−1.39	3.58	−8.42
10.0	−3.20	25.20	−233.33

The parameters λ_{10}, $x_2(0)$, J corresponding to various choices of t_f

Fig. 4.5 Problem 4.3. Plots
of the performance indexes:
J (*solid line*), J_u
(*dash-single-dotted line*), J_{tx}
(*dash-double-dotted line*)

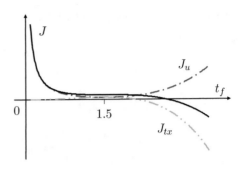

When $t_f = 1.5$ (one of the above chosen values), the function $J(\cdot)$ exhibits a flex. ∎

4.2 $x(t_f) = $ **Not Given, $(t_0, t_f) = $ Given**

The control variable is not constrained in the three problems of this section. The function $l(\cdot, \cdot, \cdot)$ in the performance index depends on the state, control and time in the first of them, on the control only in the second one and on the state and control in the third one.

Problem 4.4
Let

$$J = \int_0^{t_f} l(x(t), u(t), t)\, dt, \quad l(x, u, t) = \frac{u^2}{2(1 + \mu t)} + x_1, \quad u(t) \in R, \ \forall t, \quad t_f = 1$$

$$x(0) \in \overline{S}_0 = \{x | \alpha_0(x) = 0\}, \quad \alpha_0(x) = x_1 - 1$$

$$x(t_f) \in \overline{S}_f = \{x | \alpha_f(x) = 0\}, \quad \alpha_f(x) = x_1 - 2$$

where the integer and nonnegative parameter μ is given. The two sets $\overline{S}_0$ and $\overline{S}_f$ of admissible initial and final states are shown in Fig. 4.6. They are regular varieties. In fact for the first one we have

$$\Sigma_0(x) = \frac{d\alpha_0(x)}{dx} = \begin{bmatrix} 1 & 0 \end{bmatrix}$$

and rank$(\Sigma_0(x)) = 1, \forall x \in \overline{S}_0$. Analogously, for the second one we have

$$\Sigma_f(x) = \frac{d\alpha_f(x)}{dx} = \begin{bmatrix} 1 & 0 \end{bmatrix}$$

and rank$(\Sigma_f(x)) = 1, \forall x \in \overline{S}_f$.

Fig. 4.6 Problem 4.4. State trajectories when $\mu = 0$ (*solid line*), when $\mu = 5$ (*dotted line*), when $\mu = 30$ (*dash-single-dotted line*)

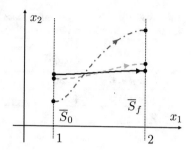

The hamiltonian function and the H−minimizing control are

$$H = \frac{u^2}{2(1 + \mu t)} + x_1 + \lambda_1 x_2 + \lambda_2 u, \quad u_h = -(1 + \mu t)\lambda_2$$

so that Eq. (2.5a) and its general solution over an interval beginning at time 0 are

$$\dot{\lambda}_1(t) = -1 \qquad \lambda_1(t) = \lambda_{10} - t$$
$$\dot{\lambda}_2(t) = -\lambda_1 \qquad \lambda_2(t) = \lambda_{20} - \lambda_{10}t + \frac{t^2}{2}$$

The orthogonality condition at the initial time

$$\begin{bmatrix} \lambda_1(0) \\ \lambda_2(0) \end{bmatrix} = \begin{bmatrix} \lambda_{10} \\ \lambda_{20} \end{bmatrix} = \vartheta_0 \left.\frac{d\alpha_0(x)}{dx}\right|'_{x=x(0)} = \vartheta_0 \begin{bmatrix} 1 \\ 0 \end{bmatrix}$$

implies $\lambda_{20} = 0$. In view of this outcome, the orthogonality condition at the final time

$$\begin{bmatrix} \lambda_1(t_f) \\ \lambda_2(t_f) \end{bmatrix} = \begin{bmatrix} \lambda_{10} - t_f \\ \lambda_{20} - \lambda_{10}t_f + \frac{t_f^2}{2} \end{bmatrix} = \vartheta_f \left.\frac{d\alpha_f(x)}{dx}\right|'_{x=x(t_f)} = \vartheta_f \begin{bmatrix} 1 \\ 0 \end{bmatrix}$$

entails $\lambda_{10} = 0.5$. Thus the system state motion is

$$x_1(t) = 1 + x_2(0)t + \frac{1}{12}t^3 + \frac{\mu - 1}{24}t^4 - \frac{\mu}{40}t^5$$
$$x_2(t) = x_2(0) + \frac{1}{4}t^2 + \frac{\mu - 1}{6}t^3 - \frac{\mu}{8}t^4$$

By enforcing feasibility ($x_1(t_f) = 2$), we get $x_2(0)$ and, successively, $x_2(t_f)$. More precisely

$$x_2(0) = \frac{23}{24} - \frac{\mu}{60}, \quad x_2(t_f) = \frac{25}{24} + \frac{\mu}{40}$$

Table 4.3 The values of the performance index resulting from the solutions of Problems 3.2, 3.13, 4.4 and corresponding to various choices of the parameters $\beta = \varphi = \mu$

	J		
	Problem 3.2	Problem 3.13	Problem 4.4
$\beta = \varphi = \mu = 0$	7.50	2.87	1.50
$\beta = \varphi = \mu = 5$	3.51	2.00	1.49
$\beta = \varphi = \mu = 30$	1.93	1.47	1.43

The state trajectories corresponding to some values of μ are reported in Fig. 4.6.

We note that the problem under consideration is quite similar to Problems 3.2 and 3.13. As a matter of fact, the performance index, the length of the control interval and the constraint on the first state variable at the initial and final time are the same for all of them. The problems differ because of the values of $x_2(0)$ and $x_2(t_f)$. They are given in Problem 3.2, $x_2(t_f)$ is free in Problem 3.13, while both $x_2(0)$ and $x_2(t_f)$ are free in the present problem. Therefore when $\beta = \varphi = \mu$ we expect that the values of the performance index decrease when passing from the first to the third problem. This clearly results from Table 4.3. ∎

Problem 4.5
Let

$$J = \int_0^{t_f} l(x(t), u(t), t) \, dt, \quad l(x, u, t) = \frac{u^2}{2}, \quad u(t) \in R, \ \forall t, \quad t_f = 1$$

$$x(0) \in \overline{S}_0 = \{x | \alpha_0(x) = 0\}, \quad \alpha_0(x) = x_2 - x_{20}$$

$$x(t_f) \in \overline{S}_f = \{x | \alpha_f(x) = 0\}, \quad \alpha_f(x) = x_1^2 + (x_2 - 2)^2 - 1$$

where $x_{20} \geq 0$ is the given initial value of the second state variable.

The two sets $\overline{S}_0$ and $\overline{S}_f$ of the admissible initial and final states are regular varieties. In fact for the first one we have

$$\Sigma_0(x) = \frac{d\alpha_0(x)}{dx} = \begin{bmatrix} 0 & 1 \end{bmatrix}$$

and $\text{rank}(\Sigma_0(x)) = 1$, $\forall x \in \overline{S}_0$. Analogously, for the second one we have

$$\Sigma_f(x) = \frac{d\alpha_f(x)}{dx} = \begin{bmatrix} 2x_1 & 2(x_2 - 2) \end{bmatrix}$$

and $\text{rank}(\Sigma_f(x)) = 1$, $\forall x \in \overline{S}_f$ because, if $x_1 = 0$, then $x_2 \neq 2$. The set $\overline{S}_f$ is shown in Fig. 4.7.

Fig. 4.7 Problem 4.5. State trajectories corresponding to various x_{20}s

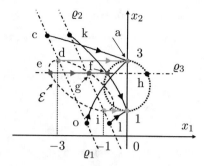

The hamiltonian function and the $H-$minimizing control are

$$H = \frac{u^2}{2} + \lambda_1 x_2 + \frac{\lambda_2}{2} u^2, \quad u_h = -\lambda_2$$

Therefore Eq. (2.5a) and its general solution over a time interval beginning at time 0 are

$$\dot{\lambda}_1(t) = 0 \qquad \lambda_1(t) = \lambda_{10}$$
$$\dot{\lambda}_2(t) = -\lambda_1(t) \qquad \lambda_2(t) = \lambda_{20} - \lambda_{10}t$$

and the orthogonality condition at the initial time

$$\begin{bmatrix} \lambda_1(0) \\ \lambda_2(0) \end{bmatrix} = \begin{bmatrix} \lambda_{10} \\ \lambda_{20} \end{bmatrix} = \vartheta_0 \frac{d\alpha_0(x)}{dx} \bigg|'_{x=x(0)} = \vartheta_0 \begin{bmatrix} 0 \\ 1 \end{bmatrix}$$

implies $\lambda_{10} = 0$, so that the control $u(\cdot)$ is constant, equal to $-\lambda_{20}$ and

$$x_1(t) = x_1(0) + x_{20}t - \frac{\lambda_{20}}{2}t^2, \qquad x_2(t) = x_{20} - \lambda_{20}t$$

The orthogonality condition at the final time is

$$\begin{bmatrix} \lambda_1(t_f) \\ \lambda_2(t_f) \end{bmatrix} = \begin{bmatrix} \lambda_{10} \\ \lambda_{20} - \lambda_{10}t_f \end{bmatrix} = \begin{bmatrix} 0 \\ \lambda_{20} \end{bmatrix} = \vartheta_f \frac{d\alpha_f(x)}{dx} \bigg|'_{x=x(t_f)} = \vartheta_f \begin{bmatrix} 2x_1(t_f) \\ 2(x_2(t_f) - 2) \end{bmatrix}$$

where the obtained value for λ_{10} has been taken into account. This relation is consistent with three cases:

Case 1. $\vartheta_f = 0, \quad x_1(t_f) = 0$

Case 2. $\vartheta_f = 0, \quad x_1(t_f) \neq 0$

Case 3. $\vartheta_f \neq 0, \quad x_1(t_f) = 0$

which will be sequentially examined.

Case 1. $\vartheta_f = 0$, $x_1(t_f) = 0$

We have the following consequences

$$\vartheta_f = 0 \Rightarrow \lambda_{20} = 0 \Rightarrow u(\cdot) = 0 \Rightarrow \begin{cases} x_2(t_f) = x_{20} \\ x_1(t_f) = x_{20}t_f + x_1(0) \end{cases}$$

$$x_1(t_f) = 0 \Rightarrow x_2(t_f) = \begin{cases} 1 \\ 3 \end{cases}$$

Thus either $x_{20} = 1$ and $x_1(0) = -1$ or $x_{20} = 3$ and $x_1(0) = -3$. This solution is surely optimal because $J = 0$, the smallest possible value.

Case 2. $\vartheta_f = 0$, $x_1(t_f) \neq 0$

In the light of previous achievements (in particular $u(\cdot) = 0$), we have $x_2(t_f) = x_{20}$, $x_1(t_f) = x_1(0)t_f + x_{20}$. Then, by enforcing feasibility $(x_1^2(t_f) = 1 - (x_2(t_f) - 2)^2)$, we get

$$x_1(0) = -x_{20} \pm \sqrt{1 - (x_{20} - 2)^2}$$

It is obvious that the case under consideration can occur only if $1 - (x_{20} - 2)^2 \geq 0$, that is only if $1 \leq x_{20} \leq 3$. Within this interval, the above equation for $x_1(0)$ is the equation of the ellipsis $\mathcal{E}$ shown in Fig. 4.7 (dotted line). When $1 \leq x_{20} \leq 3$, it is easy to ascertain that we have two solutions. Again, they are certainly optimal because $J = 0$. They are specified by the values of $x_1(0)$ which correspond to the intersection of $\mathcal{E}$ with the straight line ϱ_3 defined by the equation $x_2 = x_{20}$.

Case 3. $x_1(t_f) = 0$, $\vartheta_f \neq 0$

First of all observe that feasibility $(x_1^2(t_f) = 1 - (x_2(t_f) - 2)^2)$ requires either $x_2(t_f) = 3$ or $x_2(t_f) = 1$. Furthermore by recalling that $u(t) = -\lambda_2(t) = -\lambda_{20}$, we get the following implications

$$x_2(t_f) = 3 \Rightarrow \begin{cases} \lambda_{20} = x_{20} - 3 \\ x_1(0) = -\dfrac{x_{20} + 3}{2} \end{cases}$$

$$x_2(t_f) = 1 \Rightarrow \begin{cases} \lambda_{20} = x_{20} - 1 \\ x_1(0) = -\dfrac{x_{20} + 1}{2} \end{cases}$$

The above relations between x_{20} and $x_1(0)$ define the two straight lines ϱ_1 and ϱ_2 shown in Fig. 4.7. Thus, corresponding to any given value of x_{20}, $x_{20} > 3$ or $x_{20} < 1$, two solutions there exist. They are specified, by the values of $x_1(0)$ resulting from the intersections of the straight line $x_2 = x_{20}$ with ϱ_1 and ϱ_2. By computing the value of the performance index, we conclude that the best solution is the one where $x_2(t_f) = 1$, if $x_{20} < 2$, while is the one where $x_2(t_f) = 3$, if $x_{20} > 2$.

As a result of the preceding discussion we claim that the solution of the problem depends on x_{20} in the following way

$$x_{20} < 1 \qquad u(\cdot) = 1 - x_{20} \quad \begin{cases} x_1(0) = -\dfrac{x_{20} + 1}{2} \\ x_1(t_f) = 0 \\ x_2(t_f) = 1 \end{cases}$$

$$x_{20} = 1 \qquad u(\cdot) = 0 \quad \begin{cases} x_1(0) = -1 \\ x_1(t_f) = 0 \\ x_2(t_f) = 1 \end{cases}$$

$$1 < x_{20} < 3 \quad u(\cdot) = 0 \quad \begin{cases} x_1(0) = -x_{20} \pm \sqrt{1 - (x_{20} - 2)^2} \\ x_1(t_f) = \pm\sqrt{1 - (x_{20} - 2)^2} \\ x_2(t_f) = x_{20} \end{cases}$$

$$x_{20} = 3 \qquad u(\cdot) = 0 \quad \begin{cases} x_1(0) = -3 \\ x_1(t_f) = 0 \\ x_2(t_f) = 3 \end{cases}$$

$$x_{20} > 3 \qquad u(\cdot) = 3 - x_{20} \quad \begin{cases} x_1(0) = -\dfrac{x_{20} + 3}{2} \\ x_1(t_f) = 0 \\ x_2(t_f) = 3 \end{cases}$$

The shapes of the corresponding trajectories are shown in Fig. 4.7. More in detail:
If $x_{20} > 3$ we have trajectories of the kind (c–a).
If $x_{20} = 3$ we have the trajectory (d–a).
If $1 < x_{20} < 3$ we have trajectories of the kind (e–f) or (g–h).
If $x_{20} = 1$ we have the trajectory (i–b).
If $x_{20} < 1$ we have trajectories of the kind (l–b).

The number of solutions which satisfy the NC are as follows:
If either $x_{20} < 1$ or $x_{20} > 3$, then we have two solutions which are characterized by

$$\begin{cases} x_1(0) = -\dfrac{x_{20} + 1}{2} \\ x_2(t_f) = 1 \end{cases} \qquad \begin{cases} x_1(0) = -\dfrac{x_{20} + 3}{2} \\ x_2(t_f) = 3 \end{cases}$$

If $x_{20} = 1$, then we have one solution which is characterized by $x_1(0) = -1$, $x_2(t_f) = 1$.

If $1 < x_{20} < 3$, then we have four solutions which are characterized by

$$\begin{cases} x_1(0) = -x_{20} \pm \sqrt{1 - (x_{20} - 2)^2} \\ x_2(t_f) = x_{20} \end{cases}$$

$$\begin{cases} x_1(0) = -\dfrac{x_{20} + 1}{2} \\ x_2(t_f) = 1 \end{cases} \qquad \begin{cases} x_1(0) = -\dfrac{x_{20} + 3}{2} \\ x_2(t_f) = 3 \end{cases}$$

If $x_{20} = 3$, then we have one solution which is characterized by $x_1(0) = -3$, $x_2(t_f) = 3$.

Problem 4.6

Let

$$x(0) \in \overline{S}_0 = \{x | \alpha_0(x) = 0\}, \quad \alpha_0(x) = x_1 + 1$$

$$x(t_f) \in \overline{S}_f = \{x | \alpha_f(x) = 0\}, \quad \alpha_f(x) = x_1 - 1$$

$$J = \int_0^{t_f} l(x(t), u(t), t)\, dt, \quad l(x, u, t) = \frac{u^2}{2} + \beta x_1 + \gamma x_2, \ u(t) \in R, \ \forall t, \quad t_f = 1$$

where β and γ are two given parameters.

The sets $\overline{S}_0$ and $\overline{S}_f$ of the admissible initial and final states are shown in Fig. 4.8. They are regular varieties. In fact we have for the first one

$$\Sigma_0(x) = \frac{d\alpha_0(x)}{dx} = \begin{bmatrix} 1 & 0 \end{bmatrix}$$

Fig. 4.8 Problem 4.6. State trajectories corresponding to some values of β

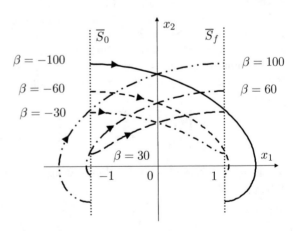

and rank$(\Sigma_0(x)) = 1, \forall x \in \overline{S}_0$. Analogously, we have for the second one

$$\Sigma_f(x) = \frac{d\alpha_f(x)}{dx} = \begin{bmatrix} 1 & 0 \end{bmatrix}$$

and rank$(\Sigma_f(x)) = 1, \forall x \in \overline{S}_f$.

The hamiltonian function and the $H-$minimizing control are

$$H = \frac{u^2}{2} + \beta x_1 + \gamma x_2 + \lambda_1 x_2 + \lambda_2 u, \quad u_h = -\lambda_2$$

Then Eq. (2.5a) and its general solution over an interval beginning at time 0 are

$$\dot{\lambda}_1(t) = -\beta \qquad\qquad \lambda_1(t) = \lambda_{10} - \beta t$$
$$\dot{\lambda}_2(t) = -\gamma - \lambda_1(t) \qquad\qquad \lambda_2(t) = \lambda_{20} - (\lambda_{10} + \gamma)t + \frac{\beta}{2}t^2$$

and the orthogonality condition at the initial time

$$\begin{bmatrix} \lambda_1(0) \\ \lambda_2(0) \end{bmatrix} = \begin{bmatrix} \lambda_{10} \\ \lambda_{20} \end{bmatrix} = \vartheta_0 \left. \frac{d\alpha_0(x)}{dx} \right|'_{x=x(0)} = \vartheta_0 \begin{bmatrix} 1 \\ 0 \end{bmatrix}$$

implies $\lambda_{20} = 0$, while the one at the final time

$$\begin{bmatrix} \lambda_1(t_f) \\ \lambda_2(t_f) \end{bmatrix} = \begin{bmatrix} \lambda_{10} \\ \lambda_{20} - (\lambda_{10} + \gamma)t_f + \frac{\beta}{2}t_f^2 \end{bmatrix} = \begin{bmatrix} \lambda_{10} \\ -(\lambda_{10} + \gamma) + \frac{\beta}{2} \end{bmatrix} = $$
$$= \vartheta_f \left. \frac{d\alpha_f(x)}{dx} \right|'_{x=x(t_f)} = \vartheta_f \begin{bmatrix} 1 \\ 0 \end{bmatrix}$$

entails $\lambda_2(t_f) = 0$. From the two conditions we have

$$\lambda_{10} = \frac{\beta}{2} - \gamma$$

so that

$$\lambda_2(t) = \frac{\beta}{2}(t - 1)t$$

In view of the form of the $H-$minimizing control, the state motion of the system is

$$x_1(t) = -1 + x_2(0)t + \frac{\beta}{12}t^3 - \frac{\beta}{24}t^4$$
$$x_2(t) = x_2(0) + \frac{\beta}{4}t^2 - \frac{\beta}{6}t^3$$

By enforcing feasibility $(x_1(t_f) = 1)$ we obtain

$$x_2(0) = 2 - \frac{\beta}{24}, \quad x_2(t_f) = 2 + \frac{\beta}{24}$$

Note that both the initial and final state are functions of the parameter β only. As this parameter increases, the second component of the state decreases at $t = 0$ and increases at $t = t_f$. The solution of the problem is not affected by the value of the second parameter γ. This is a consequence of the fact that the integral of $\gamma x_2(t) = \gamma \dot{x}_1(t)$ does not depend on $x_2(\cdot)$, because its value over the interval $[0, t_f]$ is $\gamma(x_1(1) - x_1(0)) = 2\gamma$, whatever $u(\cdot)$ is.

The state trajectories corresponding to some values of β are reported in Fig. 4.8.

∎

4.3 $x(t_f) = $ Not Given, $(t_0, t_f) = $ Free

In this section we present four problems. In all of them the function $l(\cdot, \cdot, \cdot)$ which defines the integral term of the performance index depends on time. It depends on time in the first and third problem only, also on the state in the second problem and, finally, also on the control in the last problem. Furthermore an infinite number of solutions which satisfy the NC can be found in Problem 4.7, while one or more solutions are possible for Problem 4.8, according to the value of the parameter β in the performance index. Finally, the solution is not unique also in the first of the two subproblems dealt with in Problem 4.10.

Problem 4.7
Let

$$x(t_0) \in \overline{S}_0 = \{x \mid \alpha_0(x) = 0\}, \quad \alpha_0(x) = x_2 - 1$$
$$x(t_f) \in \overline{S}_f = \{x \mid \alpha_f(x) = 0\}, \quad \alpha_f = x_2 + 1$$
$$J = \int_{t_0}^{t_f} l(x(t), u(t), t) \, dt, \quad l(x, u, t) = t^2$$
$$u(t) \in \overline{U}, \ \forall t, \quad \overline{U} = \{u \mid -1 \le u \le 1\}$$

where the initial (t_0) and final (t_f) times are free.

The sets $\overline{S}_0$ and $\overline{S}_f$ of the admissible initial and final states are shown in Fig. 4.9. They are regular varieties. In fact, with reference to $\overline{S}_0$, we have

$$\Sigma_0(x) = \frac{d\alpha_0(x)}{dx} = [0 \ 1]$$

Fig. 4.9 Problem 4.7. The regular varieties $\overline{S}_0$ and $\overline{S}_f$ (*dotted lines*) and some state trajectories (*solid lines*)

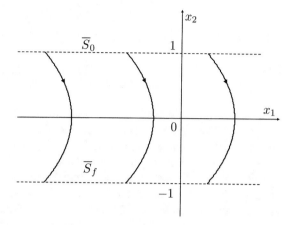

and $\mathrm{rank}(\Sigma_0(x)) = 1$, $\forall x \in \overline{S}_0$. Analogously, with reference to $\overline{S}_f$, it results

$$\Sigma_f(x) = \frac{d\alpha_f(x)}{dx} = \begin{bmatrix} 0 & 1 \end{bmatrix}$$

and $\mathrm{rank}(\Sigma_f(x)) = 1$, $\forall x \in \overline{S}_f$.

The hamiltonian function and the H−minimizing control are

$$H = t^2 + \lambda_1 x_2 + \lambda_2 u, \quad u_h = -\mathrm{sign}\,(\lambda_2)$$

so that Eq. (2.5a) and its general solution over a time interval beginning at t_0 are

$$\begin{aligned} \dot{\lambda}_1(t) &= 0 & \lambda_1(t) &= \lambda_1(t_0) \\ \dot{\lambda}_2(t) &= -\lambda_1(t) & \lambda_2(t_0) &- \lambda_1(t_0)(t - t_0) \end{aligned}$$

The orthogonality condition at the initial time

$$\begin{bmatrix} \lambda_1(t_0) \\ \lambda_2(t_0) \end{bmatrix} = \vartheta_0 \left. \frac{d\alpha_0(x)}{dx} \right|'_{x=x(t_0)} = \vartheta_0 \begin{bmatrix} 0 \\ 1 \end{bmatrix}$$

implies $\lambda_1(t_0) = 0$, whereas the orthogonality condition at the final time

$$\begin{bmatrix} \lambda_1(t_f) \\ \lambda_2(t_f) \end{bmatrix} = \vartheta_f \left. \frac{d\alpha_f(x)}{dx} \right|'_{x=x(t_f)} = \vartheta_f \begin{bmatrix} 0 \\ 1 \end{bmatrix}$$

entails $\lambda_1(t_f) = 0$.

Thus $\lambda_2(\cdot) = \lambda_2(t_0)$ and the control is constant.

In the light of these results, the transversality conditions at the initial and final times are

$$H(t_0) = t_0^2 + \lambda_1(t_0)x_2(t_0) + \lambda_2(t_0)u(t_0) = t_0^2 - |\lambda_2(t_0)| = 0$$
$$H(t_f) = t_f^2 + \lambda_1(t_f)x_2(t_f) + \lambda_2(t_f)u(t_f) = t_f^2 - |\lambda_2(t_f)| = t_f^2 - |\lambda_2(t_0)| = 0$$

so that

$$t_0 = -\sqrt{|\lambda_2(t_0)|}, \quad t_f = \sqrt{|\lambda_2(t_0)|}$$

Note that $\lambda_2(t_0) \neq 0$ because, otherwise, the two transversality conditions imply $t_0 = t_f = 0$ and the constraints on the initial and final states can not be satisfied. When $\lambda_2(t_0) = 0$ we have $t_0 \neq 0$ and $t_f \neq 0$ only if $\lambda_0 = 0$ (pathologic problem) and condition (v) of Theorem 2.1 is violated.

Also note that the control $u(\cdot) = 1$ is not feasible as it implies $x_2(t) = 1 + (t - t_0)$ and $x_2(t_f) \neq -1$, because $t_f \geq t_0$. Thus $u(\cdot) = -1$, so that $x_2(t) = 1 - (t - t_0)$ and, enforcing feasibility, we obtain $t_f - t_0 = 2$. By recalling the expressions for t_f and t_0, it follows that $2\sqrt{|\lambda_2(t_0)|} = 2$. Therefore

$$t_f = 1, \quad t_0 = -1$$

Furthermore note that any value can be given to $x_1(t_0)$: we conclude that an infinite number of solutions which satisfy the NC can be found. This result is quite obvious because both the performance index and the constraints on the initial and final states do not depend, either directly or indirectly, on the first state variable.

Some of the possible state trajectories (lines (a–b), (c–d), (e–f)) are reported in Fig. 4.9.

Finally, we observe that when the term $\bar{l}(\cdot, \cdot, \cdot), \bar{l}(x, u, t) = \beta x_1, \beta \neq 0$ is added to $l(\cdot, \cdot, \cdot)$ in the performance index, no solution exists. In fact the hamiltonian function and the equation for $\lambda_1(\cdot)$ are now

$$H = t^2 + \beta x_1 + \lambda_1 x_2 + \lambda_2 u, \quad \dot{\lambda}_1(t) = -\beta$$

and it is not possible to satisfy the two orthogonality conditions which require $\lambda_1(t_0) = \lambda_1(t_f) = 0$ because $\beta \neq 0$. These conditions are not violated if $\dot{\lambda}_1(t) = 0$ and this can happen only if $\lambda_0 = 0$, that is only if the problem is pathological. In this case, however, the transversality conditions require $\lambda_2(\cdot) = 0$, against point (v) of Theorem 2.1.   ∎

Problem 4.8
Let

$$x(t_0) \in \overline{S}_0 = \{x \,|\, \alpha_0(x) = 0\}, \quad \alpha_0(x) = x_2$$

$$x(t_f) \in \overline{S}_f = \{x \,|\, \alpha_f(x) = 0\}, \quad \alpha_f(x) = \frac{x_1^2}{2} - x_2 + 1$$

$$J = \int_{t_0}^{t_f} l(x(t), u(t), t)\, dt, \quad l(x, u, t) = t^2 + \beta x_2$$

$$u(t) \in \overline{U}, \ \forall t, \quad \overline{U} = \{u \,|\, -1 \leq u \leq 1\}$$

Fig. 4.10 Problem 4.8. State trajectories if $\beta = -3$. The trajectory is (a–f) when $u(\cdot) = u_1(\cdot)$, (c–d–f) when $u(\cdot) = u_3(\cdot)$ and the solution #1 is enforced, (b–f) when $u(\cdot) = u_3(\cdot)$ and the solution #2 is enforced

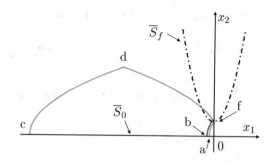

where both the initial (t_0) and final (t_f) times are free. The parameter β is given.

The sets $\overline{S}_0$ and $\overline{S}_f$ of the admissible initial and final states are shown both in Figs. 4.10 and 4.11. They are regular varieties. In fact, as for the first set, we have

$$\Sigma_0(x) = \frac{d\alpha_0(x)}{dx} = \begin{bmatrix} 0 & 1 \end{bmatrix}$$

and rank$(\Sigma_0(x)) = 1, \forall x \in \overline{S}_0$. In a similar way, we have, for the second set,

$$\Sigma_f(x) = \frac{d\alpha_f(x)}{dx} = \begin{bmatrix} x_1 & -1 \end{bmatrix}$$

and rank$(\Sigma_f(x)) = 1, \forall x \in \overline{S}_f$.

The hamiltonian function and the H−minimizing control are

$$H = t^2 + \beta x_2 + \lambda_1 x_2 + \lambda_2 u, \quad u_h = -\text{sign}(\lambda_2)$$

Fig. 4.11 Problem 4.8. State trajectories when $\beta = 3$. If $u(\cdot) = u_1(\cdot)$ the trajectory is (a–f), if $u(\cdot) = u_4(\cdot)$ and the solution #1 is enforced the trajectory is (c–d–f)

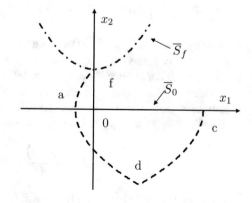

so that Eq. (2.5a) and its general solution over an interval beginning at time t_0 are

$$\dot{\lambda}_1(t) = 0 \qquad\qquad \lambda_1(t) = \lambda_1(t_0)$$
$$\dot{\lambda}_2(t) = -\lambda_1(t) - \beta \qquad \lambda_2(t) = \lambda_2(t_0) - (\lambda_1(t_0) + \beta)(t - t_0)$$

The orthogonality condition at the initial time

$$\begin{bmatrix} \lambda_1(t_0) \\ \lambda_2(t_0) \end{bmatrix} = \vartheta_0 \left. \frac{d\alpha_0(x)}{dx} \right|'_{x=x(t_0)} = \vartheta_0 \begin{bmatrix} 0 \\ 1 \end{bmatrix}$$

entails $\lambda_1(t_0) = 0$, so that $\lambda_1(\cdot) = 0$ and $\lambda_2(t) = \lambda_2(t_0) - \beta(t - t_0)$. Thus $\lambda_2(\cdot)$ is linear with respect to the time t and, as far as the control $u(\cdot)$ is concerned, we have to consider the following four *cases* only.

Case 1
$$u(t) = u_1(t) \qquad\qquad u_1(t) := 1 \qquad\qquad t_0 \le t \le t_f$$
Case 2
$$u(t) = u_2(t) \qquad\qquad u_2(t) := -u_1(t) \qquad\quad t_0 \le t \le t_f$$
Case 3
$$u(t) = u_3(t) \qquad\qquad u_3(t) := \begin{cases} 1 \\ -1 \end{cases} \qquad \begin{matrix} t_0 \le t < \tau \\ \tau < t \le t_f \end{matrix}$$
Case 4
$$u(t) = u_4(t) \qquad\qquad u_4(t) := -u_3(t) \qquad\quad t_0 \le t \le t_f$$

In the above expressions τ, $t_0 < \tau < t_f$, is a suitable time to be found. The control switches at this time only if the sign of the function $\lambda_2(\cdot)$ changes, so that $\lambda_2(\tau) = 0$.

No matter what the control action is, the orthogonality condition at the final time is

$$\begin{bmatrix} \lambda_1(t_f) \\ \lambda_2(t_f) \end{bmatrix} = \begin{bmatrix} 0 \\ \lambda_2(t_0) - \beta(t_f - t_0) \end{bmatrix} = \vartheta_f \left. \frac{d\alpha_f(x)}{dx} \right|'_{x=x(t_f)} = \vartheta_f \begin{bmatrix} x_1(t_f) \\ -1 \end{bmatrix}$$

where the result concerning $\lambda_1(\cdot)$ has been taken into account. It follows that $x_1(t_f) = 0$ and $x_2(t_f) = 1$, in view of the request $x(t_f) \in \bar{S}_f$. Observe that we do not need to consider the occurrence $\vartheta_f = 0$ which entails $\lambda_2(t_f) = 0$. In fact, if $\lambda_2(t_f) = 0$, no switching of the control can occur and we have to discuss *Case 1* or *Case 2* only. Finally, note that we can easily find the value of the performance index. Indeed by recalling that $x_2(t) = \dot{x}_1(t)$ and $x_1(t_f) = 0$, we have

$$J = \frac{t_f^3 - t_0^3}{3} - \beta x_1(t_0)$$

We now proceed to examine the four possible control functions. The objective is to find which one among them is consistent with the NC.

Case 1. $u(t) = u_1(t)$, $u_1(t) := 1$, $t_0 \leq t \leq t_f$

First, we observe that $x_2(t) = x_2(t_0) + t - t_0 = t - t_0$. Then the transversality condition at the initial time is

$$H(t_0) = t_0^2 + \beta x_2(t_0) + \lambda_1(t_0)x_2(t_0) + \lambda_2(t_0)u(t_0) = t_0^2 + \lambda_2(t_0) = 0$$

In writing this equation we have taken into account that $\lambda_1(t_0) = x_2(t_0) = 0$ and $u(t_0) = 1$. It implies $t_0^2 = -\lambda_2(t_0)$, because $\lambda_2(t_0) \leq 0$, in accordance with $u(t_0) = 1$. Analogously, the transversality condition at the final time is

$$H(t_f) = t_f^2 + \beta x_2(t_f) + \lambda_1(t_f)x_2(t_f) + \lambda_2(t_f)u(t_f) =$$
$$= t_f^2 + \beta(t_f - t_0) + \lambda_2(t_0) - \beta(t_f - t_0) = t_f^2 + \lambda_2(t_0) = 0$$

provided that what has been found for $\lambda_1(\cdot), \lambda_2(\cdot), x_2(\cdot)$ is taken into account. Therefore $t_f^2 = -\lambda_2(t_0)$. Finally, by enforcing feasibility $(x_2(t_f) = 1)$ and writing down the expression for $x_1(\cdot)$, that is

$$x_1(t) = x_1(t_0) + \frac{(t - t_0)^2}{2}$$

all the parameters of the solution can easily be determined. We get

$$t_0 = -\frac{1}{2}, \quad t_f = \frac{1}{2}, \quad \lambda_2(t_0) = -\frac{1}{4}, \quad x_1(t_0) = -\frac{1}{2}$$

and, for the sake of completeness,

$$J = \frac{t_f^3 - t_0^3}{3} - \beta x_1(t_0) = \frac{1}{12} + \frac{\beta}{2}$$

Note that the solution does not depend on β, whereas the value of the performance index does. This outcome can also be verified in Table 4.4 when β is negative and in Table 4.5 when β is positive. The state trajectories (a–f) are plotted in Figs. 4.10 and 4.11 when $\beta = -3$ and $\beta = 3$, respectively.

Case 2. $u(t) = u_2(t)$, $u_2(t) := -u_1(t)$

The control $u(\cdot) = u_2(\cdot)$ is not feasible, because $x_2(t) \leq 0, t \geq t_0$.

Case 3. $u(t) = u_3(t)$, $u_3(t) := \begin{cases} 1, & t_0 \leq t < \tau \\ -1, & \tau < t \leq t_f \end{cases}$

The transversality condition at the initial time is the same of *Case 1*, so that

$$t_0^2 = -\lambda_2(t_0)$$

Table 4.4 Problem 4.8

		$u(\cdot) = u_1(\cdot)$	$u(\cdot) = u_3(\cdot)$	
			Sol. #1	Sol. #2
$\beta = -5$	t_0	-0.50	-5.78	2.51
	t_f	0.50	6.59	4.04
	τ	–	0.90	3.78
	$x_1(t_0)$	-0.5	-44.20	-1.10
	J	-2.42	-61.18	11.20
$\beta = -3$	t_0	-0.50	-3.70	1.83
	t_f	0.50	4.44	3.06
	τ	–	0.87	2.94
	$x_1(t_0)$	-0.50	-20.40	-0.74
	J	-1.42	-15.09	5.26
$\beta = -2$	t_0	-0.50	-2.63	1.42
	t_f	0.50	3.31	2.46
	τ	–	0.84	2.44
	$x_1(t_0)$	-0.50	-11.55	-0.53
	J	-0.92	-4.94	2.91
$\beta = -1$	t_0	-0.50	-1.51	–
	t_f	0.50	2.07	–
	τ	–	0.78	–
	$x_1(t_0)$	-0.50	-4.76	–
	J	-0.42	-0.64	–

The parameters which specify the solution when negative values of β are chosen

As already mentioned, if the control switches at time τ then

$$\lambda_2(\tau) = \lambda_2(t_0) - \beta(\tau - t_0) = 0$$

Furthermore we have

$$x_2(t_f) = \tau - t_0 - (t_f - \tau)$$

$$x_1(t_f) = x_1(t_0) + \frac{(\tau - t_0)^2}{2} + (\tau - t_0)(t_f - \tau) - \frac{(t_f - \tau)^2}{2}$$

$$\lambda_2(t_f) = -\beta(t_f - \tau)$$

The transversality condition at the final time is

$$H(t_f) = t_f^2 + \beta x_2(t_f) + \lambda_1(t_f)x_2(t_f) + \lambda_2(t_f)u(t_f) = t_f^2 + \beta + \beta(t_f - \tau) = 0$$

Table 4.5 Problem 4.8

		$u(\cdot) = u_1(\cdot)$	$u(\cdot) = u_4(\cdot)$ Sol. #1	Sol. #2
$\beta = 1$	t_0	-0.50	$-$	$-$
	t_f	0.50	$-$	$-$
	τ	$-$	$-$	$-$
	$x_1(t_0)$	-0.50	$-$	$-$
	J	1.08	$-$	$-$
$\beta = 2$	t_0	-0.50	-1.00	-1.00
	t_f	0.50	1.00	1.00
	τ	$-$	-0.50	-0.50
	$x_1(t_0)$	-0.50	-0.25	-0.25
	J	1.08	1.17	1.17
$\beta = 3$	t_0	-0.50	-2.37	-0.63
	t_f	0.50	2.37	0.63
	τ	$-$	-0.50	-0.50
	$x_1(t_0)$	-0.50	2.98	-0.48
	J	1.58	-0.12	1.62
$\beta = 5$	t_0	-0.50	-4.44	-0.56
	t_f	0.50	4.44	0.56
	τ	$-$	-0.50	-0.50
	$x_1(t_0)$	-0.50	14.00	-0.50
	J	2.58	-16.77	2.60

The parameters which specify the solution when positive values of β are chosen

where we have taken into account what has been found for $\lambda_1(\cdot)$, $\lambda_2(t_f)$ and the fact that $u(t_f) = -1$, $x_2(t_f) = 1$, as required by the orthogonality condition at the final time.

From the four equations

$$x_2(t_f) = \tau - t_0 - (t_f - \tau) = 1$$
$$H(t_f) = t_f^2 + \beta + \beta(t_f - \tau) = 0$$
$$t_0^2 = -\lambda_2(t_0)$$
$$\lambda_2(\tau) = \lambda_2(t_0) - \beta(\tau - t_0) = 0$$

we find the four parameters $\lambda_2(t_0)$, τ, t_0, t_f.

First, we observe that β can not be zero because, if $\beta = 0$, then $t_f = 0$, $\lambda_2(t_0) = 0$, $t_0 = 0$. As a matter of fact, β must be negative as, otherwise, the forth equation is not satisfied because $\tau - t_0 > 0$ and $u(t_0) = 1$ imply $\lambda_2(t_0) \leq 0$. The value of the first state variable at the initial time follows from the equation $x_1(t_f) = 0$.

Two distinct solutions are (numerically) found when $\beta < 0$.

The parameters which characterize the two solutions are reported in Table 4.4. If $\beta = -3$ and the solution #1 is chosen, we obtain the state trajectory (c–d–f) plotted in Fig. 4.10. On the contrary, if the solution #2 is adopted, we get the state trajectory (b–f) plotted in the same figure.

The trajectory which results when the solution #2 is chosen is very similar to the one corresponding to $u(\cdot) = u_1(\cdot)$.

Case 4. $u(t) = u_4(t)$, $u_4(t) := -u_3(t)$

Observe that the request $x_1(t_f) = 0$ implies that the final part of the trajectories must coincide with the final part of the trajectory which results when $u(\cdot) = u_1(\cdot)$. The transversality condition at the initial time is

$$H(t_0) = t_0^2 + \beta x_2(t_0) + \lambda_1(t_0)x_2(t_0) + \lambda_2(t_0)u(t_0) = t_0^2 - \lambda_2(t_0) = 0$$

In writing down this equation we have recalled that $\lambda_1(t_0) = x_2(t_0) = 0$ and $u(t_0) = -1$. It implies $t_0^2 = \lambda_2(t_0)$, whereas the switching of the control at time τ requires, as when discussing *Case 3*,

$$\lambda_2(\tau) = \lambda_2(t_0) - \beta(\tau - t_0) = 0$$

Furthermore we have

$$x_2(t_f) = -(\tau - t_0) + (t_f - \tau)$$

$$x_1(t_f) = x_1(t_0) - \frac{(\tau - t_0)^2}{2} - (\tau - t_0)(t_f - \tau) + \frac{(t_f - \tau)^2}{2}$$

$$\lambda_2(t_f) = -\beta(t_f - \tau)$$

The transversality condition at the final time is

$$H(t_f) = t_f^2 + \beta x_2(t_f) + \lambda_1(t_f)x_2(t_f) + \lambda_2(t_f)u(t_f) = t_f^2 + \beta - \beta(t_f - \tau) = 0$$

where we have taken into account what has been found for $\lambda_1(\cdot)$, $\lambda_2(t_f)$ and the fact that $u(t_f) = 1$, $x_2(t_f) = 1$, as it is required by the orthogonality condition at the final time.

From the four equations

$$x_2(t_f) = -(\tau - t_0) + (t_f - \tau) = 1$$
$$H(t_f) = t_f^2 + \beta - \beta(t_f - \tau) = 0$$
$$t_0^2 = \lambda_2(t_0)$$
$$\lambda_2(\tau) = \lambda_2(t_0) - \beta(\tau - t_0) = 0$$

we compute the four parameters $\lambda_2(t_0)$, τ, t_0, t_f. Remarks similar to those made with reference to *Case 3* lead to the conclusion that β must be positive.

Fig. 4.12 Problem 4.8. The
initial part (b–e–a) of the
state trajectory when $\beta = 3$,
$u(\cdot) = u_4(\cdot)$, and the
solution #2 is enforced

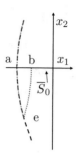

Again, the value of the first state variable at the initial time follows from the
equation $x_1(t_f) = 0$.

Numerically, we find that the four equations above admit a solution only if $\beta \geq 2$.
Two distinct solutions exist if $\beta > 2$, whereas the solution is unique when $\beta = 2$.
The parameters which characterize the two solutions are reported in Table 4.5. If
$\beta = 3$ and the solution #1 is enforced, we obtain the state trajectory (c–d–f) reported
in Fig. 4.11. It is apparent that the final part of the trajectory coincides with the
(whole) trajectory resulting from the choice $u(\cdot) = u_1(\cdot)$. Because of scale reasons
we can not distinguish the trajectory resulting from the choice of the other solution.
Again, the final part of this trajectory coincides with the final part of the trajectory
corresponding to $u(\cdot) = u_1(\cdot)$. If the solution #2 is chosen, the initial part (b–e–a) of
the trajectory is plotted in Fig. 4.12 with a different scale.

By summarizing, we are able to find three solutions if either $\beta > 2$ or $\beta < 0$,
a single solution if $0 \leq \beta < 2$, two solutions if $\beta = 2$. Obviously, whenever
$\beta \notin [0, 2)$ the solution to be adopted is the one characterized by the smallest value
of the performance index. ∎

Problem 4.9
Let

$$x(t_0) = \in \overline{S}_0 = \{x \mid \alpha_0(x) = 0\}, \quad \alpha_0(x) = x_1^2 - x_2 + 1$$
$$x(t_f) = \in \overline{S}_f = \{x \mid \alpha_f(x) = 0\}, \quad \alpha_f(x) = x_2$$
$$J = \int_{t_0}^{t_f} l(x(t), u(t), t) \, dt, \quad l(x, u, t) = t^2$$
$$u(t) \in \overline{U}, \ \forall t, \ \overline{U} = \{u \mid -1 \leq u \leq 1\}$$

where the initial (t_0) and final (t_f) times are free.

The sets $\overline{S}_0$ and $\overline{S}_f$ of the admissible initial and final states are shown in Fig. 4.13.
They are regular varieties. In fact we have, for the first one,

$$\Sigma_0(x) = \frac{d\alpha_0(x)}{dx} = \begin{bmatrix} 2x_1 & -1 \end{bmatrix}$$

Fig. 4.13 Problem 4.9. The regular varieties $\overline{S}_0$ (*dotted line*), $\overline{S}_f$ (coinciding with the x_1 axis) and the state trajectory (a–b) (*solid line*)

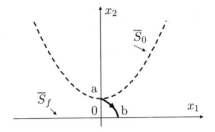

and rank$(\Sigma_0(x)) = 1$, $\forall x \in \overline{S}_0$, whereas, for the second set,

$$\Sigma_f(x) = \frac{d\alpha_f(x)}{dx} = \begin{bmatrix} 0 & 1 \end{bmatrix}$$

and rank$(\Sigma_f(x)) = 1$, $\forall x \in \overline{S}_f$.

The hamiltonian function and the H−minimizing control are

$$H = t^2 + \lambda_1 x_2 + \lambda_2 u, \quad u_h = -\text{sign}(\lambda_2)$$

As a consequence, Eq. (2.5a) and its general solution over an interval beginning at time t_0 are

$$\dot{\lambda}_1(t) = 0 \qquad \lambda_1(t) = \lambda_1(t_0)$$
$$\dot{\lambda}_2(t) = -\lambda_1(t) \quad \lambda_2(t) = \lambda_2(t_0) - \lambda_1(t_0)(t - t_0)$$

The orthogonality condition at the final time

$$\begin{bmatrix} \lambda_1(t_f) \\ \lambda_2(t_f) \end{bmatrix} = \begin{bmatrix} \lambda_1(t_0) \\ \lambda_2(t_0) - \lambda_1(t_0)(t_f - t_0) \end{bmatrix} = \vartheta_f \frac{d\alpha_f(x)}{dx}' \bigg|_{x=x(t_f)} = \vartheta_f \begin{bmatrix} 0 \\ 1 \end{bmatrix}$$

implies $\lambda_1(t_f) = 0$. Thus $\lambda_1(\cdot) = 0$, $\lambda_2(\cdot) = \lambda_2(t_0)$ and the control is constant. The orthogonality condition at the initial time is

$$\begin{bmatrix} \lambda_1(t_0) \\ \lambda_2(t_0) \end{bmatrix} = \begin{bmatrix} 0 \\ \lambda_2(t_0) \end{bmatrix} = \vartheta_0 \frac{d\alpha_0(x)}{dx}' \bigg|_{x=x(t_0)} = \vartheta_0 \begin{bmatrix} 2x_1(t_0) \\ -1 \end{bmatrix}$$

where we have taken into account that $\lambda_1(\cdot) = 0$. We obtain $x_1(t_0) = 0$ and, enforcing feasibility, $x_2(t_0) = 1$. Observe that the resulting initial state is, among those admissibles, the one closest to the set $\overline{S}_f$. The transversality conditions at the initial and final time are

$$H(t_0) = t_0^2 - \lambda_1(t_0)x_2(t_0) + \lambda_2(t_0)u(t_0) = t_0^2 - |\lambda_2(t_0)| = 0$$
$$H(t_f) = t_f^2 - \lambda_1(t_f)x_2(t_f) + \lambda_2(t_f)u(t_f) = t_f^2 - |\lambda_2(t_0)| = 0$$

where we have exploited the achievements concerning $\lambda_1(\cdot)$ and $\lambda_2(\cdot)$ and the fact that $u(t_0) = -\text{sign}(\lambda_2(t_0))$. The above conditions entail

$$t_0 = -\sqrt{|\lambda_2(t_0)|}, \quad t_f = \sqrt{|\lambda_2(t_0)|}$$

Considerations similar to those presented for Problem 4.7 lead to the conclusion that $\lambda_2(t_0) \neq 0$ and, enforcing feasibility, $u(\cdot) = -1$. Therefore $0 = x_2(t_f) = 1 - (t_f - t_0)$, because $x_2(t) = 1 - (t - t_0)$. Thus we have $t_f - t_0 = 1$ and, exploiting the two expressions for the initial and final times, we conclude that $t_f - t_0 = 2\sqrt{|\lambda_2(t_0)|} = 1$ and also

$$t_0 = -0.5, \quad t_f = 0.5, \quad x_1(t) = t - t_0 - \frac{(t - t_0)^2}{2}$$

The resulting state trajectory is plotted in Fig. 4.13.

Finally, observe that the orthogonality condition at the initial time can not be satisfied by $\vartheta_0 = 0$, because, as a consequence, $\lambda_2(t_0) = 0$ and the two transversality conditions would imply $t_0 = t_f = 0$. This result is plainly unfeasible unless the problem is pathological, that is if $\lambda_0 = 0$. However, this event violates condition (v) of Theorem 2.1. ∎

Problem 4.10
Let

$$x(t_0) = \in \overline{S}_0 = \{x \mid \alpha_0(x) = 0\}, \quad \alpha_0(x) = x_2$$

$$J = \int_{t_0}^{t_f} l(x(t), u(t), t) \, dt, \quad l(x, u, t) = t^2 + \beta \left(\frac{u^2}{2} + u \right), \quad u(t) \in R, \ \forall t$$

where the initial (t_0) and final (t_f) times are free and the positive parameter β is given.

The final state $x(t_f)$ belongs to the set

$$\overline{S}_{f_1} = \{x \mid \alpha_{f_1}(x) = 0\}, \quad \alpha_{f_1}(x) = x_2 - 1$$

in Problem 4.10-1, whereas it belongs to the set

$$\overline{S}_{f_2} = \{x \mid \alpha_{f_2}(x) = 0\}, \quad \alpha_{f_2}(x) = x_1^2 - x_2 + 1$$

in Problem 4.10-2. The set $\overline{S}_0$ of the admissible initial states is shown in Figs. 4.14 and 4.15. The sets $\overline{S}_{f_1}$ and $\overline{S}_{f_2}$ are illustrated in the first and second of these figures, respectively.

All sets are regular varieties. In fact we have

$$\Sigma_0(x) = \frac{d\alpha_0(x)}{dx} = \begin{bmatrix} 0 & 1 \end{bmatrix}$$

Fig. 4.14 Problem 4.10-1. The regular varieties $\overline{S}_0$ (coinciding with the x_1 axis) and $\overline{S}_{f_1}$ (*dotted line*). The state trajectory is (a–0) when $\beta = 1$, (b–0) when $\beta = 10$, (c–0) when $\beta = 50$

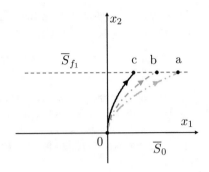

for the set $\overline{S}_0$ and rank$(\Sigma_0(x)) = 1$, $\forall x \in \overline{S}_0$. Analogously, we have

$$\Sigma_{f_1}(x) = \frac{d\alpha_{f_1}(x)}{dx} = \begin{bmatrix} 0 & 1 \end{bmatrix}$$

for the set $\overline{S}_{f_1}$ and rank$(\Sigma_{f_1}(x)) = 1$, $\forall x \in \overline{S}_{f_1}$. Finally, we have

$$\Sigma_{f_2}(x) = \frac{d\alpha_{f_2}(x)}{dx} = \begin{bmatrix} 2x_1 & -1 \end{bmatrix}$$

for the set $\overline{S}_{f_2}$ and rank$(\Sigma_{f_2}(x)) = 1$, $\forall x \in \overline{S}_{f_2}$.

The hamiltonian function and the H−minimizing control are

$$H = t^2 + \beta \left(\frac{u^2}{2} + u \right) + \lambda_1 x_2 + \lambda_2 u, \quad u_h = -\frac{\lambda_2 + \beta}{\beta}$$

for Problems 4.10-1 and 4.10-2. Therefore Eq. (2.5a) and its general solution over an interval beginning at time t_0 are the same for the two problems. More in detail,

$$\dot{\lambda}_1(t) = 0 \qquad\qquad \lambda_1(t) = \lambda_1(t_0)$$
$$\dot{\lambda}_2(t) = -\lambda_1(t) \qquad\qquad \lambda_2(t) = \lambda_2(t_0) - \lambda_1(t_0)(t - t_0)$$

Also the orthogonality condition at the initial time is the same. It is

$$\begin{bmatrix} \lambda_1(t_0) \\ \lambda_2(t_0) \end{bmatrix} = \vartheta_0 \left. \frac{d\alpha_0(x)}{dx} \right|'_{x=x(t_0)} = \vartheta_0 \begin{bmatrix} 0 \\ 1 \end{bmatrix}$$

and implies

$$\lambda_1(t_0) = 0, \quad \lambda_1(\cdot) = 0, \quad \lambda_2(\cdot) = \lambda_2(t_0)$$

In particular, the above equation, together with the expression of the H−minimizing control, entails that $u(\cdot)$ is constant. Furthermore we conclude that the transversality condition at the initial time

$$H(t_0) = t_0^2 + \beta \left(\frac{u^2(t_0)}{2} + u(t_0) \right) + \lambda_1(t_0)x_2(t_0) + \lambda_2(t_0)u(t_0) =$$

$$= t_0^2 - \frac{(\lambda_2(t_0) + \beta)^2}{2\beta} = 0$$

as well as the one at the final time

$$H(t_f) = t_f^2 + \beta \left(\frac{u^2(t_f)}{2} + u(t_f) \right) + \lambda_1(t_f)x_2(t_f) + \lambda_2(t_f)u(t_f) =$$

$$= t_f^2 - \frac{(\lambda_2(t_0) + \beta)^2}{2\beta} = 0$$

are the same for both problems. From these conditions it follows that the initial and final times are given, as functions of $\lambda_2(t_0)$, by

$$t_0 = -\frac{|\lambda_2(t_0) + \beta|}{\sqrt{2\beta}}, \quad t_f = \frac{|\lambda_2(t_0) + \beta|}{\sqrt{2\beta}}$$

Note that, according to the data, it results $t_0 = -t_f$, i.e., the control interval is centered on the time zero.

Finally, also the equations for the state motion

$$x_1(t) = x_1(t_0) - \frac{\lambda_2(t_0) + \beta}{2\beta}(t - t_0)^2$$

$$x_2(t) = -\frac{\lambda_2(t_0) + \beta}{\beta}(t - t_0)$$

which take into account that $x_2(t_0) = 0$, are equal.

We now discuss the two problems.

Problem 4.10-1 $\alpha_{f_1}(x) = x_2 - 1$.

The orthogonality condition at the final time

$$\begin{bmatrix} \lambda_1(t_f) \\ \lambda_2(t_f) \end{bmatrix} = \begin{bmatrix} \lambda_1(t_0) \\ \lambda_2(t_0) - \lambda_1(t_0)(t_f - t_0) \end{bmatrix} = \vartheta_{f_1} \left. \frac{d\alpha_{f_1}(x)}{dx} \right|'_{x=x(t_f)} = \vartheta_{f_1} \begin{bmatrix} 0 \\ 1 \end{bmatrix}$$

leads to the already found conclusions about $\lambda_1(\cdot)$ and $\lambda_2(\cdot)$.

By enforcing feasibility ($x_2(t_f) = 1$), we get

$$\lambda_2(t_0) = -\left(\beta + \sqrt[4]{\frac{\beta^3}{2}} \right)$$

Fig. 4.15 Problem 4.10-2.
The regular varieties $\overline{S}_0$
(coinciding with the x_1 axis)
and $\overline{S}_{f_2}$ (*dotted line*). The
state trajectory is (a–d) when
$\beta = 1$, (b–d) when $\beta = 10$,
(c–d) when $\beta = 50$

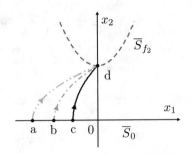

Now observe that the first state component does not appear in none of the equations
for the NC. We conclude that $x_1(t_0)$ is arbitrary, or, in other words, that the problem
admits an infinite number of solutions.

The state trajectories corresponding to various choices for the parameter β are
plotted in Fig. 4.14. In drawing such trajectories we have taken $x_1(t_0) = 0$. Different
values of $x_1(t_0)$ simply entail a shifting along the x_1 axis.

Finally, it is easy to verify that an increase of β implies an increase of the final
time t_f and a decrease of both the initial time t_0 and the (constant) value of $u(\cdot)$.

Problem 4.10-2 $\alpha_{f_2}(x) = x_1^2 - x_2 + 1$.
The orthogonality condition at the final time

$$\begin{bmatrix} \lambda_1(t_f) \\ \lambda_2(t_f) \end{bmatrix} = \begin{bmatrix} \lambda_1(t_0) \\ \lambda_2(t_0) - \lambda_1(t_0)(t_f - t_0) \end{bmatrix} = \vartheta_{f_2} \left. \frac{d\alpha_{f_2}(x)}{dx} \right|'_{x=x(t_f)} = \vartheta_{f_2} \begin{bmatrix} x_1(t_f) \\ -1 \end{bmatrix}$$

implies $x_1(t_f) = 0$, because $\lambda_1(t_0) = 0$. We have, as in Problem 4.10-1, $x_2(t_f) = 1$
because $x(t_f) \in \overline{S}_{f_2}$. It follows that the expression of $\lambda_2(t_0)$ does not change and
that $x_1(t_0)$ is computed by imposing $x_1(t_f) = 0$. We obtain

$$x_1(t_0) = \frac{(\lambda_2(t_0) + \beta)^3}{\beta^2}$$

The state trajectories corresponding to various choices of β are plotted in Fig. 4.15.

It is obvious that the state trajectories have the same shape because the values
of $\lambda_2(t_0)$, t_f, t_0 and $u(\cdot)$ are the same for the two problems. When the parameter β
changes we find that t_f, t_0 and the (constant) value of $u(\cdot)$ vary in the same way as
in Problem 4.10-1, whereas $x_1(t_0)$ decreases as β increases.

Finally, observe that the orthogonality condition at the final time can not be sat-
isfied by letting $\vartheta_{f_2} = 0$ because this event entails $\lambda_2(t_0) = 0$ and $u(\cdot) = -1$ which
is not a feasible control.   ∎

Chapter 5
Simple Constraints: $J = \int +m$, $x(t_0) =$ Given, $x(t_f) =$ Not Given

Abstract The main features of the optimal control problems considered here are: (1) the initial time is zero; (2) the performance index is not of a purely integral type, that is it always includes a function of the final event; (3) the initial state is given; (4) only simple constraints act on the system, namely we only require that (a) the final state is free or belongs to a regular variety, (b) the final time is either free or given, (c) the control variable is free or takes on values in a closed subset of the real numbers. In the considered problems we have the final time given or free.

As in Chaps. 3 and 4 we consider optimal control problems defined on a double integrator which can be handled by making reference to the material in Sect. 2.2.2. Their main features are: (1) the initial time is zero; (2) the performance index is not of a purely integral type, that is it always includes a function of the final event; (3) the initial state is given; (4) only simple constraints act on the system, namely we only require that (a) the final state is free or belongs to a regular variety, (b) the final time is either free or given, (c) the control variable is free or takes on values in a closed subset of the real numbers. In the considered problems we have the final time given (Sect. 5.1) or free (Sect. 5.2).

5.1 $(t_0, t_f) =$ Given

The performance indexes of the first, second and forth problem include both an integral type term and a function of the final state. On the contrary, in the third problem we have a performance index containing a function of the final state only. The initial time is always zero.

© Springer International Publishing Switzerland 2017
A. Locatelli, *Optimal Control of a Double Integrator*, Studies in Systems, Decision and Control 68, DOI 10.1007/978-3-319-42126-1_5

Problem 5.1

Let

$$x(0) = \begin{bmatrix} 1 \\ 1 \end{bmatrix}, \quad x(t_f) = \text{free}, \quad t_f = 1$$

$$J = m(x(t_f), t_f) + \int_0^{t_f} l(x(t), u(t), t)\, dt, \quad u(t) \in R, \quad \forall t$$

$$m(x, t) = \beta x_1, \quad l(x, u, t) = \frac{u^2}{2}$$

where $\beta > 0$ is a given parameter.

The hamiltonian function and the H−minimizing control are

$$H = \frac{u^2}{2} + \lambda_1 x_2 + \lambda_2 u, \quad u_h = -\lambda_2$$

so that Eq. (2.5a) and its general solution over an interval beginning at time 0 are

$$\begin{aligned} \dot{\lambda}_1(t) &= 0 \\ \dot{\lambda}_2(t) &= -\lambda_1(t) \end{aligned} \qquad\qquad \begin{aligned} \lambda_1(t) &= \lambda_{10} \\ \lambda_2(t) &= \lambda_{20} - \lambda_{10} t \end{aligned}$$

The orthogonality condition at the final time is

$$\begin{bmatrix} \lambda_1(t_f) \\ \lambda_2(t_f) \end{bmatrix} - \left.\frac{\partial m(x, t)}{\partial x}\right|'_{x=x(t_f)} = \begin{bmatrix} \lambda_{10} \\ \lambda_{20} - \lambda_{10} t_f \end{bmatrix} - \begin{bmatrix} \beta \\ 0 \end{bmatrix} = \begin{bmatrix} 0 \\ 0 \end{bmatrix}$$

and implies

$$\lambda_{10} = \lambda_{20} = \beta, \quad u(t) = \beta(t - 1)$$

because $t_f = 1$. Therefore

$$x_1(t) = 1 + t - \frac{\beta}{2} t^2 + \frac{\beta}{6} t^3$$

$$x_2(t) = 1 - \beta t + \frac{\beta}{2} t^2$$

The state trajectories corresponding to some values of β are plotted in Fig. 5.1. They show that, as β increases, it is more convenient to make the values of $x_1(\cdot)$ smaller, even negative.   ∎

Fig. 5.1 Problem 5.1. State trajectories when $\beta = 0.5$ (*solid line*), $\beta = 1$ (*dashed line*), $\beta = 5$ (*dash-single-dotted line*)

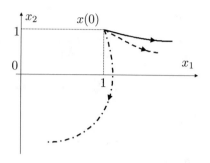

Problem 5.2

Let

$$x(0) = \begin{bmatrix} 0 \\ 0 \end{bmatrix}, \quad x(t_f) = \text{free}, \quad u(t) \in \overline{U}, \; \forall t, \; \overline{U} = \{u| -1 \le u \le 1\}, \quad t_f = 1$$

$$J = m(x(t_f), t_f) + \int_0^{t_f} l(x(t), u(t), t)\,dt, \; m(x, t) = \frac{\beta}{2}x_1^2, \; l(x, u, t) = tu$$

where the value of the parameter β is either 10 or -10.

The hamiltonian function and the H−minimizing control are

$$H = tu + \lambda_1 x_2 + \lambda_2 u, \quad u_h = -\text{sign}(t + \lambda_2)$$

Thus Eq. (2.5a) and its general solution over an interval beginning at time 0 are

$$\dot{\lambda}_1(t) = 0 \qquad\qquad \lambda_1(t) = \lambda_{10}$$
$$\dot{\lambda}_2(t) = -\lambda_1(t) \qquad\qquad \lambda_2(t) = \lambda_{20} - \lambda_{10}t$$

The orthogonality condition at the final time is

$$\begin{bmatrix} \lambda_1(t_f) \\ \lambda_2(t_f) \end{bmatrix} - \frac{\partial m(x, t)}{\partial x}\Bigg|'_{x=x(t_f)} = \begin{bmatrix} \lambda_{10} \\ \lambda_{20} - \lambda_{10}t_f \end{bmatrix} - \begin{bmatrix} \beta x_1(t_f) \\ 0 \end{bmatrix} = \begin{bmatrix} 0 \\ 0 \end{bmatrix}$$

and implies

$$\lambda_2(t_f) = \lambda_{20} - \lambda_{10}t_f = 0, \quad \lambda_{10} = \beta x_1(t_f)$$

The function $\lambda_2(\cdot)$ is linear with respect to t and zero at the final time, thus we only need to consider the three following *cases*. They refer to the initial value λ_{20} of such function.

Case 1. $\lambda_{20} = 0$.

Thanks to the orthogonality condition at the final time, we have $\lambda_{10} = 0$, $\lambda_2(\cdot) = 0$ e $x_1(t_f) = 0$. This is a contradiction, because

$$\lambda_2(t) + t > 0, \ 0 < t \leq t_f \ \Rightarrow \ u(t) = -1, \ 0 < t \leq t_f \ \Rightarrow \ x_1(t_f) < 0$$

Case 2. $\lambda_{20} > 0$.

This case occurs only if $\lambda_{10} > 0$ as well, because the orthogonality condition at the final time requires $\lambda_{20} - \lambda_{10}t_f = 0$. Therefore $\lambda_2(t) + t > 0$, $0 \leq t < t_f$, so that $x_1(t_f) < 0$ and the orthogonality condition can be satisfied only if $\beta = -10$. Correspondingly, we get

$$x_1(t) = -\frac{t^2}{2}, \quad x_2(t) = -t, \quad J = -1.75$$

Case 3. $\lambda_{20} < 0$.

This case occurs only if $\lambda_{10} < 0$ as well (see the orthogonality condition at the final time). Thus $\lambda_2(\tau) + \tau = 0$ for a suitable $\tau \in (0, t_f)$. In fact when $t = 0$ we have $\lambda_2(t) + t = \lambda_{20} + 0 < 0$, whereas, when $t = t_f$, we have $\lambda_2(t) + t = 0 + t_f > 0$. Therefore the control switches from $+1$ to -1 and

$$
\begin{array}{ll}
0 \leq t < \tau & \tau < t \leq t_f \\
u(t) = 1 & u(t) = -1 \\
x_1(t) = \dfrac{t^2}{2} & x_1(t) = \dfrac{\tau^2}{2} + \tau(t - \tau) - \dfrac{(t - \tau)^2}{2} \\
x_2(t) = t & x_2(t) = \tau - (t - \tau)
\end{array}
$$

The switching condition $(\lambda_2(\tau) + \tau = 0)$ together with the above relations among λ_{10}, λ_{20} and $x_1(t_f)$ constitute the system of three equations

$$\lambda_{20} - \lambda_{10}\tau + \tau = 0$$
$$\lambda_{20} - \lambda_{10}t_f = 0$$
$$\lambda_{10} = \beta \left(\frac{\tau^2}{2} + \tau(t_f - \tau) - \frac{(t_f - \tau)^2}{2} \right)$$

We deduce that τ must be a root of the polynomial

$$p(z) = 2\beta z^3 - 6\beta z^2 + (5\beta + 2)z - \beta$$

This root must belong to the interval $(0, t_f)$ and be such that $\beta x_1(t_f) < 0$. If $\beta = -10$ we have two solutions

$$\tau = 0.82, \quad x_1(t_f) = 0.47, \quad J = -0.92$$
$$\tau = 0.33, \quad x_1(t_f) = 0.05, \quad J = -0.40$$

whereas, if $\beta = 10$, we have the unique solution

$$\tau = 0.27, \quad x_1(t_f) = -0.04, \quad J = -0.42$$

In summary:
if $\beta = -10$ three different solutions exist and the best control is $u(\cdot) = -1$;
if $\beta = 10$ only one solution exists and the control is

$$u(t) = 1, \ 0 \le t < \tau, \quad u(t) = -1, \ \tau < t \le t_f, \quad \tau = 0.27.$$

■

Problem 5.3

Let

$$x(0) = \begin{bmatrix} 0 \\ 0 \end{bmatrix}, \quad x(t_f) = \text{free}, \quad t_f = 1, \quad J = m(x(t_f), t_f)$$

$$m(x, t) = \beta \frac{(x_1 - 1)^2}{2} + \frac{x_2^2}{2}, \quad u(t) \in \overline{U}, \ \forall t, \ \overline{U} = \{u| -1 \le u \le 1\}$$

where the positive parameter β is given.
    The hamiltonian function and the H−minimizing control are

$$H = \lambda_1 x_2 + \lambda_2 u, \quad u_h = -\text{sign}\,(\lambda_2)$$

Therefore Eq. (2.5a) and its general solution over an interval beginning at time 0 are

$$\dot{\lambda}_1(t) = 0 \qquad\qquad \lambda_1(t) = \lambda_{10}$$
$$\dot{\lambda}_2(t) = -\lambda_1(t) \qquad\qquad \lambda_2(t) = \lambda_{20} - \lambda_{10}t$$

The function $\lambda_2(\cdot)$ is linear with respect to time, thus the control is piecewise constant and switches at most once at time τ from 1 to −1 (or vice versa). Consistently, only the four following *cases* can occur:

Case 1. $\lambda_2(t) \le 0, \quad 0 \le t \le t_f \Rightarrow u(t) = 1, \ 0 \le t < t_f.$

Case 2. $\lambda_2(t) \ge 0, \quad 0 \le t \le t_f \Rightarrow u(t) = -1, \ 0 \le t < t_f.$

Case 3. $\lambda_2(t) \begin{cases} < 0, & 0 \le t < \tau \\ > 0, & \tau < t \le t_f \end{cases} \Rightarrow u(t) = \begin{cases} 1, & 0 \le t < \tau \\ -1, & \tau < t \le t_f \end{cases}, \tau$ to be found.

Case 4. $\lambda_2(t) \begin{cases} > 0, & 0 \le t < \tau \\ < 0, & \tau < t \le t_f \end{cases} \Rightarrow u(t) = \begin{cases} -1, & 0 \le t < \tau \\ 1, & \tau < t \le t_f \end{cases}, \tau$ to be found.

No matter what the considered case is, the orthogonality condition at the final time is

$$\begin{bmatrix} \lambda_1(t_f) \\ \lambda_2(t_f) \end{bmatrix} - \frac{\partial m(x, t_f)}{\partial x}\Bigg|'_{x=x(t_f)} = \begin{bmatrix} \lambda_{10} \\ \lambda_{20} - \lambda_{10}t_f \end{bmatrix} - \begin{bmatrix} \beta(x_1(t_f) - 1) \\ x_2(t_f) \end{bmatrix} = \begin{bmatrix} 0 \\ 0 \end{bmatrix}$$

and implies

$$\lambda_{10} = \beta(x_1(t_f) - 1), \quad \lambda_{20} = \lambda_{10}t_f + x_2(t_f)$$

We now discuss the four cases above.

Case 1. $\lambda_2(t) \leq 0$, $0 \leq t \leq t_f \Rightarrow u(t) = 1$, $0 \leq t < t_f$.

We have

$$x_1(t) = \frac{t^2}{2}, \quad x_2(t) = t$$

so that, in view of the orthogonality condition, it results

$$x_1(t_f) = \frac{1}{2}, \quad x_2(t_f) = 1, \quad \lambda_{10} = -\frac{\beta}{2}, \quad \lambda_{20} = 1 - \frac{\beta}{2}, \quad \lambda_2(t_f) = 1$$

This is a contradiction because $\lambda_2(t_f) > 0$.

Case 2. $\lambda_2(t) \geq 0$, $0 \leq t \leq t_f \Rightarrow u(t) = -1$, $0 \leq t < t_f$.

We have

$$x_1(t) = -\frac{t^2}{2}, \quad x_2(t) = -t$$

so that, in view of the orthogonality condition, it results

$$x_1(t_f) = -\frac{1}{2}, \quad x_2(t_f) = -1, \quad \lambda_{10} = -\frac{3\beta}{2}, \quad \lambda_{20} = -1 - \frac{3\beta}{2}$$

This is a contradiction because $\lambda_{20} < 0$.

As for the two remaining cases, it is convenient to recall that, if the control switches at time τ, then the switching condition $\lambda_2(\tau) = 0$ must be verified. It implies

$$\lambda_2(\tau) = \lambda_{20} - \lambda_{10}\tau = 0 \Rightarrow \tau = \frac{\lambda_{20}}{\lambda_{10}}$$

Case 3. $\lambda_2(t) \begin{cases} < 0, & 0 \leq t < \tau \\ > 0, & \tau < t \leq t_f \end{cases} \Rightarrow u(t) = \begin{cases} 1, & 0 \leq t < \tau \\ -1, & \tau < t \leq t_f \end{cases}$, τ to be found.

Corresponding to this control function, the state motion is

$$\begin{aligned} &0 \leq t \leq \tau & &\tau \leq t \leq t_f \\ &x_1(t) = \frac{t^2}{2} & &x_1(t) = -\tau^2 + 2\tau t - \frac{t^2}{2} \\ &x_2(t) = t & &x_2(t) = 2\tau - t \end{aligned}$$

Table 5.1 Problem 5.3

β	τ	$x_1(t_f)$	$x_2(t_f)$	λ_{10}	λ_{20}	J
1	0.62	0.36	0.24	−0.64	−0.40	0.24
5	0.79	0.46	0.58	−2.72	−2.15	0.91
20	0.92	0.49	0.84	−10.14	−9.30	2.92

The values of the parameters which characterize the solution as functions of β

so that

$$x(t_f) = -\begin{bmatrix} \dfrac{1}{2} - 2\tau + \tau^2 \\ 1 - 2\tau \end{bmatrix}$$

If we exploit these results into the orthogonality condition and recall the switching condition $(\lambda_2(\tau) = 0)$ we end up to the system of three equations

$$\lambda_{10} = -\beta\left(\tau^2 - 2\tau + \frac{3}{2}\right)$$
$$\lambda_{20} = \lambda_{10} + 2\tau - 1$$
$$\lambda_{20} - \lambda_{10}\tau = 0$$

for the three unknowns $\lambda_{10}, \lambda_{20}, \tau$.

When $\beta = 1, \beta = 5, \beta = 20$ we obtain the results collected in Table 5.1.

Case 4. $\lambda_2(t) \begin{cases} > 0, & 0 \leq t < \tau \\ < 0, & \tau < t \leq t_f \end{cases} \Rightarrow u(t) = \begin{cases} -1, & 0 \leq t < \tau \\ 1, & \tau < t \leq t_f \end{cases}$, τ to be found.

Corresponding to this control function, the state motion is

$$0 \leq t \leq \tau$$
$$x_1(t) = -\frac{t^2}{2}$$
$$x_2(t) = -t$$

$$\tau \leq t \leq t_f$$
$$x_1(t) = \tau^2 - 2\tau t + \frac{t^2}{2}$$
$$x_2(t) = -2\tau + t$$

so that

$$x(t_f) = \begin{bmatrix} \dfrac{1}{2} - 2\tau + \tau^2 \\ 1 - 2\tau \end{bmatrix}$$

If we exploit these results into the orthogonality condition and recall the switching condition $(\lambda_2(\tau) = 0)$, we end up to the system of three equations

$$\lambda_{10} = \beta\left(\tau^2 - 2\tau - \frac{1}{2}\right)$$
$$\lambda_{20} = \lambda_{10} - 2\tau + 1$$
$$\lambda_{20} - \lambda_{10}\tau = 0$$

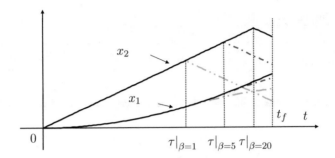

Fig. 5.2 Problem 5.3. State motion when $\beta = 1$ (*dash-double-dotted line*), $\beta = 5$ (*dash-single-dotted line*), $\beta = 20$ (*solid line*)

for the three unknowns λ_{10}, λ_{20}, τ.

When $\beta = 1$ we obtain

$$\tau = 0.16, \quad x_1(t_f) = 0.20, \quad x_2(t_f) = 0.67, \quad \lambda_{10} = -0.80, \quad \lambda_{20} = -0.13, \quad J = 0.55$$

whereas, when $\beta = 5$ or $\beta = 20$, no real solutions exist.

When $\beta = 1$, the value of the performance index for *Case 4* is greater than that for *Case 3*, so that the control of the latter case has to be preferred.

In summary, we ascertain that when the value of β increases we prefer $x_1(t_f)$ is closer to 1 rather than $x_2(t_f)$ is closer to 0. To this regard, see Fig. 5.2 where $x_1(\cdot)$ and $x_2(\cdot)$ are plotted.   ∎

Problem 5.4

Let

$$x(0) = \begin{bmatrix} 0 \\ 0 \end{bmatrix}, \quad x(t_f) = \text{free}, \quad t_f = 1$$

$$J = m(x(t_f), t_f) + \int_0^{t_f} l(x(t), u(t), t)\, dt$$

$$m(x, t) = \frac{(x_1 - 1)^2}{2} + \frac{x_2^2}{2}, \quad l(x, u, t) = \beta \frac{u^2}{2}, \quad u(t) \in R, \ \forall t$$

where the positive parameter β is given.

The hamiltonian function and the H−minimizing control are

$$H = \beta \frac{u^2}{2} + \lambda_1 x_2 + \lambda_2 u, \quad u_h = -\frac{\lambda_2}{\beta}$$

so that Eq. (2.5a) and its general solution over an interval beginning at time 0 are

$$\dot{\lambda}_1(t) = 0 \qquad\qquad \lambda_1(t) = \lambda_{10}$$
$$\dot{\lambda}_2(t) = -\lambda_1(t) \qquad \lambda_2(t) = \lambda_{20} - \lambda_{10}t$$

As a consequence, the state motion and the orthogonality condition at the final time are

$$x_1(t) = -\frac{\lambda_{20}}{2\beta}t^2 + \frac{\lambda_{10}}{6\beta}t^3, \quad x_2(t) = -\frac{\lambda_{20}}{\beta}t + \frac{\lambda_{10}}{2\beta}t^2$$

and

$$\begin{bmatrix} \lambda_1(t_f) \\ \lambda_2(t_f) \end{bmatrix} - \left.\frac{\partial m(x, t_f)}{\partial x}\right|'_{x=x(t_f)} = \begin{bmatrix} \lambda_{10} - (x_1(t_f) - 1) \\ \lambda_{20} - \lambda_{10}t_f - x_2(t_f) \end{bmatrix} =$$

$$= \begin{bmatrix} \lambda_{10} + \dfrac{\lambda_{20}}{2\beta} - \dfrac{\lambda_{10}}{6\beta} + 1 \\[2mm] \lambda_{20} - \lambda_{10} + \dfrac{\lambda_{20}}{\beta} - \dfrac{\lambda_{10}}{2\beta} \end{bmatrix} = \begin{bmatrix} 0 \\ 0 \end{bmatrix}$$

respectively, if we exploit the expressions of $x_1(\cdot)$ and $x_2(\cdot)$. From these equations we get

$$\lambda_{10} = -\frac{12\beta(\beta + 1)}{12\beta^2 + 16\beta + 1}, \quad \lambda_{20} = \frac{\lambda_{10}(2\beta + 1)}{2(\beta + 1)}$$

The state motions corresponding to four values of β are plotted in Fig. 5.3. We notice that, as β increases, the first state variable moves away from 1 at the final time, more and more. On the contrary, the effect on the second state variable is different. At the beginning it moves away from 0, whereas it comes close to this value later on. This is a consequence of the fact that when β increases it is more convenient not to control the system. In other words, we tend to have $u(\cdot) = 0$. On the contrary, when

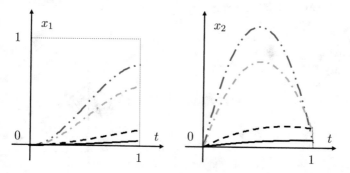

Fig. 5.3 Problem 5.4. State motions corresponding to $\beta = 5$ (*solid line*), $\beta = 1$ (*dashed line*), $\beta = 0.05$ (*dash-single-dotted line*), $\beta = 0.01$ (*dash-double-dotted line*)

β tends to 0 (that is, when the use of the control tends to be not penalized) we can easily check that $x_1(t_f)$ and $x_2(t_f)$ tend to 1 and 0, respectively. ∎

5.2 $t_0 = $ Given, $t_f = $ Free

The function $m(\cdot, \cdot)$ is always included in the performance indexes of the four forth-coming problems where the initial time is zero. Moreover, the integral term is present in Problems 5.5 and 5.7 only, where $m(\cdot, \cdot)$ depends on the final state and on the final event, respectively. The solution is not unique in Problem 5.6, where $m(\cdot, \cdot)$ depends on the final event. Finally, the last problem is worth mentioning because it seems to lead to a paradox. Indeed no solution satisfying the NC of Theorem 2.2 exists, in spite of the fact that an optimal solution does actually exist.

Problem 5.5

Let

$$x(0) = \begin{bmatrix} 0 \\ 0 \end{bmatrix}, \quad x(t_f) \in \overline{S}_f = \{x | \alpha_f(x) = 0\}, \quad \alpha_f(x) = x_1 + x_2 - 1$$

$$J = m(x(t_f), t_f) + \int_0^{t_f} l(x(t), u(t), t)\, dt,$$

$$m(x, t) = x_1 - x_2, \quad l(x, u, t) = 1 + \frac{u^2}{2}, \quad u(t) \in R, \ \forall t$$

where the final time t_f is free.

The set $\overline{S}_f$ of the admissible final states is shown in Fig. 5.4. It is a regular variety because we have

$$\Sigma_f(x) = \frac{d\alpha_f(x)}{dx} = \begin{bmatrix} 1 & 1 \end{bmatrix}$$

and rank$(\Sigma_f(x)) = 1, \forall x \in \overline{S}_f$.

Fig. 5.4 Problem 5.5. The regular variety $\overline{S}_f$ (*dashed line*) and the state trajectory (0–a)

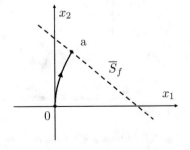

The hamiltonian function and the H−minimizing control are

$$H = 1 + \frac{u^2}{2} + \lambda_1 x_2 + \lambda_2 u, \quad u_h = -\lambda_2$$

Therefore Eq. (2.5a) and its general solution over an interval beginning at time 0 are

$$\dot{\lambda}_1(t) = 0 \qquad\qquad \lambda_1(t) = \lambda_{10}$$
$$\dot{\lambda}_2(t) = -\lambda_1(t) \qquad\qquad \lambda_2(t) = \lambda_{20} - \lambda_{10}t$$

Correspondingly, the state motion is

$$x_1(t) = -\frac{\lambda_{20}}{2}t^2 + \frac{\lambda_{10}}{6}t^3, \quad x_2(t) = -\lambda_{20}t + \frac{\lambda_{10}}{2}t^2$$

so the the control is feasible $(x_1(t_f) + x_2(t_f) = 1)$ if

$$\frac{\lambda_{10}t_f^2 \, (t_f + 3)}{6} - \frac{\lambda_{20}t_f \, (t_f + 2)}{2} = 1$$

The transversality and orthogonality conditions at the final time, the former written at the initial time because the problem is time-invariant, are

$$H(0) = 1 + \frac{u^2(0)}{2} + \lambda_1(0)x_2(0) + \lambda_2(0)u(0) = 1 - \frac{\lambda_{20}^2}{2} = 0$$

and

$$\begin{bmatrix} \lambda_1(t_f) \\ \lambda_2(t_f) \end{bmatrix} - \left.\frac{\partial m(x, t_f)}{\partial x}\right|'_{x=x(t_f)} = \begin{bmatrix} \lambda_{10} \\ \lambda_{20} - \lambda_{10}t_f \end{bmatrix} - \begin{bmatrix} 1 \\ -1 \end{bmatrix} =$$
$$= \vartheta_f \left.\frac{d\alpha_f(x)}{dx}\right|'_{x=x(t_f)} = \vartheta_f \begin{bmatrix} 1 \\ 1 \end{bmatrix}$$

respectively. We find

$$\lambda_{20} = \pm\sqrt{2}, \quad \frac{\lambda_{10} - 1}{\lambda_{20} - \lambda_{10}t_f + 1} = 1$$

and also

$$\lambda_{10} = \frac{\lambda_{20} + 2}{1 + t_f}$$

By substituting these expressions into the one relevant to control feasibility, we conclude that t_f is a root, obviously positive, of the polynomial

$$p(z) := (2 - 2\lambda_{20})z^3 + (6 - 6\lambda_{20})z^2 - (6 + 6\lambda_{20})z - 6$$

When $\lambda_{20} = \sqrt{2}$, $p(\cdot)$ does not possess positive roots, whereas, when $\lambda_{20} = -\sqrt{2}$, a positive root $\bar{z} = 0.53$ exists. By letting $t_f = \bar{z}$ we obtain $\lambda_{10} = 0.38$ and the solution is completely specified. The resulting state trajectory is shown in Fig. 5.4. ∎

Problem 5.6

Let

$$x(0) = \begin{bmatrix} 0 \\ 0 \end{bmatrix}, \quad x(t_f) = \text{free}, \quad J = m(x(t_f), t_f)$$

$$m(x, t) = \frac{1}{2}\left(x_1^2 - x_2^2\right) + t, \quad u(t) \in \overline{U}, \ \forall t, \ \overline{U} = \{u | -1 \le u \le 1\}$$

where the final time t_f is free. Observe that, if the problem admits a solution, then it is certainly not unique. In fact the nature of the performance index is such that if the final time $\bar{t}_f$ and the control $\bar{u}(\cdot)$ which originates the state motion $\bar{x}(\cdot)$ are a (possibly optimal) solution, the same final time and the control $-\bar{u}(\cdot)$ which originates the state motion $-\bar{x}(\cdot)$ constitute a (possibly optimal) solution as well.

The hamiltonian function and the H−minimizing control are

$$H = \lambda_1 x_2 + \lambda_2 u, \quad u_h = -\text{sign}\,(\lambda_2)$$

Therefore Eq. (2.5a) and its general solution over an interval beginning at time 0 are

$$\dot{\lambda}_1(t) = 0 \qquad\qquad\qquad \lambda_1(t) = \lambda_{10}$$
$$\dot{\lambda}_2(t) = -\lambda_1(t) \qquad\qquad \lambda_2(t) = \lambda_{20} - \lambda_{10}t$$

The transversality and orthogonality conditions at the final time are

$$H(t_f) + \left.\frac{\partial m(x(t_f), t)}{\partial t}\right|_{t=t_f} = \lambda_1(t_f)x_2(t_f) + \lambda_2(t_f)u(t_f) + 1 =$$
$$= \lambda_{10}x_2(t_f) + (\lambda_{20} - \lambda_{10}t_f)u(t_f) + 1 = 0$$

and

$$\begin{bmatrix} \lambda_1(t_f) \\ \lambda_2(t_f) \end{bmatrix} - \left.\frac{\partial m(x, t_f)}{\partial x}\right|'_{x=x(t_f)} = \begin{bmatrix} \lambda_{10} \\ \lambda_{20} - \lambda_{10}t_f \end{bmatrix} - \begin{bmatrix} x_1(t_f) \\ -x_2(t_f) \end{bmatrix} = \begin{bmatrix} 0 \\ 0 \end{bmatrix}$$

respectively.

Thanks to what has been noticed about non-uniqueness of the solution and by taking into account that $\lambda_2(\cdot)$ is a linear function of time, we must consider only the two following *cases*.

Case 1. $u(\cdot) = 1$.

Case 2. $u(t) = -1, 0 \le t < \tau, u(t) = 1, \tau < t \le t_f, 0 < \tau < t_f, \tau$ to be found.

In the second case the control switches at time τ where the sign of $\lambda_2(\cdot)$ must change, so that $\lambda_2(\tau) = 0$.

The two cases above are now discussed.

Case 1. $u(\cdot) = 1$.

We have

$$x_1(t) = \frac{t^2}{2}, \quad x_2(t) = t$$

and the above transversality and orthogonality conditions imply

$$\lambda_{10}t_f + \lambda_{20} - \lambda_{10}t_f + 1 = 0$$
$$\lambda_{10} - \frac{t_f^2}{2} = 0$$
$$\lambda_{20} - \lambda_{10}t_f + t_f = 0$$

From this system of equations we get

$$\lambda_{10} = \frac{t_f^2}{2}, \quad \lambda_{20} = \frac{t_f^3}{2} - t_f$$

where t_f is a root of the polynomial

$$p(z) := \frac{z^3}{2} - z + 1$$

No root of $p(\cdot)$ is positive. Therefore the control $u(\cdot) = 1$ does not satisfy Theorem 2.2.

Case 2. $u(t) = -1, 0 \le t < \tau, u(t) = 1, \tau < t \le t_f, \tau$ to be found.

We have

$$0 \le t < \tau, \qquad\qquad \tau < t \le t_f$$
$$u(t) = -1 \qquad\qquad u(t) = 1$$
$$0 \le t \le \tau \qquad\qquad \tau \le t \le t_f$$
$$x_1(t) = -\frac{t^2}{2} \qquad x_1(t) = -\frac{\tau^2}{2} - \tau(t - \tau) + \frac{(t - \tau)^2}{2}$$
$$x_2(t) = -t \qquad\qquad x_2(t) = -\tau + (t - \tau)$$

and the switching condition $(\lambda_2(\tau) = 0)$ must be satisfied. The switching, transversality and orthogonality conditions constitute the system of four equations

$$\lambda_2(\tau) = \lambda_{20} - \lambda_{10}\tau = 0$$
$$H(t_f) + 1 = \lambda_{10}x_2(t_f) + \lambda_2(t_f)u(t_f) + 1 = 0$$
$$\lambda_1(t_f) - x_1(t_f) = 0$$
$$\lambda_2(t_f) + x_2(t_f) = 0$$

for the four unknowns $\lambda_{10}, \lambda_{20}, \tau, t_f$. If, in the above equations, we take into account that

$$\lambda_1(t_f) = \lambda_{10}, \quad \lambda_2(t_f) = -\lambda_{10}(t_f - \tau), \quad u(t_f) = 1$$
$$x_1(t_f) = -\frac{\tau^2}{2} - \tau(t_f - \tau) + \frac{(t_f - \tau)^2}{2}, \quad x_2(t_f) = -\tau + (t_f - \tau)$$

we obtain

$$\lambda_{10} = \frac{1}{\tau}, \quad \lambda_{20} = 1, \quad t_f = \tau\frac{1 - 2\tau}{1 - \tau}$$

where τ is a root of the polynomial

$$q(z) := 2z^5 - 4z^4 + z^3 + 2z^2 - 4z + 2$$

Observe that $\lambda_{20} > 0$ in accordance with $u(0) = -1$. The root $\bar{\tau}$ of $q(\cdot)$ to be found must be such that $\bar{\tau} > 1$. In fact

$$t_f - \bar{\tau} = \frac{\bar{\tau}^2}{\bar{\tau} - 1} > 0 \Rightarrow \bar{\tau} > 1$$

The only root possessing this property is $\bar{\tau} = 1.64$ which in turn gives

$$t_f = 5.84, \quad \lambda_{10} = 0.61$$

Finally, we have $\lambda_2(t_f) = -2.57 < 0$, in accordance with $u(t_f) = 1$. ∎

Problem 5.7

Let

$$x(0) = \begin{bmatrix} 0 \\ 0 \end{bmatrix}, \quad x(t_f) = \text{free}, \quad u(t) \in \overline{U}, \ \forall t, \ \overline{U} = \{u | -1 \le u \le 1\}$$

$$J = m(x(t_f), t_f) + \int_0^{t_f} l(x(t), u(t), t)\, dt$$

$$m(x, t) = \beta t - x_1, \quad l(x, u, t) = \frac{u^2(t)}{2}$$

where the final time t_f is free and $\beta > 0$ is a given parameter.

The hamiltonian function and the H−minimizing control are

$$H = \frac{u^2}{2} + \lambda_1 x_2 + \lambda_2 u, \quad u_h = \begin{cases} -1, & \lambda_2 \geq 1 \\ -\lambda_2, & |\lambda_2| \leq 1 \\ 1, & \lambda_2 \leq -1 \end{cases}$$

Therefore Eq. (2.5a) and its general solution over an interval beginning at time 0 are

$$\dot{\lambda}_1(t) = 0 \qquad\qquad\qquad \lambda_1(t) = \lambda_{10}$$
$$\dot{\lambda}_2(t) = -\lambda_1(t) \qquad\qquad \lambda_2(t) = \lambda_{20} - \lambda_{10} t$$

The orthogonality condition at the final time is

$$\begin{bmatrix} \lambda_1(t_f) \\ \lambda_2(t_f) \end{bmatrix} - \frac{\partial m(x, t_f)}{\partial x}\Bigg|_{x=x(t_f)}^{\prime} = \begin{bmatrix} \lambda_{10} \\ \lambda_{20} - \lambda_{10} t_f \end{bmatrix} + \begin{bmatrix} 1 \\ 0 \end{bmatrix} = \begin{bmatrix} 0 \\ 0 \end{bmatrix}$$

Thus

$$\lambda_{10} = -1, \quad \lambda_2(t_f) = 0, \quad \lambda_{20} = -t_f, \quad \lambda_2(t) = t - t_f$$

These results make the transversality condition at the final time to be

$$H(t_f) + \frac{\partial m(x(t_f), t)}{\partial t}\Bigg|_{t=t_f} = \frac{u^2(t_f)}{2} + \lambda_1(t_f) x_2(t_f) + \lambda_2(t_f) u(t_f) + \beta =$$
$$= -x_2(t_f) + \beta = 0$$

so that $x_2(t_f) = \beta$. The function $\lambda_2(\cdot)$ is linear with respect to t, negative at the initial time and zero at the final one: thus the control can be either

$$u(t) = -\lambda_2(t), \ 0 \leq t \leq t_f$$

or

$$u(t) = 1, \ 0 \leq t \leq \tau; \quad u(t) = -\lambda_2(t), \ \tau \leq t \leq t_f$$

where τ is a suitable time. We now discuss these two cases.

Case 1. $u(t) = -\lambda_2(t) = t_f - t, 0 \leq t \leq t_f$.

We have

$$x_1(t) = \frac{t_f}{2} t^2 - \frac{1}{6} t^3$$
$$x_2(t) = t_f t - \frac{1}{2} t^2$$

from which it follows

$$t_f = \sqrt{2\beta}$$

because $x_2(t_f) = \beta$. The considered control requires that $-1 \le \lambda_2(t) \le 0, 0 \le t \le t_f$. Therefore $\beta \le 0.5$, because $\lambda_{20} = -t_f$.

Case 2. $u(t) = 1, 0 \le t \le \tau, u(t) = -\lambda_2(t) = t_f - t, \tau \le t \le t_f$.

We have

$$0 \le t \le \tau$$
$$u(t) = 1$$
$$x_1(t) = \frac{t^2}{2}$$

$$\tau \le t \le t_f$$
$$u(t) = t_f - t$$
$$x_1(t) = \frac{\tau^2}{2} + \tau(t - \tau) + t_f \frac{(t - \tau)^2}{2} + $$
$$-\frac{t^3 - \tau^3}{6} + \frac{\tau^2(t - \tau)}{2}$$

$$x_2(t) = t$$

$$x_2(t) = \tau + t_f(t - \tau) - \frac{t^2 - \tau^2}{2}$$

By enforcing the switching condition $(\lambda_2(\tau) = t_f - \tau = -1)$ and recalling that $x_2(t_f) = \beta$, we get

$$t_f = \beta + \frac{1}{2}, \quad \tau = \beta - \frac{1}{2}$$

The considered control requires that $\lambda_{20} \le -1$, so that $\beta \ge 0.5$.

In summary, the control which satisfies the NC is

$$\text{if } \beta \le \frac{1}{2}$$

$$u(t) = \sqrt{2\beta} - t, \quad 0 \le t \le \sqrt{2\beta} = t_f$$

otherwise

$$\text{if } \beta \ge \frac{1}{2}$$

$$u(t) = \begin{cases} 1, & 0 \le t \le \beta - \frac{1}{2} \\ \beta + \frac{1}{2} - t, & \beta - \frac{1}{2} \le t \le \beta + \frac{1}{2} = t_f \end{cases}$$

Observe that the same conclusions hold if the given performance index is replaced by

$$J = \int_0^{t_f} \left(\frac{u^2(t)}{2} + \beta - x_2(t) \right) dt$$

because

$$\beta t_f = \int_0^{t_f} \beta \, dt, \quad x_1(t_f) = \int_0^{t_f} x_2(t) \, dt$$

∎

As stated at the beginning of this section, the next problem is of particular interest because no solution which satisfies the NC can be found, although an optimal solution exists.

Problem 5.8

Let

$$x(0) = \begin{bmatrix} 0 \\ 0 \end{bmatrix}, \quad x(t_f) = \text{free}, \quad u(t) \in \overline{U}, \ \forall t, \ \overline{U} = \{u | -1 \le u \le 1\}$$

$$J = m(x(t_f), t_f), \quad m(x, t) = \frac{1}{2}(x_1 - 1)^2 + t$$

where the final time t_f is free.

The hamiltonian function and the $H-$minimizing control are

$$H = \lambda_1 x_2 + \lambda_2 u, \quad u_h = -\text{sign}(\lambda_2)$$

Therefore Eq. (2.5a) and its general solution over an interval beginning at time 0 are

$$\dot{\lambda}_1(t) = 0 \qquad\qquad \lambda_1(t) = \lambda_{10}$$
$$\dot{\lambda}_2(t) = -\lambda_1(t) \qquad\qquad \lambda_2(t) = \lambda_{20} - \lambda_{10} t$$

The orthogonality condition at the final time is

$$\begin{bmatrix} \lambda_1(t_f) \\ \lambda_2(t_f) \end{bmatrix} - \frac{\partial m(x, t_f)}{\partial x}\Bigg|'_{x=x(t_f)} = \begin{bmatrix} \lambda_{10} \\ \lambda_{20} - \lambda_{10} t_f \end{bmatrix} - \begin{bmatrix} x_1(t_f) - 1 \\ 0 \end{bmatrix} = \begin{bmatrix} 0 \\ 0 \end{bmatrix}$$

and implies

$$\lambda_{10} = x_1(t_f) - 1, \quad \lambda_2(t_f) = \lambda_{20} - \lambda_{10} t_f = 0$$

The function $\lambda_2(\cdot)$ is linear with respect to t: thus we conclude that it does not change sign and only the following two *cases* can occur.

Case 1. $u(\cdot) = 1$.

Case 2. $u(\cdot) = -1$.

No matter what the actual case is, the transversality condition at the final time is

$$H(t_f) + \frac{\partial m(x(t_f), t)}{\partial t}\Bigg|_{t=t_f} = \lambda_1(t_f)x_2(t_f) + \lambda_2(t_f)u(t_f) + 1 = \lambda_{10} x_2(t_f) + 1 = 0$$

where we have exploited the expressions of $\lambda_1(\cdot)$ and $\lambda_2(\cdot)$. Therefore

$$\lambda_{10} = -\frac{1}{x_2(t_f)}$$

On the other hand we have already established that $\lambda_{10} = x_1(t_f) - 1$: thus we conclude that

$$x_2(t_f)(x_1(t_f) - 1) + 1 = 0$$

Now we consider the two alternatives for $u(\cdot)$.

Case 1. $u(\cdot) = 1$.

We have

$$x_1(t_f) = \frac{t_f^2}{2}, \quad x_2(t_f) = t_f$$

so that

$$x_2(t_f)(x_1(t_f) - 1) + 1 = \frac{t_f^3}{2} - t_f + 1 = 0$$

We conclude that the final time t_f must be a root of the polynomial

$$p(z) := z^3 - 2z + 2$$

This polynomial does not possess real and non negative roots. Therefore the control $u(\cdot) = 1$ does not satisfy Theorem 2.2.

Case 2. $u(\cdot) = -1$.

We have

$$x_1(t_f) = -\frac{t_f^2}{2}, \quad x_2(t_f) = -t_f$$

so that

$$x_2(t_f)(x_1(t_f) - 1) + 1 = \frac{t_f^3}{2} + t_f + 1 = 0$$

We conclude that the final time t_f must be a root of the polynomial

$$q(z) := z^3 + 2z + 2$$

Also this polynomial does not possess real and non negative roots. Therefore also the control $u(\cdot) = -1$ does not satisfy Theorem 2.2.

Fig. 5.5 Problem 5.8. The
value of the performance
index as a function of the
given final time

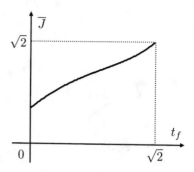

However an optimal solution exists. In order to clarify this seeming contradiction, assume that the final time is given and not negative. Then we have to minimize

$$I = \frac{(x_1(t_f) - 1)^2}{2}$$

It is obvious that the control $u(t) = 1$, $t \in [0, t_f]$, $t_f \leq \sqrt{2}$, is optimal because it makes $x_1(t_f) = 0.5 t_f^2$ which is the value as close as possible to 1. Correspondingly we have

$$I(t_f) = \frac{(t_f^2 - 2)^2}{4}$$

When $t_f > \sqrt{2}$ it is again obvious that the control $u(t) = 1$, $t \in [0, \sqrt{2}]$, $u(t) = 0$, $t < \sqrt{2}$, is optimal because $I = 0$, the least possible value. Thus the optimal value of the performance index J is

$$J(t_f) = \frac{(t_f^2 - 2)^2}{4} + t_f$$

This function is plotted in Fig. 5.5. It attains its minimum value at $t_f = 0$ where the derivative is positive. This is why the NC can not be satisfied. Indeed they are stated under the assumption that the initial and final times do not coincide. ∎

Chapter 6
Nonstandard Constraints on the Final State

Abstract In this chapter we consider control problems where the final state of the controlled system is not constrained to belong to a set with the properties of a regular variety but rather to a set which is a generalization of it. In the last problem of this chapter the final time is not given but must be smaller than an assigned value.

In this chapter we consider control problems where the final state of the double integrator is not constrained to belong to a set with the properties of a regular variety, but rather to a set which is a generalization of it. We refer to the material presented in Sect. 2.2.3 and begin by discussing five problems which exhibit the following characteristics.

(1) In the first four of them only one component of the state at the final time must belong to a given interval, whereas such a constraint is imposed to both components in Problem 6.5. Such intervals are bounded in Problems 6.1–6.3 and unbounded in Problems 6.4 and 6.5.
(2) The performance index of Problem 6.3 is not purely integral, whereas in Problems 6.2 and 6.4 it is simply the length of the control interval.
(3) The control variable is not constrained in Problems 6.1, 6.3 and 6.5.

Finally, in the last problem of this chapter, namely Problem 6.6, the final time is not given. It must be smaller than an assigned value. By exploiting Remark 2.4 it is possible to restate the problem in terms of an *enlarged* third order system where, at the final time, two out of the three state variables are constrained to belong to a bounded interval.

Problem 6.1
Let

$$x(0) = \begin{bmatrix} \beta \\ \gamma \end{bmatrix}, \quad J = \int_0^{t_f} l(x(t), u(t), t)\, dt, \quad l(x, u, t) = \frac{u^2}{2}, \quad u(t) \in R, \ \forall t, \quad t_f = 1$$

$$x(t_f) \in \widehat{S}_f = \{x \mid x \in S_f, \ a_2 \le x_2 \le b_2\}, \quad a_2 = -1, \ b_2 = 1$$

$$S_f = \overline{S}_f = \{x \mid \alpha_f(x) = 0\}, \quad \alpha_f(x) = x_1$$

© Springer International Publishing Switzerland 2017
A. Locatelli, *Optimal Control of a Double Integrator*, Studies in Systems,
Decision and Control 68, DOI 10.1007/978-3-319-42126-1_6

Fig. 6.1 Problem 6.1. The
set $\widehat{S}_f$ and the *straight lines*
ϱ_1 and ϱ_2. They bound the
set $\mathcal{R}$ of the initial states
where the constraint on
$x_2(t_f)$ is not binding

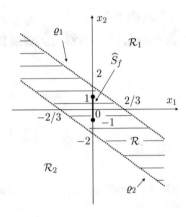

where the parameters β and γ are given. Therefore the set $\widehat{S}_f$ of the admissible final
states is the line segment of the x_2 axis lying between the points $x_2 = -1$ and $x_2 = 1$
(see Fig. 6.1). The set $\overline{S}_f$ is a regular variety because

$$\Sigma_f(x) = \frac{d\alpha_f(x)}{dx} = \begin{bmatrix} 1 & 0 \end{bmatrix}$$

and $\text{rank}(\Sigma_f(x)) = 1, \forall x \in \overline{S}_f$.

The hamiltonian function and the H-minimizing control are

$$H = \frac{u^2}{2} + \lambda_1 x_2 + \lambda_2 u, \quad u_h = -\lambda_2$$

and Eq. (2.5a) and its general solution over an interval beginning at time 0 are

$$\dot{\lambda}_1(t) = 0 \qquad\qquad \lambda_1(t) = \lambda_{10}$$
$$\dot{\lambda}_2(t) = -\lambda_1(t) \qquad \lambda_2(t) = \lambda_{20} - \lambda_{10}t$$

First, we discuss the case where the inequality constraints on $x_2(t_f)$ are not present,
namely the case where $x(t_f) \in \overline{S}_f$. Then the orthogonality condition at the final time
is

$$\begin{bmatrix} \lambda_1(t_f) \\ \lambda_2(t_f) \end{bmatrix} = \begin{bmatrix} \lambda_{10} \\ \lambda_{20} - \lambda_{10}t_f \end{bmatrix} = \vartheta_f \left. \frac{d\alpha_f(x)}{dx} \right|_{x=x(t_f)}' = \vartheta_f \begin{bmatrix} 1 \\ 0 \end{bmatrix}$$

and implies $\lambda_2(t_f) = \lambda_2(1) = 0$, that is $\lambda_{20} = \lambda_{10}$. Consequently, the state motion
is

$$x_1(t) = \beta + \gamma t - \frac{\lambda_{10}}{2}t^2 + \frac{\lambda_{10}}{6}t^3$$
$$x_2(t) = \gamma - \lambda_{10}t + \frac{\lambda_{10}}{2}t^2$$

Fig. 6.2 Problem 6.1. State trajectories corresponding to the initial state
$x(0) = \bar{x}_0 := [0 \;\; 1.5]'$ (not binding constraint),
$x(0) = \hat{x}_0 := [1 \;\; -1]' \in \varrho_1$ (not binding constraint),
$x(0) = \tilde{x}_0 := [0 \;\; 3]'$ (binding constraint),
$x(0) = x_0^* := [2 \;\; 3]'$ (binding constraint)

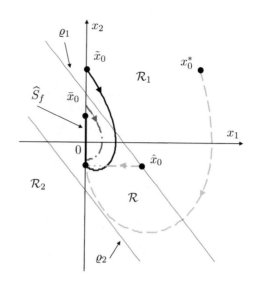

and, by enforcing feasibility ($x_1(t_f) = 0$), we obtain

$$\lambda_{10} = 3\,(\beta + \gamma), \quad x_2(t_f) = -\frac{(\gamma + 3\beta)}{2}$$

Thus the inequality constraints on $x_2(t_f)$ are not binding whenever the parameters β and γ are such that the initial state belongs to the region $\mathcal{R}$ of the state plane which is shown in Fig. 6.1. It is bounded by the two straight lines ϱ_1, ϱ_2 and includes them. The two straight lines are defined, respectively, by the equations $r_1(x_1, x_2) = 0$ and $r_2(x_1, x_2) = 0$ where $r_1(x_1, x_2) := 3x_1 + x_2 - 2$ and $r_2(x_1, x_2) := 3x_1 + x_2 + 2$. Two state trajectories corresponding to initial states belonging to the region $\mathcal{R}$ are plotted in Fig. 6.2.

The solution of the problem entails $x_2(t_f) = -1$ whenever the initial state is located above the straight line ϱ_1. In the light of what has been previously reported, this result is completely consistent with $x_1(t_f) = 0$, $x_2(t_f) = -1$ if the initial state is a point of ϱ_1.

By following a more systematic approach, we first enforce feasibility, namely

$$x(t_f) \in S_{fM} := \left\{ \begin{bmatrix} 0 \\ 1 \end{bmatrix} \right\}$$

or

$$x(t_f) \in S_{fm} := \left\{ \begin{bmatrix} 0 \\ -1 \end{bmatrix} \right\}$$

which are equivalent to requiring that

$$0 = x_1(t_f) = \beta + \gamma - \frac{\lambda_{20}}{2} + \frac{\lambda_{10}}{6}$$

$$1 = x_2(t_f) = \gamma - \lambda_{20} + \frac{\lambda_{10}}{2}$$

or

$$0 = x_1(t_f) = \beta + \gamma - \frac{\lambda_{20}}{2} + \frac{\lambda_{10}}{6}$$

$$-1 = x_2(t_f) = \gamma - \lambda_{20} + \frac{\lambda_{10}}{2}$$

respectively. When $x(t_f) \in S_{fM}$ we get

$$\lambda_{10} = 6(2\beta + \gamma + 1), \quad \lambda_{20} = 2(3\beta + 2\gamma + 1)$$

whereas, when $x(t_f) \in S_{fm}$, we obtain

$$\lambda_{10} = 6(2\beta + \gamma - 1), \quad \lambda_{20} = 2(3\beta + 2\gamma - 1)$$

The value of the performance index is

$$J = \frac{(\lambda_{10}^2 - 3\lambda_{10}\lambda_{20} + 3\lambda_{20}^2)}{6}$$

and it is easy to conclude that

$$J|_{x(t_f) \in S_{fm}} < J|_{x(t_f) \in S_{fM}}$$

above the straight line ϱ_1.

Two state trajectories corresponding to initial states belonging to the region $\mathcal{R}_1$ are plotted in Fig. 6.2. We have plotted trajectories of this kind only in view of obvious symmetry considerations. Indeed the state trajectory relevant to an initial state $\check{x}_0$ located below the straight line ϱ_2 is symmetric, with respect to the origin, to the trajectory relevant to the initial state $-\check{x}_0$, which is located above the straight line ϱ_1.

By summarizing, the solution of the problem is $u(t) = \lambda_{10}t - \lambda_{20}$, where the parameters λ_{10} and λ_{20} depend on the initial state in the following way

$$x(0) \in \mathcal{R} := \left\{ x \left| \frac{|3x_1 + x_2|}{2} \le 1 \right. \right\} \Rightarrow \left\{ \begin{array}{l} \lambda_{10} = 3(x_1(0) + x_2(0)) \\ \lambda_{20} = \lambda_{10} \end{array} \right.$$

$$x(0) \in \mathcal{R}_1 := \left\{ x \left| \frac{3x_1 + x_2}{2} > 1 \right. \right\} \Rightarrow \left\{ \begin{array}{l} \lambda_{10} = 6(2x_1(0) + x_2(0) - 1) \\ \lambda_{20} = 2(3x_1(0) + 2x_2(0) - 1) \end{array} \right.$$

$$x(0) \in \mathcal{R}_2 := \left\{ x \left| \frac{3x_1 + x_2}{2} < 1 \right. \right\} \Rightarrow \left\{ \begin{array}{l} \lambda_{10} = 6(2x_1(0) + x_2(0) + 1) \\ \lambda_{20} = 2(3x_1(0) + 2x_2(0) + 1) \end{array} \right.$$

■

Fig. 6.3 Problem 6.2. The set $\widehat{S}_f$ and the curves Γ_1 and Γ_2. They bound the set $\mathcal{R}$ of the initial states where the constraint on $x_1\left(t_f\right)$ is not binding

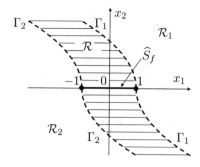

Problem 6.2

Let

$$x(0) = \begin{bmatrix} \beta \\ \gamma \end{bmatrix}, \quad u(t) \in \overline{U}, \ \forall t, \ \overline{U} = \{u \mid -1 \le u \le 1\}$$

$$J = \int_0^{t_f} l(x(t), u(t), t) \, dt, \ l(x, u, t) = 1$$

$$x(t_f) \in \widehat{S}_f = \{x \mid x \in S_f, \ a_1 \le x_1 \le b_1\} \ a_1 = -1, \ b_1 = 1$$

$$S_f = \overline{S}_f = \{x \mid \alpha_f(x) = 0\}, \ \alpha_f(x) = x_2$$

where β and γ are given parameters such that $x(0) \notin \widehat{S}_f$. The final time t_f is free.

Thus the set $\widehat{S}_f$ of the admissible final states is the line segment of the x_1 axis lying between the points $x_1 = -1$ and $x_1 = 1$ (see Fig. 6.3). The set $\overline{S}_f$ is a regular variety because

$$\Sigma_f(x) = \frac{d\alpha_f(x)}{dx} = \begin{bmatrix} 0 & 1 \end{bmatrix}$$

and $\text{rank}(\Sigma_f(x)) = 1, \ \forall x \in \overline{S}_f$.

The hamiltonian function and the H-minimizing control are

$$H = 1 + \lambda_1 x_2 + \lambda_2 u, \quad u_h = -\text{sign}(\lambda_2)$$

Therefore Eq. (2.5a) and its general solution over an interval beginning at time 0 are

$$\begin{aligned} \dot{\lambda}_1(t) &= 0 & \lambda_1(t) &= \lambda_{10} \\ \dot{\lambda}_2(t) &= -\lambda_1(t) & \lambda_2(t) &= \lambda_{20} - \lambda_{10} t \end{aligned}$$

We notice that the control is either constant (and equal to ± 1) or is piecewise constant with a single switching (from 1 to -1 or viceversa). In fact the function $\lambda_2(\cdot)$ is linear with respect to time and its sign can change at most once.

First, we discuss the case where the inequality constraints on $x_1(t_f)$ are not present, namely the case where $x(t_f) \in \overline{S}_f$. Correspondingly, the orthogonality condition at the final time is

$$\begin{bmatrix} \lambda_1(t_f) \\ \lambda_2(t_f) \end{bmatrix} = \begin{bmatrix} \lambda_{10} \\ \lambda_{20} - \lambda_{10}t_f \end{bmatrix} = \vartheta_f \left. \frac{d\alpha_f(x)}{dx} \right|'_{x=x(t_f)} = \vartheta_f \begin{bmatrix} 0 \\ 1 \end{bmatrix}$$

It implies $\lambda_1(t_f) = 0$, $\lambda_1(\cdot) = 0$ and $\lambda_2(\cdot) = \lambda_{20}$, so that the control is constant and equal to 1 or -1. The transversality condition at the final time, written at the initial time because the problem is time-invariant, is

$$H(0) = 1 + \lambda_1(0)x_2(0) + \lambda_2(0)u(0) = 1 + \lambda_{10}x_2(0) + \lambda_{20}u(0) = 1 + \lambda_{20}u(0) = 0$$

and entails $\lambda_{20} = -\text{sign}(u(0))$.

Therefore the inequality constraints on $x_1(t_f)$ are not binding whenever the initial state is located inside the region $\mathcal{R}$ which is shown in Fig. 6.3 and bounded by the two curves Γ_1 and Γ_2. They are defined by the equations $g_1(x) = 0$ and $g_2(x) = 0$, respectively, where

$$g_1(x) := \begin{cases} x_1 + \dfrac{x_2^2}{2} - 1, & x_2 \geq 0 \\ x_1 - \dfrac{x_2^2}{2} - 1, & x_2 \leq 0 \end{cases} \qquad g_2(x) := \begin{cases} x_1 + \dfrac{x_2^2}{2} + 1, & x_2 \geq 0 \\ x_1 - \dfrac{x_2^2}{2} + 1, & x_2 \leq 0 \end{cases}$$

The region $\mathcal{R}$ includes the two curves as well and is given by

$$\mathcal{R} := \left\{ x \,\middle|\, -1 \leq x_1 + \frac{x_2^2}{2} \leq 1, \ x_2 \geq 0 \right\} \cup \left\{ x \,\middle|\, -1 \leq x_1 - \frac{x_2^2}{2} \leq 1, \ x_2 \leq 0 \right\}$$

If $x(0) \in \mathcal{R}$ and $\gamma < 0$, we have

$$u(\cdot) = 1, \quad t_f = -\gamma$$

whereas, if $\gamma > 0$, we have

$$u(\cdot) = -1, \quad t_f = \gamma$$

Observe that the parameter γ can not be zero, because, $x(0) \in \mathcal{R}$ and $\gamma = 0$ imply $x(0) \in \widehat{S}_f$. When the initial state is located inside the region $\mathcal{R}_1$ which lies to the right of the curve Γ_1 and is defined by

$$\mathcal{R}_1 := \left\{ x \,\middle|\, x_1 + \frac{x_2^2}{2} > 1, \ x_2 \geq 0 \right\} \cup \left\{ x \,\middle|\, x_1 - \frac{x_2^2}{2} > 1, \ x_2 \leq 0 \right\}$$

the solution can be found as follows. We notice that a constant control equal to 1 or -1 is not feasible. Therefore the constraint is binding and it must be either

$$x(t_f) \in S_{fm} := \left\{ \begin{bmatrix} -1 \\ 0 \end{bmatrix} \right\}$$

or

$$x(t_f) \in S_{fM} := \left\{ \begin{bmatrix} 1 \\ 0 \end{bmatrix} \right\}$$

In both cases the control switches from 1 to -1 (or viceversa). Only one switch can occur because of the expressions of $\lambda_2(\cdot)$ and the H-minimizing control.

Thanks to the fact that the performance index is the time required to reach the set $\widehat{S}_f$, it is straightforward to verify that the solution is constituted by a control which is equal to -1 at the beginning and switches to 1 when the state becomes a point of the curve Γ_1. This solution gives $x(t_f) \in S_{fM}$.

When the initial state is located inside the region $\mathcal{R}_2$ which lies to the left of the curve Γ_2 and is defined by

$$\mathcal{R}_2 := \left\{ x \left| x_1 - \frac{x_2^2}{2} < -1, \ x_2 \le 0 \right\} \cup \left\{ x \left| x_1 + \frac{x_2^2}{2} < -1, \ x_2 \ge 0 \right\} \right.$$

the solution can be computed in a completely similar way. We find that it is constituted by a control which switches from 1 to -1 when the state becomes a point of the curve Γ_2. This solution gives $x(t_f) \in S_{fm}$.

Figure 6.4 reports some state trajectories corresponding to various initial conditions, all of them located inside the region $\mathcal{R}_1 \cup \mathcal{R}$.

We have plotted trajectories of this kind only in view of obvious symmetry considerations. Indeed the state trajectory relevant to an initial state $\check{x}_0$ belonging to the region $\mathcal{R}_2$ is symmetric, with respect to the origin, to the trajectory relevant to the initial state $-\check{x}_0$ which belongs to $\mathcal{R}_1$.

If we look at Fig. 6.4, we note that the control is uniquely determined, at any time, by the position of the state in the $x_1 - x_2$ plane at the same time.

Fig. 6.4 Problem 6.2. State trajectories corresponding to $x(0) = \bar{x}_0 = [1 \ -1]'$ (not binding constraint), $x(0) = \hat{x}_0 = [2 \ 1]'$ (binding constraint), $x(0) = \check{x}_0 = [3 \ -1]'$ (binding constraint). In the last two cases the final part of the trajectory coincides with that piece of the curve Γ_1 which is relevant to negative values of x_2

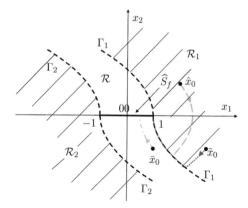

Indeed for all $t \in [0, t_f]$, we have

$$
\begin{aligned}
x(t) \in \mathcal{R}_1 && \Rightarrow u(t) = -1 \\
x(t) \in \mathcal{R}_2 && \Rightarrow u(t) = 1 \\
x(t) \in \mathcal{R}, \; x_2(t) < 0 &\Rightarrow u(t) = 1 \\
x(t) \in \mathcal{R}, \; x_2(t) > 0 &\Rightarrow u(t) = -1 \\
x(t) \in \widehat{S}_f && \Rightarrow u(t) = 0
\end{aligned}
$$

Therefore the solution of the problem can be implemented in a closed-loop form by resorting to a regulator characterized by the control law $u(x)$ which results from the above relations.   ∎

Problem 6.3
Let

$$
x(0) = \begin{bmatrix} \beta \\ \gamma \end{bmatrix}, \quad x(t_f) \in \widehat{S}_f = \left\{ x \mid x \in S_f, \; a_1 \le x_1 \le b_1 \right\}, \; a_1 = -1, \; b_1 = 1
$$

$$
S_f = R^2, \quad J = m(x(t_f), t_f) + \int_0^{t_f} l(x(t), u(t), t)\, dt
$$

$$
m(x, t) = \frac{x_2^2}{2}, \quad l(x, u, t) = \frac{u^2}{2}, \; u(t) \in R, \; \forall t, \quad t_f = 1
$$

where β and γ are given parameters. The set $\widehat{S}_f$ of the admissible final states is constituted by the vertical strip of width 2 which is centered on the x_2 axis and shown in Fig. 6.5.

The hamiltonian function and the H-minimizing control are

$$
H = \frac{u^2}{2} + \lambda_1 x_2 + \lambda_2 u, \quad u_h = \lambda_2
$$

Fig. 6.5 Problem 6.3. The set $\widehat{S}_f$ and the *straight lines* ϱ_1 and ϱ_2 which bound the set $\mathcal{R}$ of the initial states where the constraint on $x_1(t_f)$ is not binding

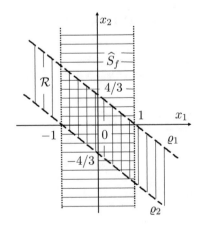

so that Eq. (2.5a) and its general solution over an interval beginning at time 0 are

$$\dot{\lambda}_1(t) = 0 \qquad\qquad \lambda_1(t) = \lambda_{10}$$
$$\dot{\lambda}_2(t) = -\lambda_1(t) \qquad \lambda_2(t) = \lambda_{20} - \lambda_{10}t$$

and the state motion is

$$x_1(t) = \beta + \gamma t - \frac{\lambda_{20}}{2}t^2 + \frac{\lambda_{10}}{6}t^3$$
$$x_2(t) = \gamma - \lambda_{20}t + \frac{\lambda_{10}}{2}t^2$$

First, we discuss the case where the inequality constraints on $x_1(t_f)$ are not present, namely the case where $x(t_f) = $ free. Correspondingly, the orthogonality condition at the final time implies

$$\begin{bmatrix} \lambda_1(t_f) \\ \lambda_2(t_f) \end{bmatrix} - \frac{\partial m(x, t_f)}{\partial x}\bigg|'_{x=x(t_f)} = \begin{bmatrix} \lambda_{10} \\ \lambda_{20} - \lambda_{10}t_f \end{bmatrix} - \begin{bmatrix} 0 \\ x_2(t_f) \end{bmatrix} = \begin{bmatrix} 0 \\ 0 \end{bmatrix}$$

From these equations and those relevant to the state motion we get

$$x_1(t_f) = \beta + \frac{3\gamma}{4}, \quad x_2(t_f) = \frac{\gamma}{2}$$

Therefore the inequality constraints on $x_1(t_f)$ are not binding whenever the initial state is located inside the region

$$\mathcal{R} := \left\{ x \,\middle|\, -1 \leq \frac{4x_1 + 3x_2}{4} \leq 1 \right\}$$

It is constituted by that part of the state plane which is bounded by the straight lines ϱ_1 and ϱ_2. These lines are shown in Fig. 6.5 and are defined by the equations $r_1(x) = 0$ and $r_2(x) = 0$, respectively, where

$$r_1(x) := \frac{4(x_1 - 1)}{3} + x_2, \qquad r_2(x) := \frac{(4x_1 + 1)}{3} + x_2$$

The region $\mathcal{R}$ incorporates the two straight lines as well.

Whenever the initial state does not belong to this region, we can find the solution in the following way. First, we consider the case where the initial state is located into the region

$$\mathcal{R}_1 := \left\{ x \,\middle|\, x_2 > \frac{4(1 - x_1)}{3} \right\}$$

that is to the right of the straight line ϱ_1. Then the final state must belong to one of the two sets

$$S_{fm} := \{x \mid \alpha_{fm}(x) = 0\}, \quad \alpha_{fm}(x) = x_1 + 1$$
$$S_{fM} := \{x \mid \alpha_{fM}(x) = 0\}, \quad \alpha_{fM}(x) = x_1 - 1$$

Both sets are regular varieties because

$$\Sigma_{fm}(x) = \frac{d\alpha_{fm}(x)}{dx} = \Sigma_{fM}(x) = \frac{d\alpha_{fM}(x)}{dx} = \begin{bmatrix} 1 & 0 \end{bmatrix}$$

and $\text{rank}(\Sigma_{fm}(x)) = 1, \forall x \in S_{fm}, \text{rank}(\Sigma_{fM}(x)) = 1, \forall x \in S_{fM}$. The orthogonality conditions at the final time do coincide, as they are

$$\begin{bmatrix} \lambda_1(t_f) \\ \lambda_2(t_f) \end{bmatrix} - \frac{\partial m(x, t_f)}{\partial x}\bigg|'_{x=x(t_f)} = \begin{bmatrix} \lambda_{10} \\ \lambda_{20} - \lambda_{10} t_f \end{bmatrix} - \begin{bmatrix} 0 \\ x_2(t_f) \end{bmatrix} =$$
$$= \vartheta_{fm} \frac{d\alpha_{fm}}{dx}\bigg|'_{x=x(t_f)} = \vartheta_{fm} \begin{bmatrix} 1 \\ 0 \end{bmatrix}$$

and

$$\begin{bmatrix} \lambda_1(t_f) \\ \lambda_2(t_f) \end{bmatrix} - \frac{\partial m(x, t_f)}{\partial x}\bigg|'_{x=x(t_f)} = \begin{bmatrix} \lambda_{10} \\ \lambda_{20} - \lambda_{10} t_f \end{bmatrix} - \begin{bmatrix} 0 \\ x_2(t_f) \end{bmatrix} =$$
$$= \vartheta_{fM} \frac{d\alpha_{fM}}{dx}\bigg|'_{x=x(t_f)} = \vartheta_{fM} \begin{bmatrix} 1 \\ 0 \end{bmatrix}$$

respectively. Therefore we have $\lambda_2(t_f) - x_2(t_f) = 0$ in both cases. This relation together with the expressions of $x_2(t_f)$ and $\lambda_2(t_f)$ gives

$$\lambda_{20} = \frac{\gamma}{2} + \frac{3\lambda_{10}}{4}$$

If we require that either $x_1(t_f) \in S_{fm}$ or $x_1(t_f) \in S_{fM}$, we obtain

$$\lambda_{10} = \frac{24}{5}\left(\beta + \frac{3\gamma}{4} + 1\right)$$

or

$$\lambda_{10} = \frac{24}{5}\left(\beta + \frac{3\gamma}{4} - 1\right)$$

It is easy to check that

$$J|_{x_1(t_f)=-1} > J|_{x_1(t_f)=-1}, \quad \forall x(0) \in \mathcal{R}_1$$

Table 6.1 Problem 6.3

β	γ	$J_{x_1(t_f)=1}$	$J_{x_1(t_f)=-1}$
3.0	6	110.40	182.40
3.0	-2	1.60	16.00
0.5	6	47.40	95.40
-0.5	6	30.60	69.00
-3.0	6	9.60	24.00

The values of the performance index corresponding to $x_1(t_f) = \pm 1$

so that the control to be used is

$$u(t) = -\lambda_{20} + \lambda_{10} t, \quad \lambda_{10} = \frac{24}{5}\left(\beta + \frac{3\gamma}{4} + 1\right), \quad \lambda_{20} = \frac{\gamma}{2} + \frac{3\lambda_{10}}{4}$$

Obvious symmetry considerations lead to the conclusion that, whenever the initial state is located inside the region

$$\mathcal{R}_2 := \left\{ x \,\middle|\, x_2 < -\frac{4(1 + x_1)}{3} \right\}$$

namely to the left of the straight line ϱ_2, the solution is the one which entails $x_1(t_f) = -1$.

Table 6.1 shows the values of the performance index corresponding to five different initial states. All of them are located to the right of the straight line ϱ_1 because, thanks to the already mentioned symmetry considerations, no further insight is gained if initial states located to the left of the straight line ϱ_2 are considered.

By summarizing, the solution of the problem takes on the form $u(t) = \lambda_{10}t - \lambda_{20}$, where the parameters λ_{10} and λ_{20} depend on the initial state as follows

$$x(0) = \begin{bmatrix} \beta \\ \gamma \end{bmatrix} \in \mathcal{R} \Rightarrow \lambda_{10} = 0, \ \lambda_{20} = \frac{\gamma}{2}$$

$$x(0) = \begin{bmatrix} \beta \\ \gamma \end{bmatrix} \in \mathcal{R}_1 \Rightarrow \lambda_{10} = \frac{6(4\beta + 3\gamma - 4)}{5}, \ \lambda_{20} = \frac{2\gamma + 3\lambda_{10}}{4}$$

$$x(0) = \begin{bmatrix} \beta \\ \gamma \end{bmatrix} \in \mathcal{R}_2 \Rightarrow \lambda_{10} = \frac{6(4\beta + 3\gamma + 4)}{5}, \ \lambda_{20} = \frac{2\gamma + 3\lambda_{10}}{4}$$

■

Problem 6.4

Let

$$x(0) = \begin{bmatrix} \beta \\ \gamma \end{bmatrix}, \quad x\left(t_f\right) \in \widehat{S}_f = \left\{x \,|\, x \in S_f,\; x_1 \leq b_1\right\}, \quad b_1 = 0$$

$$S_f = \overline{S}_f = \left\{x \,|\, \alpha_f(x) = 0\right\}, \quad \alpha_f(x) = x_2$$

$$J := \int_0^{t_f} l(x(t), u(t), t)\, dt, \quad l(x, u, t) = 1, \quad u(t) \in \overline{U}, \; \forall t, \; \overline{U} = \{u \,|\, -1 \leq u \leq 1\}$$

where β and γ are given parameters such that $x(0) \notin \widehat{S}_f$. Obviously, the final time t_f is free. The set $\widehat{S}_f$ of the admissible final states is constituted by the whole negative part of the x_1 axis, as shown in Fig. 6.6. The set $\overline{S}_f$ is a regular variety because

$$\Sigma_f(x) = \frac{d\alpha_f(x)}{dx} = \begin{bmatrix} 0 & 1 \end{bmatrix}$$

and $\mathrm{rank}(\Sigma_f(x)) = 1$, $\forall x \in \overline{S}_f$.

The hamiltonian function and the H-minimizing control are

$$H = 1 + \lambda_1 x_2 + \lambda_2 u, \quad u_h = -\mathrm{sign}(\lambda_2)$$

and Eq. (2.5a) and its general solution over an interval beginning at time 0 are

$$\begin{aligned} \dot{\lambda}_1(t) &= 0 & \lambda_1(t) &= \lambda_{10} \\ \dot{\lambda}_2(t) &= -\lambda_1(t) & \lambda_2(t) &= \lambda_{20} - \lambda_{10}t \end{aligned}$$

First, we discuss the case where the inequality constraints on $x_1\left(t_f\right)$ are not present, namely the case where $x(t_f) \in \overline{S}_f$. The orthogonality condition at the final time

$$\begin{bmatrix} \lambda_1(t_f) \\ \lambda_2(t_f) \end{bmatrix} = \begin{bmatrix} \lambda_{10} \\ \lambda_{20} - \lambda_{10}t_f \end{bmatrix} = \vartheta_f \left.\frac{d\alpha_f(x)}{dx}\right|'_{x=x(t_f)} = \vartheta_f \begin{bmatrix} 0 \\ 1 \end{bmatrix}$$

implies $\lambda_1\left(t_f\right) = \lambda_{10} = 0$, so that $\lambda_2(\cdot)$ is constant, equal to λ_{20} and $u(\cdot) = \pm 1$. When $u(\cdot) = 1$, the state trajectories are the curves defined by the equations $x_2^2 - 2x_1 = \mathrm{const.}$ (see also Fig. 7.1a). They intersect the negative part of the x_1 axis (so that the control is feasible) only if the initial state belongs to the region $\mathcal{R}_1$ which is shown in Fig. 6.6 and lies below the set $\widehat{S}_f$ and the curve Γ_1 defined by the equations

$$g_1(x) = 0, \quad g_1(x) := x_1 - \frac{x_2^2}{2}, \quad x_2 \leq 0$$

The region $\mathcal{R}_1$ includes Γ_1, but does not include the set $\widehat{S}_f$ and is formally given by

$$\mathcal{R}_1 := \left\{x \,|\, 2x_1 - x_2^2 \leq 0,\; x_2 < 0\right\}$$

Fig. 6.6 Problem 6.4. The set $\widehat{S}_f$, the curves Γ_1, Γ_2 and some state trajectories. The curves Γ_1, Γ_2 bound the sets $\mathcal{R}_1$ and $\mathcal{R}_2$ of the initial conditions where the constraint on $x_1\left(t_f\right)$ is not binding

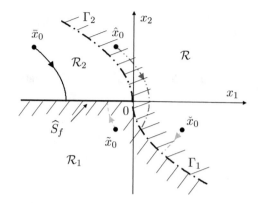

In a similar way, when $u(\cdot) = -1$, the state trajectories are the curves defined by the equations $x_2^2 + 2x_1 = $ const. (again see Fig. 7.1a). They intersect the negative part of the x_1 axis (so that the control is feasible) only if the initial state belongs to the region $\mathcal{R}_2$ which is shown in Fig. 6.6 and lies above the set $\widehat{S}_f$ and below the curve Γ_2 defined by the equations

$$g_2(x) = 0, \quad g_2(x) := x_1 - \frac{x_2^2}{2}, \quad x_2 \geq 0$$

The region $\mathcal{R}_2$ includes Γ_2, but does not include the set $\widehat{S}_f$ and is formally given by

$$\mathcal{R}_2 := \left\{x \,\middle|\, 2x_1 + x_2^2 \leq 0, \; x_2 > 0\right\}$$

From the transversality condition at the final time

$$H(t_f) = 1 + \lambda_1(t_f)x_2(t_f) + \lambda_2(t_f)u(t_f) = 1 + \lambda_{20}u(t_f) = 1 + \lambda_{20}u(0) = 0$$

we get λ_{20}, which turns out to be equal to either -1 or $+1$, according to whether the initial state belongs to the region $\mathcal{R}_1$ or $\mathcal{R}_2$.

The inequality constraint on $x_1\left(t_f\right)$ can not be ignored when the initial state is located inside the region $\mathcal{R}$ which lies above the two curves Γ_1 and Γ_2, but does not include them. This region is formally specified by

$$\mathcal{R} := \left\{x \,\middle|\, 2x_1 + x_2^2 > 0, \; x_2 \geq 0\right\} \cup \left\{x \,\middle|\, 2x_1 - x_2^2 > 0, \; x_2 \leq 0\right\}$$

When $x(0) \in \mathcal{R}$, we have

$$x(t_f) \in S_{fM} := \left\{\begin{bmatrix} 0 \\ 0 \end{bmatrix}\right\}$$

and the control switches at a suitable time τ. This is consistent with the fact that $\lambda_2(\cdot)$ is a linear function of t. In view of the shapes of the state trajectories corresponding to $u(\cdot) = \pm 1$ (once more see Fig. 7.1a), we conclude that the control switches from -1 to 1 at the time τ when the state belongs to the curve Γ_1. Simple computations lead to the following values of τ and t_f, namely

$$\tau = \gamma + \sqrt{\beta + \frac{\gamma^2}{2}}, \quad t_f = \gamma + 2\sqrt{\beta + \frac{\gamma^2}{2}}$$

The transversality condition at the final time, written at the initial time because the problem is time-invariant,

$$H(0) = 1 + \lambda_1(0)x_2(0) + \lambda_2(0)u(0) = 1 + \lambda_{10}x_2(0) + \lambda_{20}u(0) = 0$$

together with the constraint on $\lambda_2(\tau)$ deriving from control switching

$$\lambda_2(\tau) = \lambda_{20} - \lambda_{10}\tau = 0$$

implies

$$\lambda_{10} = \frac{1}{\sqrt{\beta + \frac{\gamma^2}{2}}}, \quad \lambda_{20} = 1 + \frac{\gamma}{\sqrt{\beta + \frac{\gamma^2}{2}}}$$

and we conclude that the NC are satisfied. Some state trajectories are plotted in Fig. 6.6: the final part of the trajectories resulting from control switching lies on the curve Γ_1.

Finally, if we make again reference to this figure, we conclude that the control to be applied uniquely depends on the actual position of the state in the $x_1 - x_2$ plane. More in detail, for $t \in [0, t_f]$ we have

$$x(t) \in \begin{cases} \mathcal{R}_1 & \Rightarrow u(t) = 1 \\ \mathcal{R}_2 \cup \mathcal{R} & \Rightarrow u(t) = -1 \\ \widehat{S}_f & \Rightarrow u(t) = 0 \end{cases}$$

Thus, as for Problem 6.2, the solution can be implemented in closed-loop form, the control law $u(x)$ being specified by the above equation. ∎

Problem 6.5

Let

$$x(0) = \begin{bmatrix} 1 \\ 1 \end{bmatrix}, \quad J = \int_0^{t_f} l(x(t), u(t), t)\, dt, \quad l(x, u, t) = \frac{u^2}{2}, \quad u(t) \in R, \; \forall t, \quad t_f = 1$$

$$x(t_f) \in \widehat{S}_f = \{x | x \in S_f, \; x_1 \leq b_1, \; x_2 \leq b_2\}, \quad b_1 = 0, \; b_2 = 0, \; S_f = R^2$$

Fig. 6.7 Problem 6.5. The set $\widehat{S}_f$ and state trajectories relevant to the considered cases

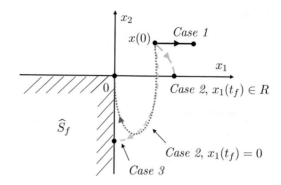

The set $\widehat{S}_f$ of the admissible final states is the region of the state plane below the negative part of the x_1 axis and to left of the negative part of the x_2 axis. It includes the negative parts of these axis as well and is shown in Fig. 6.7.

The hamiltonian function and the H-minimizing control are

$$H = \frac{u^2}{2} + \lambda_1 x_2 + \lambda_2 u, \quad u_h = -\lambda_2$$

Therefore Eq. (2.5a) and its general solution over an interval beginning at time 0 are

$$\dot{\lambda}_1(t) = 0 \qquad\qquad \lambda_1(t) = \lambda_{10}$$
$$\dot{\lambda}_2(t) = -\lambda_1(t) \qquad \lambda_2(t) = \lambda_{20} - \lambda_{10}t$$

Consistently with the material in Sect. 2.2.3, we consider the four distinct cases which correspond to the two inequality constraints on the final state being binding or not.

Case 1. $x(t_f) \in R^2$

Case 2. $x(t_f) \in \widehat{S}_{fM_1} := \{x | x \in S_{fM_1}, \; x_1 \le b_1\}, \; b_1 = 0$
            $S_{fM_1} = \overline{S}_{fM_1} := \{x | \alpha_{fM_1}(x) = 0\}, \; \alpha_{fM_1}(x) = x_2$

Case 3. $x(t_f) \in \widehat{S}_{fM_2} := \{x | x \in S_{fM_2}, \; x_2 \le b_2\}, \; b_2 = 0$
            $S_{fM_2} = \overline{S}_{fM_2} := \{x | \alpha_{fM_2}(x) = 0\}, \; \alpha_{fM_2}(x) = x_1$

Case 4. $x(t_f) \in \overline{S}_{fM_{12}} := \{x | \alpha_{fM_1}(x) = 0, \; \alpha_{fM_2}(x) = 0\}$
            $\alpha_{fM_1}(x) = x_2, \quad \alpha_{fM_2}(x) = x_1$

Case 1. $x(t_f) \in R^2$

The orthogonality condition at the final time

$$\begin{bmatrix} \lambda_1(t_f) \\ \lambda_2(t_f) \end{bmatrix} = \begin{bmatrix} \lambda_{10} \\ \lambda_{20} - \lambda_{10}t_f \end{bmatrix} = \begin{bmatrix} 0 \\ 0 \end{bmatrix}$$

implies

$$\lambda_{10} = 0, \quad \lambda_{20} = 0, \quad u(\cdot) = 0$$

so that

$$x_1(t) = 1 + t, \quad x_2(t) = 1$$

The resulting trajectory is plotted in Fig. 6.7 and is plainly not feasible.

Case 2. $x(t_f) \in \widehat{S}_{fM_1}$

The set $\overline{S}_{fM_1}$ is a regular variety because

$$\Sigma_{fM_1}(x) = \frac{d\alpha_{fM_1}}{dx} = \begin{bmatrix} 0 & 1 \end{bmatrix}$$

and rank$(\Sigma_{fM_1}(x)) = 1, \forall x \in \overline{S}_{fM_1}$.

First, we suppose that the constraint on $x_1(t_f)$ is not present, so that $x(t_f) \in \overline{S}_{fM_1}$.
The orthogonality condition at the final time

$$\begin{bmatrix} \lambda_1(t_f) \\ \lambda_2(t_f) \end{bmatrix} = \begin{bmatrix} \lambda_{10} \\ \lambda_{20} - \lambda_{10}t_f \end{bmatrix} = \vartheta_{fM_1} \frac{d\alpha_{fM_1}(x)}{dx}' \bigg|_{x=x(t_f)} = \vartheta_{fM_1} \begin{bmatrix} 0 \\ 1 \end{bmatrix}$$

implies $\lambda_{10} = 0$ and $\lambda_2(t) = \lambda_{20}$. Therefore the control is $u(\cdot) = -\lambda_{20}$ and the resulting state motion is

$$x_1(t) = 1 + t - \lambda_{20}\frac{t^2}{2}, \quad x_2(t) = 1 - \lambda_{20}t$$

By enforcing feasibility ($x_2(t_f) = 0$), we get

$$\lambda_{20} = 1, \quad x_1(t_f) = 1.5$$

Therefore the constraint can not be ignored, because $x(t_f) \notin \widehat{S}_{fM_1}$. See Fig. 6.7.

It follows that $x_1(t_f) = 0$. Then the control is $u(t) = \lambda_{10}t - \lambda_{20}$ and the resulting state motion is

$$x_1(t) = 1 + t - \lambda_{20}\frac{t^2}{2} + \frac{\lambda_{10}}{6}t^3, \quad x_2(t) = 1 - \lambda_{20}t + \frac{\lambda_{10}}{2}t^2$$

By enforcing feasibility ($x(t_f) = 0$), we get

$$\lambda_{10} = 18, \quad \lambda_{20} = 10$$

The corresponding state trajectory is plotted in Fig. 6.7 and the value of the performance index is $J_{fM_1} = 14$.

Case 3. $x(t_f) \in \widehat{S}_{fM_2}$

The set $\overline{S}_{fM_2}$ is a regular variety because

$$\Sigma_{fM_2}(x) = \frac{d\alpha_{fM_2}}{dx} = \begin{bmatrix} 1 & 0 \end{bmatrix}$$

and $\text{rank}(\Sigma_{fM_2}(x)) = 1, \forall x \in \overline{S}_{fM_2}$.

First, we suppose that the constraint on $x_2(t_f)$ is not present, so that $x(t_f) \in \overline{S}_{fM_2}$. The orthogonality condition at the final time

$$\begin{bmatrix} \lambda_1(t_f) \\ \lambda_2(t_f) \end{bmatrix} = \begin{bmatrix} \lambda_{10} \\ \lambda_{20} - \lambda_{10} t_f \end{bmatrix} = \vartheta_{fM_2} \left. \frac{d\alpha_{fM_2}(x)}{dx} \right|'_{x=x(t_f)} = \vartheta_{fM_2} \begin{bmatrix} 1 \\ 0 \end{bmatrix}$$

implies $\lambda_{10} = \lambda_{20}$ and $\lambda_2(t) = \lambda_{10}(1-t)$. Therefore the control is $u(t) = \lambda_{10}(t-1)$ and the resulting state motion is

$$x_1(t) = 1 + t - \lambda_{10}\frac{t^2}{2} + \lambda_{10}\frac{t^3}{6}, \quad x_2(t) = 1 - \lambda_{10}t + \lambda_{10}\frac{t^2}{2}$$

By enforcing feasibility ($x_1(t_f) = 0$), we get

$$\lambda_{10} = \lambda_{20} = 6, \quad x_2(t_f) = -2$$

and conclude that the constraint on $x_2(t_f)$ is not binding because $x(t_f) \in \widehat{S}_{fM_2}$. The corresponding state trajectory is plotted in Fig. 6.7.

The value of the performance index is $J_{fM_2} = 6$ which is lower than J_{fM_1}. This outcome is easily predictable because the case under consideration is less constrained than the previous one.

Case 4. $x(t_f) \in \overline{S}_{fM_{12}}$

The discussion of this case is substantially similar to the one relevant to *Case 2*, after having ascertained that the constraint on $x_1(t_f)$ can not be ignored. Thus $x(t_f) = 0$ and the control and the value of the performance index are those of *Case 2*.

In summary, we have found two controls which satisfy the NC: the control of *Case 3*, namely $u(t) = 6(t-1)$ is the best one.   ∎

Problem 6.6

Let

$$x(0) = \begin{bmatrix} 0 \\ 0 \end{bmatrix}, \quad J = \int_0^{t_f} l(x(t), u(t), t)dt, \quad l(x, u, t) = \frac{u^2}{2}, \quad u(t) \in R, \ \forall t, \quad 0 \le t_f \le 1$$

$$x(t_f) \in \widehat{S}_f = \{x | x \in S_f, a_1 \le x_1 \le b_1\}, \quad a_1 = 1, \ b_1 = 3$$

$$S_f = \overline{S}_f = \{x | \alpha_f(x) = 0\}, \quad \alpha_f(x) = x_1 - x_2 - 2$$

The set $\overline{S}_f$ is a regular variety because

$$\Sigma_f(x) = \frac{d\alpha_f(x)}{dx} = \begin{bmatrix} 1 & -1 \end{bmatrix}$$

and rank($\Sigma_f(x)$) $= 1$, $\forall x \in \overline{S}_f$.

An interesting feature of this problem is the presence of an upper bound on the final time: this constraint can be dealt with by resorting to Remark 2.4. Consistently, we introduce a new state variable x_3 which must satisfy the following equations

$$\dot{x}_3(t) := 1, \quad x_3(0) = 0$$

Therefore $x_3(t) = t$ and this state variable can legitimately be identified with the time t. In so doing the system is *enlarged* because its state $x_a := \begin{bmatrix} x_1 & x_2 & x_3 \end{bmatrix}'$ is now a three-dimensional vector. The state is constrained at the final time, namely

$$x_a(t_f) \in \widehat{S}_{af} = \{ x_a | x_a \in S_{af}, a_1 \le x_1 \le b_1, a_3 \le x_3 \le b_3 \}, a_3 = 0, b_3 = 1$$
$$S_{af} = \overline{S}_{af} = \{ x_a | \alpha_{af1}(x_a) = 0 \}, \quad \alpha_{af1}(x_a) = x_1 - x_2 - 2$$

The set $\widehat{S}_{af}$ is shown in Fig. 6.8.

First, we observe that the final time t_f can not be zero, so that the occurrence $x_3(t_f) = 0$ must not be discussed.

The set $\overline{S}_{af}$ is a regular variety. In fact

$$\Sigma_{af}(x_a) = \frac{d\alpha_{af1}(x_a)}{dx_a} = \begin{bmatrix} 1 & -1 & 0 \end{bmatrix}$$

and rank($\Sigma_{af}(x_a)$) $= 1$, $\forall x_a \in \overline{S}_{af}$.

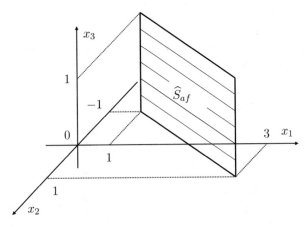

Fig. 6.8 Problem 6.6. The set $\widehat{S}_{af}$

The hamiltonian function and the H-minimizing control are

$$H = \frac{u^2}{2} + \lambda_1 x_2 + \lambda_2 u + \lambda_3, \quad u_h = -\lambda_2$$

Thus Eq. (2.5a) and its general solution over an interval beginning at time 0 are

$$\begin{aligned}
\dot{\lambda}_1(t) &= 0 & \lambda_1(t) &= \lambda_{10} \\
\dot{\lambda}_2(t) &= -\lambda_1(t) & \lambda_2(t) &= \lambda_{20} - \lambda_{10} t \\
\dot{\lambda}_3(t) &= 0 & \lambda_3(t) &= \lambda_{30}
\end{aligned}$$

Consistently with the material in Sect. 2.2.3, we consider the following six cases

Case 1. $x_a(t_f) \in \overline{S}_{af}$

Case 2. $x_a(t_f) \in \overline{S}_{afm_1} := \{x_a | x_a \in \overline{S}_{af}, \; \alpha_{af2}(x_a) = 0\}, \; \alpha_{af2}(x_a) = x_1 - a_1$

Case 3. $x_a(t_f) \in \overline{S}_{afM_1} := \{x_a | x_a \in \overline{S}_{af}, \; \alpha_{af3}(x_a) = 0\}, \; \alpha_{af3}(x_a) = x_1 - b_1$

Case 4. $x_a(t_f) \in \overline{S}_{afM_2} := \{x_a | x_a \in \overline{S}_{af}, \; \alpha_{af4}(x_a) = 0\}, \; \alpha_{af4}(x_a) = x_3 - 1$

Case 5. $x_a(t_f) \in \overline{S}_{afm_1 M_2} := \{x_a | x_a \in \overline{S}_{afm_1}, \; \alpha_{af4}(x_a) = 0\}$

Case 6. $x_a(t_f) \in \overline{S}_{afM_1 M_2} := \{x_a | x_a \in \overline{S}_{afM_1}, \; \alpha_{af4}(x_a) = 0\}$

It is easy to verify that all sets $\overline{S}_{afm_1}, \overline{S}_{afM_1}, \overline{S}_{afM_2}, \overline{S}_{afm_1 M_2}, \overline{S}_{afM_1 M_2}$ are regular varieties. In fact

$$\Sigma_{afm_1}(x_a) = \begin{bmatrix} \dfrac{d\alpha_{af1}(x_a)}{dx_a} \\[2mm] \dfrac{d\alpha_{af2}(x_a)}{dx_a} \end{bmatrix} = \begin{bmatrix} 1 & -1 & 0 \\ 1 & 0 & 0 \end{bmatrix}$$

$$\text{rank}(\Sigma_{afm_1}(x_a)) = 2, \; \forall x_a \in \overline{S}_{afm_1}$$

$$\Sigma_{afM_1}(x_a) = \begin{bmatrix} \dfrac{d\alpha_{af1}(x_a)}{dx_a} \\[2mm] \dfrac{d\alpha_{af3}(x_a)}{dx_a} \end{bmatrix} = \begin{bmatrix} 1 & -1 & 0 \\ 1 & 0 & 0 \end{bmatrix}$$

$$\text{rank}(\Sigma_{afM_1}(x_a)) = 2, \; \forall x_a \in \overline{S}_{afM_1}$$

$$\Sigma_{afM_2}(x_a) = \begin{bmatrix} \dfrac{d\alpha_{af1}(x_a)}{dx_a} \\[2mm] \dfrac{d\alpha_{af4}(x_a)}{dx_a} \end{bmatrix} = \begin{bmatrix} 1 & -1 & 0 \\ 0 & 0 & 1 \end{bmatrix}$$

$$\text{rank}(\Sigma_{afM_2}(x_a)) = 2, \; \forall x_a \in \overline{S}_{afM_2}$$

$$\Sigma_{afm_1M_2}(x_a) = \begin{bmatrix} \dfrac{d\alpha_{af1}(x_a)}{dx_a} \\ \dfrac{d\alpha_{af2}(x_a)}{dx_a} \\ \dfrac{d\alpha_{af4}(x_a)}{dx_a} \end{bmatrix} = \begin{bmatrix} 1 & -1 & 0 \\ 1 & 0 & 0 \\ 0 & 0 & 1 \end{bmatrix}$$

$$\text{rank}(\Sigma_{afm_1M_2}(x_a)) = 3, \ \forall x_a \in \overline{S}_{afm_1M_2}$$

$$\Sigma_{afM_1M_2}(x_a) = \begin{bmatrix} \dfrac{d\alpha_{af1}(x_a)}{dx_a} \\ \dfrac{d\alpha_{af3}(x_a)}{dx_a} \\ \dfrac{d\alpha_{af4}(x_a)}{dx_a} \end{bmatrix} = \begin{bmatrix} 1 & -1 & 0 \\ 1 & 0 & 0 \\ 0 & 0 & 1 \end{bmatrix}$$

$$\text{rank}(\Sigma_{afM_1M_2}(x_a)) = 3, \ \forall x_a \in \overline{S}_{afM_1M_2}$$

We now discuss in detail the six cases above.

Case 1. $x_a(t_f) \in \overline{S}_{af}$

The final time t_f is free because no constraints are imposed on $x_3(t_f)$. Thus both the orthogonality and transversality conditions at the final time must be satisfied. The latter can be written at the initial time because the problem is time-invariant.

If the expression of the H-minimizing control and the solution of Eq. (2.5a) are taken into account, we get

$$\begin{bmatrix} \lambda_1(t_f) \\ \lambda_2(t_f) \\ \lambda_3(t_f) \end{bmatrix} = \begin{bmatrix} \lambda_{10} \\ \lambda_{20} - \lambda_{10}t_f \\ \lambda_{30} \end{bmatrix} = \vartheta_f \left. \frac{d\alpha_{af1}(x_a)}{dx_a} \right|'_{x_a = x_a(t_f)} = \vartheta_f \begin{bmatrix} 1 \\ -1 \\ 0 \end{bmatrix}$$

$$H(0) = \frac{u^2(0)}{2} + \lambda_1(0)x_2(0) + \lambda_2(0)u(0) + \lambda_3(0) = -\frac{\lambda_{20}^2}{2} + \lambda_{30} = 0$$

and

$$\lambda_{20} = 0, \quad \lambda_{30} = 0, \quad t_f = 1$$

The motion of the first two state variables of the enlarged system is

$$x_1(t) = \frac{\lambda_{10}}{6}t^3, \quad x_2(t) = \frac{\lambda_{10}}{2}t^2$$

because $u(t) = \lambda_{10}t$, and, by requiring that $x(t_f) \in \overline{S}_f$ ($\alpha_f(x) = 0$), we obtain $\lambda_{10} = -6$ which in turn implies $x_1(t_f) = -1 \notin [1, 3]$.

Case 2. $x_a(t_f) \in \overline{S}_{afm_1}$

First, we observe that the constraint $x(t_f) \in \overline{S}_f$ requires $x_2(t_f) = -1$, so that only the third component of the state of the enlarged system is free. Therefore the admissible final states $x_a(t_f)$ must satisfy the constraint

$$x_a(t_f) \in \{x_a | \alpha_{afm1}(x_a) = 0, \ \alpha_{afm2}(x_a) = 0\}$$
$$\alpha_{afm1}(x_a) = x_1 - 1, \quad \alpha_{afm2}(x_a) = x_2 + 1$$

As in *Case 1*, the orthogonality and transversality conditions at the final time must be verified. The latter can be written at the initial time because the problem is time-invariant.

If the expression of the H-minimizing control and the solution of Eq. (2.5a) are taken into account, we get

$$\begin{bmatrix} \lambda_1(t_f) \\ \lambda_2(t_f) \\ \lambda_3(t_f) \end{bmatrix} = \begin{bmatrix} \lambda_{10} \\ \lambda_{20} - \lambda_{10}t_f \\ \lambda_{30} \end{bmatrix} = \sum_{i=1}^{2} \vartheta_{fmi} \frac{d\alpha_{afmi}(x_a)}{dx_a} \Bigg|'_{x_a = x_a(t_f)} = \begin{bmatrix} \vartheta_{fm1} \\ -\vartheta_{fm2} \\ 0 \end{bmatrix}$$

$$H(0) = \frac{u^2(0)}{2} + \lambda_1(0)x_2(0) + \lambda_2(0)u(0) + \lambda_3(0) = -\frac{\lambda_{20}^2}{2} + \lambda_{30} = 0$$

so that

$$\lambda_{20} = 0, \quad \lambda_{30} = 0$$

The motion of the first two state variables of the enlarged system is

$$x_1(t) = \frac{\lambda_{10}}{6}t^3, \quad x_2(t) = \frac{\lambda_{10}}{2}t^2$$

because $u(t) = \lambda_{10}t$, and, by requiring that $x(t_f) \in \overline{S}_f$ ($x_1(t_f) = 1$ and $x_2(t_f) = -1$), we obtain

$$\lambda_{10} = -\frac{2}{9}, \quad t_f = -3$$

Obviously, this result must be discarded.

Case 3. $x_a(t_f) \in \overline{S}_{afM_1}$

We proceed in a completely similar way to the one followed for *Case 2* and obtain

$$\lambda_{20} = 0, \quad \lambda_{30} = 0$$

The motion of the first two state variables of the enlarged system is the same of the previous case because $u(t) = \lambda_{10}t$, and, by requiring that $x(t_f) \in \overline{S}_f$ ($x_1(t_f) = 3$ and $x_2(t_f) = 1$), we find

$$\lambda_{10} = \frac{2}{81} \quad t_f = 9$$

Again this result must be discarded because the constraint on the final time is violated.

Case 4. $x_a(t_f) \in \overline{S}_{afM_2}$

Observe that the final time is now given and equal to 1, consequently only the orthogonality condition at the final time must be imposed. In view of the solution of Eq. (2.5a), such a condition is

$$
\begin{bmatrix} \lambda_1(t_f) \\ \lambda_2(t_f) \\ \lambda_3(t_f) \end{bmatrix} = \begin{bmatrix} \lambda_{10} \\ \lambda_{20} - \lambda_{10}t_f \\ \lambda_{30} \end{bmatrix} = \sum_{i=1}^{2} \vartheta_{fMi} \left. \frac{d\alpha_{afMi}(x_a)}{dx_a} \right|'_{x_a=x_a(t_f)} = \begin{bmatrix} \vartheta_{fM1} \\ -\vartheta_{fM1} \\ \vartheta_{fM2} \end{bmatrix}
$$

and $\lambda_{20} = 0$. The motion of the first two state variables of the enlarged system is the same of the previous cases because $u(t) = \lambda_{10}t$, and, by requiring that $x_a(t_f) \in \overline{S}_{af}$ ($\alpha_{af1}(x_a) = 0$), it results, as for *Case 1*, $\lambda_{10} = -6$ which, in turn, implies $x_1(t_f) = -1 \notin [1, 3]$.

Case 5. $x_a(t_f) \in \overline{S}_{afm_1M_2}$

First, we observe that the request $x(t_f) \in \overline{S}_f$ implies $x_2(t_f) = -1$, so that the final state of the enlarged system is given. The motion of the first two components of x_a is

$$
x_1(t) = \frac{\lambda_{10}}{6}t^3 - \frac{\lambda_{20}}{2}t^2, \quad x_2(t) = \frac{\lambda_{10}}{2}t^2 - \lambda_{20}t
$$

because $u(t) = \lambda_{10}t - \lambda_{20}$. Then condition $x_1(t_f) = 1$, $x_2(t_f) = -1$ supplies

$$
\lambda_{10} = -18, \quad \lambda_{20} = -8, \quad J = 14
$$

Case 6. $x_a(t_f) \in \overline{S}_{afM_1M_2}$

First, we observe that the request $x(t_f) \in \overline{S}_f$ implies $x_2(t_f) = 1$, so that the final state of the enlarged system is given. The motion of the first two components of x_a is given by the same equations of *Case 5* because $u(t) = \lambda_{10}t - \lambda_{20}$. Thus the condition $x_1(t_f) = 3$, $x_2(t_f) = 1$ supplies

$$
\lambda_{10} = -30, \quad \lambda_{20} = -16, \quad J = 38
$$

In summary, we have found two solutions which satisfy the NC. On the basis of the values of the performance index, the solution to be preferred is the one of *Case 5*. ∎

Chapter 7
Minimum Time Problems

Abstract The performance index of the seven problems considered in this chapter is simply the time required to transfer the state of the system from a regular variety to a set which is always a regular variety but in the fifth problem. The amplitude of the control variable is constrained.

The performance index of the seven problems considered in this chapter is the time required to transfer the state of the double integrator from a regular variety $\overline{S}_0$ to a set S_f which is always a regular variety but in the fifth problem. As we noticed in Sect. 2.2.4, to which reference is extensively made, the amplitude of the control variable is constrained. More specifically,

$$J = \int_0^{t_f} l(x(t), u(t), t) \, dt, \quad l(x, u, t) = 1$$
$$u(t) \in \overline{U} \; \forall t, \quad \overline{U} = \{u \mid -1 \le u \le 1\}$$

It is straightforward to verify that Assumption 2.3, presented in Sect. 2.2.4, is satisfied. As a consequence, the optimal control, if it exists, is piecewise constant (Theorem 2.3). Furthermore the number of switchings of any extremal control is not greater than one (Theorem 2.4). In fact the eigenvalues of the dynamic matrix A are zero and the set $\overline{U}$ is a parallelepiped. Therefore any extremal control can be either constant over all the control interval and equal to ± 1 or piecewise constant with a single switching from 1 to -1 or viceversa.

The seven problems of this chapter have the following features.

(1) The initial state x_0 is given in Problem 7.1 where the final state is the origin of the state space. We ascertain that the optimal control exists whatever x_0 is.

(2) Particular initial and final states are chosen in Problem 7.2, the latter being different from the origin. We verify that two extremal controls exist, one of which is indeed optimal.

(3) The final state in Problems 7.3 and 7.4 is not completely specified and we show that the optimal control exists, though not unique.

© Springer International Publishing Switzerland 2017

A. Locatelli, *Optimal Control of a Double Integrator*, Studies in Systems, Decision and Control 68, DOI 10.1007/978-3-319-42126-1_7

149

(4) The final state must belong to a set which is not a regular variety in Problem 7.5. Also in this case we find the optimal control for each initial state.

(5) The last two problems (Problems 7.6 and 7.7) are somehow atypical because the initial state is not given: indeed it must belong to a set which is a regular variety. Their discussion is anyhow not complex and we can compute an optimal solution.

Problem 7.1

Let

$$x(0) = x_0, \quad x(t_f) = \begin{bmatrix} 0 \\ 0 \end{bmatrix},$$

where x_0 is given.

First, we consider the state trajectories corresponding to $u(\cdot) = \pm 1$. Their equations are

$$x_1 = \frac{x_2^2}{2} + k, \quad \text{if} \quad u(\cdot) = 1$$

$$x_1 = -\frac{x_2^2}{2} + k, \quad \text{if} \quad u(\cdot) = -1$$

where k is a constant. Along these trajectories x_2 increases when $u(\cdot) = 1$, whereas x_2 decreases when $u(\cdot) = -1$ (see Fig. 7.1a). It is straightforward to ascertain that, no matter what the initial state x_0 is, we have an infinite number of piecewise constant controls which take on the values ± 1 only and generate trajectories eventually ending at the origin. In particular, it is easy to verify that there always exists a control which switches only once. In view of the material presented in Sect. 2.2.4 we can conclude: (i) the optimal control exists (Theorem 2.7); (ii) the optimal control is unique (Theorem 2.5); (iii) the optimal control coincides with the unique feasible extremal (Theorem 2.6) which switches only once between the values ± 1 (Theorem 2.4). Therefore the optimal control exists for all initial states and is a piecewise constant function which switches at most once between the values ± 1. The shape of the state trajectories (see Fig. 7.1a) shows that the value taken on by the optimal control at any time only depends on the location, at the same time, of the system state in the $x_1 - x_2$ plane. More precisely, when at time t the state $x(t)$ is located above the curve $A - 0 - B$ or on the piece $A - 0$ of it (see Fig. 7.1b) we have $u(t) = -1$,

Fig. 7.1 Problem 7.1. **a** The state trajectories when $u(\cdot) = 1$ (*solid lines*) and when $u(\cdot) = -1$ (*dashed lines*). **b** The switching curve $A - 0 - B$ and the two optimal trajectories corresponding to the initial states $\hat{x}_0$ and $\bar{x}_0$

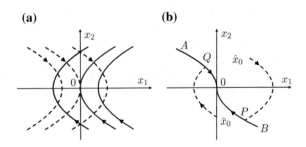

whereas when the state $x(t)$ is located below the curve $A - 0 - B$ or on the piece $0 - B$ of it (see Fig. 7.1b) we have $u(t) = 1$. The curve $A - 0 - B$, formally defined by

$$x_1 = \begin{cases} -\dfrac{x_2^2}{2} & x_2 \geq 0 \\[2mm] \dfrac{x_2^2}{2} & x_2 \leq 0 \end{cases}$$

can legitimately be referred to as the *switching curve* because the control switches when the state reaches it. Notice that the switching curve allows to compute the optimal control by resorting to a closed-loop structure of the control system. At each time the regulator supplies the value of the control variable as a function of the value of the state at that time only. Two state trajectories of the optimally controlled system are plotted (dashed lines) in Fig. 7.1b: the trajectory $\hat{x}_0$-P-0 corresponds to the initial state $\hat{x}_0$, whereas the trajectory $\bar{x}_0$-Q-0 corresponds to the initial state $\bar{x}_0$. ∎

Problem 7.2

Let

$$x(0) = x_0 := \begin{bmatrix} 1 \\ 1 \end{bmatrix}, \quad x(t_f) = x_f := \begin{bmatrix} 2 \\ 1 \end{bmatrix}$$

We can easily find two feasible extremals by looking at the trajectories corresponding to $u(\cdot) = 1$ and $u(\cdot) = -1$ which pass through x_0 and x_f (see Fig. 7.1 (a)). Consistently, we note that Theorem 2.6 can not be exploited because $x_f \neq 0$. The first extremal, $u^{(1)}(\cdot)$, switches from 1 to -1 at time $\tau = \sqrt{2} - 1$, whereas the second one $u^{(2)}(\cdot)$, switches from -1 to 1 at time $\tau = 1$. The relevant two state trajectories are plotted in Fig. 7.2. The optimal control exists thanks to Theorem 2.7 and, in view of Theorem 2.5, coincides with the extremal to which the smaller performance index corresponds. When $u(\cdot) = u^{(1)}(\cdot)$ the performance index is $t_{f_1} = 2(\sqrt{2} - 1)$, whereas when $u(\cdot) = u^{(2)}(\cdot)$ the performance index is $t_{f_2} = 2$. The optimal control coincides with the first extremal, namely

Fig. 7.2 Problem 7.2. The two state trajectories corresponding to the two extremals $u^{(1)}(\cdot)$ (*solid line*) and $u^{(2)}(\cdot)$ (*dotted line*)

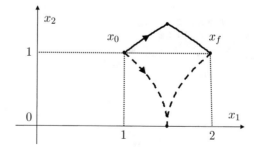

$$u(t) = 1, \quad 0 \le t < \sqrt{2} - 1$$
$$u(t) = -1, \quad \sqrt{2} - 1 < t \le 2(\sqrt{2} - 1)$$

because $t_{f_1} < t_{f_2}$.   ■

Problem 7.3

Let

$$x(0) = x_0 := \begin{bmatrix} 0 \\ 0 \end{bmatrix}, \quad x(t_f) = x_f \in \overline{S}_f := \{x | \alpha_f(x) = 0\}, \quad \alpha_f(x) := x_1^2 - 1$$

The set $\overline{S}_f$ of the admissible final states is shown in Fig. 7.3 and is a regular variety.
In fact

$$\Sigma_f(x) = \frac{d\alpha_f(x)}{dx} = \begin{bmatrix} 2x_1 & 0 \end{bmatrix}$$

and rank$(\Sigma_f(x)) = 1, \forall x \in \overline{S}_f$, because $x_1 \ne 0$ if $x \in \overline{S}_f$.

We can not resort to Theorem 2.5. As a matter of fact, the admissible final states
are all the points of the two straight lines defined by the equations $x_1 = 1$ and
$x_1 = -1$. If we take into account the trajectories corresponding to $u(\cdot) = 1$ and
$u(\cdot) = -1$ (see Fig. 7.1a), we easily ascertain that an infinite number of feasible
extremals exists. However, a simple thought leads to the conclusion that the minimum
value of the performance index is attained by the two extremals $u^{(1)}(\cdot) = 1$ and
$u^{(2)}(\cdot) = -1$ which entail $x_2(t_f) = \sqrt{2}$ and $x_2(t_f) = -\sqrt{2}$, respectively. In both
cases we have $t_f = \sqrt{2}$ and we conclude that the two extremals are optimal controls.
The corresponding state trajectories are plotted in Fig. 7.3, together with the set $\overline{S}_f$.

The lack of uniqueness of the optimal control stems from the fact that if $\bar{u}(\cdot)$ is an
optimal control and $\bar{t}_f$ is the time when $\bar{x}(\bar{t}_f)$ belongs to $\overline{S}_f$, also the control $-\bar{u}(\cdot)$
is optimal. This statement is correct because at time $\bar{t}_f$ the control $-\bar{u}(\cdot)$ drives the
state to $-\bar{x}(\bar{t}_f)$ which belongs to $\overline{S}_f$ as well, this set being symmetric with respect
to the origin (see Fig. 7.3, once more).   ■

Fig. 7.3 Problem 7.3. The
state trajectories
corresponding to the two
extremals $u^{(1)}(\cdot) = 1$ (*solid
line*) and $u^{(2)}(\cdot) = -1$
(*dashed line*)

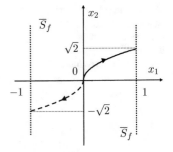

Problem 7.4

Let

$$x(0) = x_0 := \begin{bmatrix} 0 \\ 0 \end{bmatrix}, \quad x(t_f) = x_f \in \overline{S}_f = \{x | \alpha_{f1}(x) = 0, \ \alpha_{f2}(x) = 0\}$$

$$\alpha_{f1}(x) = x_1^2 - 1, \ \alpha_{f2}(x) = x_2$$

The set $\overline{S}_f$ of the admissible final states is a regular variety. In fact

$$\Sigma_f(x) = \frac{d\left(\begin{bmatrix} \alpha_{f1}(x) \\ \alpha_{f2}(x) \end{bmatrix}\right)}{dx} = \begin{bmatrix} 2x_1 & 0 \\ 0 & 1 \end{bmatrix}$$

and $\operatorname{rank}(\Sigma_f(x)) = 2, \ \forall x \in \overline{S}_f$, because $x_1 \neq 0$ if $x \in \overline{S}_f$.

Also in the present case we can not resort to Theorem 2.5 as we have two admissible final states, namely

$$x_{f_1} := \begin{bmatrix} 1 \\ 0 \end{bmatrix}, \quad x_{f_2} := \begin{bmatrix} -1 \\ 0 \end{bmatrix}$$

Once more, we can evaluate the number of feasible extremals by inspecting the state trajectories corresponding to $u(\cdot) = 1$ and $u(\cdot) = -1$ (see Fig. 7.1a). We find two extremals, $u^{(1)}(\cdot)$ and $u^{(2)}(\cdot)$ which are defined by

$$u^{(1)}(t) = 1, \ 0 \le t < \tau, \quad u(t) = -1, \ \tau < t \le t_f$$
$$u^{(2)}(t) = -1, \ 0 \le t < \tau, \ u(t) = 1, \ \tau < t \le t_f$$

Notice that $u^{(2)}(\cdot) = -u^{(1)}(\cdot)$. In both cases, we have $\tau = 1$ and $t_f = 2$, so that they are optimal controls. The two state trajectories corresponding to these extremals are plotted in Fig. 7.4. They are symmetric with respect to the origin, consistently with a simple thought on the nature of the problem at hand. ∎

Fig. 7.4 Problem 7.4. The state trajectories corresponding to the two extremals $u^{(1)}(\cdot)$ (*solid line*) and $u^{(2)}(\cdot)$ (*dashed line*)

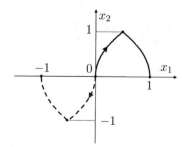

Problem 7.5

Let

$$x(0) = x_0, \quad x(t_f) \in \widehat{S}_f = \{x | x \in S_f, \ x_1 \ge a_1\}, \ a_1 = 1$$
$$S_f = \overline{S}_f = \{\alpha_f(x) = 0\}, \ \alpha_f(x) = x_2 - 1$$

where the initial state x_0 is given.

The set $\overline{S}_f$ is a regular variety. In fact

$$\Sigma_f(x) = \frac{d\alpha_f(x)}{dx} = \begin{bmatrix} 0 & 1 \end{bmatrix}$$

and rank$(\Sigma_f(x)) = 1, \ \forall x \in \overline{S}_f$.

In dealing with this problem we make also reference to the material presented in Sect. 2.2.3 and illustrated in Chap. 6, because the set $\widehat{S}_f$ of the admissible final states is not a regular variety.

Consistently, we first ignore the constraint on the state variable x_1 at the final time. Thus $x(t_f) \in \overline{S}_f$. We observe that the general solution of Eq. (2.5a) over an interval beginning at time 0 is

$$\lambda_1(t) = \lambda_1(0), \quad \lambda_2(t) = \lambda_2(0) - \lambda_1(0)t$$

so that the orthogonality condition at the final time

$$\begin{bmatrix} \lambda_1(t_f) \\ \lambda_2(t_f) \end{bmatrix} = \begin{bmatrix} \lambda_1(0) \\ \lambda_2(0) - \lambda_1(0)t_f \end{bmatrix} = \vartheta_f \frac{d\alpha_f(x)}{dx} \bigg|'_{x=x(t_f)} = \vartheta_f \begin{bmatrix} 0 \\ 1 \end{bmatrix}$$

requires $\lambda_1(0) = 0$ which in turn implies $\lambda_2(t) = \lambda_2(0)$ and any extremal is a constant function. We recall the shapes of the state trajectories corresponding to $u(\cdot) = 1$ and $u(\cdot) = -1$ by looking at Fig. 7.1a and consider, in Fig. 7.5, the curve $A - P - B$ which results from the union of the piece of trajectory corresponding to $u(\cdot) = -1$ and ending at the point P, with the piece of trajectory corresponding to

Fig. 7.5 Problem 7.5. The regions $\mathcal{R}_1, \mathcal{R}_2$ and the optimal state trajectories corresponding to the initial states $\bar{x}_0, \hat{x}_0, \check{x}_0$

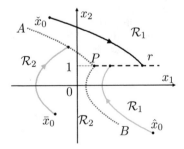

$u(\cdot) = 1$ and ending at the same point P. The coordinates of P are $x_1 = x_2 = 1$. The curve $A - P - B$ is formally defined by

$$x_1 = \begin{cases} \dfrac{3 - x_2^2}{2} & x_2 \geq 1 \\[2mm] \dfrac{1 + x_2^2}{2} & x_2 \leq 1 \end{cases}$$

We also consider the region $\mathcal{R}_1$ which is that part of the $x_1 - x_2$ plane which lies to the right of the curve $A - P - B$ and includes the curve $A - P - B$ as well (see Fig. 7.5). Then, if the initial state x_0 belongs to $\mathcal{R}_1$, we easily conclude that the optimal control coincides with the extremal which produces the trajectory starting at x_0 and ending on the half-line r defined by $x_2 = 1$, $x_1 \geq 1$ (again, see the above mentioned figure). Such an extremal is $u(\cdot) = 1$ or $u(\cdot) = -1$ according to whether x_0 is located below or above the half-line r. Now we consider the region $\mathcal{R}_2$ which is located to the left of the curve $A - P - B$ but does not include it. When $x_0 \in \mathcal{R}_2$ it is apparent that ignoring the constraint on $x_1(t_f)$ leads to violating it, so that it must be $x_1(t_f) = 1$. It is straightforward to ascertain that we have a unique feasible extremal corresponding to any initial state x_0 which belongs to $\mathcal{R}_2$. It is the piecewise constant function which takes on the value 1 over the interval $[0, \tau)$ and the value -1 over the interval $(\tau, t_f]$. The time τ is the instant when the trajectory originated at x_0 intersects the curve $A - P$.

Analogously to what has been done in Problem 7.1, the curve $A - P - B$ can legitimately be referred to as the *switching* curve because the control switches when the state reaches it. Notice that the switching curve allows to compute the optimal control by resorting to a closed-loop structure of the control system. At each time the regulator supplies the value of the control variable as a function of the value of the state at that time only. Some state trajectories and the switching curve (dotted line) are shown in Fig. 7.5. ∎

As we have anticipated the initial state is not given, but rather constrained to belong to a set $\overline{S}_0$, in the two forthcoming problems.

Problem 7.6

Let

$$x(0) \in \overline{S}_0 = \{x | \alpha_0(x) = 0\}, \quad \alpha_0(x) = x_1 + 1$$
$$x(t_f) \in \overline{S}_f = \{x | \alpha_f(x) = 0\}, \quad \alpha_f(x) = x_1 - \frac{x_2^2}{2}$$

The set $\overline{S}_0$ of the admissible initial states and the set $\overline{S}_f$ of the admissible final states are shown in Fig. 7.6. They are regular varieties. In fact we have for the first one of them

$$\Sigma_0(x) = \frac{d\alpha_0(x)}{dx} = \begin{bmatrix} 1 & 0 \end{bmatrix}$$

Fig. 7.6 Problem 7.6. The optimal state trajectory

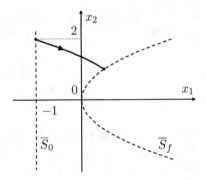

and $\mathrm{rank}(\Sigma_0(x)) = 1, \forall x \in \overline{S}_0$, whereas for the second one we have

$$\Sigma_f(x) = \frac{d\alpha_f(x)}{dx} = \begin{bmatrix} 1 & -x_2 \end{bmatrix}$$

and $\mathrm{rank}(\Sigma_f(x)) = 1, \forall x \in \overline{S}_f$.

As in the previous problems the control can switches at most once between the values ± 1, because the general solution of Eq. (2.5a) over an interval beginning at time 0 is

$$\lambda_1(t) = \lambda_1(0), \quad \lambda_2(t) = \lambda_2(0) - \lambda_1(0)t$$

The two orthogonality conditions at the initial and final time are

$$\begin{bmatrix} \lambda_1(0) \\ \lambda_2(0) \end{bmatrix} = \vartheta_0 \left. \frac{d\alpha_0(x)}{dx} \right|'_{x=x(0)} = \vartheta_0 \begin{bmatrix} 1 \\ 0 \end{bmatrix}$$

and

$$\begin{bmatrix} \lambda_1(t_f) \\ \lambda_2(t_f) \end{bmatrix} = \begin{bmatrix} \lambda_1(0) \\ \lambda_2(0) - \lambda_1(0)t_f \end{bmatrix} = \vartheta_f \left. \frac{d\alpha_f(x)}{dx} \right|'_{x=x(t_f)} = \vartheta_f \begin{bmatrix} 1 \\ -x_2(t_f) \end{bmatrix}$$

respectively. We deduce that

$$\lambda_2(0) = 0, \quad t_f = x_2(t_f)$$

Consequently, the transversality condition, written at the initial time because the problem is time-invariant,

$$H(0) = 1 + \lambda_1(0)x_2(0) + \lambda_2(0)u(0) = 1 + \lambda_1(0)x_2(0) = 0$$

supplies

$$x_2(0) = -\frac{1}{\lambda_1(0)}$$

Notice that the sign of $\lambda_2(\cdot)$ can not change, so that the control is a constant function. If we write down the expressions of $x_1(\cdot)$ and $x_2(\cdot)$ when $u(\cdot) = \pm 1$, namely

$u(t) = 1, \forall t$

$$x_1(t) = -1 - \frac{1}{\lambda_1(0)}t + \frac{t^2}{2}$$

$$x_2(t) = -\frac{1}{\lambda_1(0)} + t$$

$u(t) = -1, \forall t$

$$x_1(t) = -1 - \frac{1}{\lambda_1(0)}t - \frac{t^2}{2}$$

$$x_2(t) = -\frac{1}{\lambda_1(0)} - t$$

we can easily ascertain that the control $u(\cdot) = 1$ must be discarded. In fact, by recalling that $t_f = x_2(t_f)$, we have

$$x_2(t_f) = -\frac{1}{\lambda_1(0)} + t_f = t_f \rightarrow \frac{1}{\lambda_1(0)} = 0$$

which can not be accepted. Therefore $u(\cdot) = -1$ and, by enforcing feasibility $(x(t_f) \in \overline{S}_f)$, we obtain the equation

$$-1 - \frac{1}{\lambda_1(0)}t_f - \frac{t_f^2}{2} - \frac{\left(-\dfrac{1}{\lambda_1(0)} - t_f\right)^2}{2} = 0$$

which, together with the already mentioned relation between t_f and $x_2(t_f)$, supplies

$$\lambda_1(0) = -\frac{1}{2}, \quad t_f = 1, \quad x_2(0) = 2$$

The two regular varieties $\overline{S}_0$ and $\overline{S}_f$ are shown in Fig. 7.6 (dashed lines) together with the state trajectory (solid line). ∎

Problem 7.7

Let

$$x(0) \in \overline{S}_0 = \{x|\alpha_0(x) = 0\}, \quad \alpha_0(x) = x_1 + \frac{x_2^2}{2} + 1$$

$$x(t_f) \in \overline{S}_f = \{x|\alpha_f(x) = 0\}, \quad \alpha_f(x) = x_1 - \frac{x_2^2}{2} - 1$$

The two sets $\overline{S}_0$ and $\overline{S}_f$ of the admissible initial and final states, respectively, are shown in Fig. 7.7. They are regular varieties. In fact we have, for the first of them,

Fig. 7.7 Problem 7.7. The
optimal state trajectory

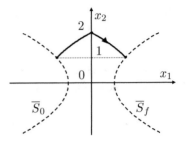

$$\Sigma_0(x) = \frac{d\alpha_0(x)}{dx} = \begin{bmatrix} 1 & x_2 \end{bmatrix}$$

and $\text{rank}(\Sigma_0(x)) = 1$, $\forall x \in \overline{S}_0$, whereas, for the second one

$$\Sigma_f(x) = \frac{d\alpha_f(x)}{dx} = \begin{bmatrix} 1 & -x_2 \end{bmatrix}$$

and $\text{rank}(\Sigma_f(x)) = 1$, $\forall x \in \overline{S}_f$.

We notice that the general solution of Eq. (2.5a) over an interval beginning at time 0 is

$$\lambda_1(t) = \lambda_1(0), \quad \lambda_2(t) = \lambda_2(0) - \lambda_1(0)t$$

so that the following four cases can occur

Case 1. $u(\cdot) = 1$

Case 2. $u(\cdot) = -1$

Case 3. $u(\cdot)$ switches at time τ and $u(0) = 1$

Case 4. $u(\cdot)$ switches at time τ and $u(0) = -1$

The two orthogonality conditions at the initial and final time are

$$\begin{bmatrix} \lambda_1(0) \\ \lambda_2(0) \end{bmatrix} = \vartheta_0 \left. \frac{d\alpha_0(x)}{dx} \right|'_{x=x(0)} = \vartheta_0 \begin{bmatrix} 1 \\ x_2(0) \end{bmatrix}$$

and

$$\begin{bmatrix} \lambda_1(t_f) \\ \lambda_2(t_f) \end{bmatrix} = \begin{bmatrix} \lambda_1(0) \\ \lambda_2(0) - \lambda_1(0)t_f \end{bmatrix} = \vartheta_f \left. \frac{d\alpha_f(x)}{dx} \right|'_{x=x(t_f)} = \vartheta_f \begin{bmatrix} 1 \\ -x_2(t_f) \end{bmatrix}$$

respectively. It follows

$$\lambda_2(0) = \lambda_1(0)x_2(0), \quad x_2(t_f) = t_f - x_2(0)$$

The transversality condition at the final time, written at the initial time because the problem is time-invariant, is

$$H(0) = 1 + \lambda_1(0)x_2(0) + \lambda_2(0)u(0) = 1 + \lambda_2(0)(1 + u(0)) = 0$$

where the above shown expression for $\lambda_2(0)$ has been exploited. In order that $1 + \lambda_2(0)(1 + u(0)) = 0$, it must be $u(0) = 1$. Therefore we disregard both *Case 2* $(u(\cdot) = -1)$ and *Case 4* (one switching of the control from -1 to 1). Also *Case 1* $(u(\cdot) = 1)$ must be disregarded. In fact, if $u(\cdot) = 1$, we have

$$x_1(t_f) = x_1(0) + x_2(0)t_f + \frac{1}{2}t_f^2$$
$$x_2(t_f) = x_2(0) + t_f$$

and, by requiring that $x(t_f) \in \overline{S}_f$, we find

$$x_1(0) = \frac{x_2^2(0)}{2} + 1$$

and

$$-x_2^2(0) - 2 = 0$$

because $x(0) \in \overline{S}_0$. The last equation does not have a real solution.

Thus only *Case 3* (one switching from 1 to -1) is left. The switching of the control at time τ requires

$$0 = \lambda_2(\tau) = \lambda_2(0) - \lambda_1(0)\tau = \lambda_1(0)(x_2(0) - \tau)$$

where we have exploited the expression for $\lambda_2(0)$ which has been deduced from the orthogonality condition at the initial time. We note that $\lambda_1(0) \neq 0$ because, otherwise, no switching takes place. Therefore we have

$$x_2(0) = \tau, \quad x_1(0) = -1 - \frac{\tau^2}{2}$$

because $x(0) \in \overline{S}_0$. These relations first imply

$$x_1(\tau) = x_1(0) + x_2(0)\tau + \frac{\tau^2}{2} = \tau^2 - 1, \quad x_2(\tau) = x_2(0) + \tau = 2\tau$$

and then

$$t_f - \tau = t_f - x_2(0) = x_2(t_f) = x_2(\tau) - (t_f - \tau) = 3\tau - t_f \rightarrow t_f = 2\tau$$

where the second equality sign is a consequence of the orthogonality condition at the final time. Furthermore we have

$$x_1(t_f) = x_1(\tau) + x_2(\tau)(t_f - \tau) - \frac{(t_f - \tau)^2}{2} = \frac{5\tau^2}{2}$$

Finally, the request $x(t_f) \in \overline{S}_f$ supplies the value of τ. As a whole, we find

$$\tau = 1, \quad t_f = 2, \quad x_1(0) = -\frac{3}{2}, \quad x_1(t_f) = \frac{3}{2}, \quad x_2(0) = x_2(t_f) = 1$$

The resulting state trajectory is shown in Fig. 7.7.   ∎

Chapter 8
Integral Constraints

Abstract In this chapter we tackle problems where the controlled system must satisfy integral-type constraints. They amount to requiring that the integral of some functions of time as well as of the state and control variables take on values which either are equal or less than given quantities. We first consider equality constraints and then inequality ones.

In this chapter we tackle problems where the controlled double integrator must also satisfy integral-type constraints. We recall that these constraints amount to requiring that the integral of some functions of time as well as of the state and control variables take on values which either are equal or less than given quantities. They have been formally presented in Sect. 2.3.1 (point (a)), whereas the relevant way of handling them has been outlined in Sect. 2.3.2. Consistently with the discussion there, we first consider equality constraints and then inequality ones.

8.1 Integral Equality Constraints

The six problems dealt with in this section are characterized by the following facts: (1) the function $w_e(\cdot, \cdot, \cdot)$ in the integral constraints of Problems 8.1–8.4 depends on the state variable x_1 only; (2) the control variable is constrained (actually its absolute value) in Problem 8.3; (3) the set of the admissible final states is a non-regular variety in Problem 8.2; (4) we consider three different scenarios in Problem 8.5 where the function $w_e(\cdot, \cdot, \cdot)$ depends on: (i) both the state variables, (ii) the first state variable and the control, (iii) the first state variable and time; (5) the function $w_e(\cdot, \cdot, \cdot)$ depends on the control variable only in Problem 8.6; (6) the final time is free in Problems 8.3 and 8.6.

The discussion needs the material presented in Sect. 2.3.2, whereas Problem 8.2 also requires Sect. 2.3.2.

© Springer International Publishing Switzerland 2017
A. Locatelli, *Optimal Control of a Double Integrator*, Studies in Systems, Decision and Control 68, DOI 10.1007/978-3-319-42126-1_8

Problem 8.1

Let

$$x(0) = \begin{bmatrix} \beta \\ \gamma \end{bmatrix}, \quad x(t_f) = \begin{bmatrix} 0 \\ 0 \end{bmatrix}, \quad J = \int_0^{t_f} l(x(t), u(t), t) \, dt, \quad l(x, u, t) = \frac{u^2}{2}$$

$$u(t) \in R, \; \forall t, \quad t_f = 1, \quad \int_0^{t_f} w_e(x(t), u(t), t) \, dt = \overline{w}_e, \quad w_e(x, u, t) = x_1, \quad \overline{w}_e = 0$$

where both the parameters β and γ are given and not contemporary zero. We take into account the integral constraint by adding a new state variable x_3 defined by the equations

$$\dot{x}_3(t) = w_e(x(t), u(t), t) = x_1(t), \quad x_3(0) = 0, \quad x_3(t_f) = \overline{w}_e$$

In so doing, the system is *enlarged* because its state $x_a := \begin{bmatrix} x_1 & x_2 & x_3 \end{bmatrix}'$ is now a three-dimensional vector which is given at the final time. Thus the hamiltonian function and the H-minimizing control are

$$H = \frac{u^2}{2} + \lambda_1 x_2 + \lambda_2 u + \lambda_3 x_1, \quad u_h = -\lambda_2$$

Consequently, Eq. (2.5a) and its general solution over an interval beginning at time 0 are

$$\dot{\lambda}_1(t) = -\lambda_3(t) \qquad\qquad \lambda_1(t) = \lambda_{10} - \lambda_{30} t$$

$$\dot{\lambda}_2(t) = -\lambda_1(t) \qquad\qquad \lambda_2(t) = \lambda_{20} - \lambda_{10} t + \frac{\lambda_{30}}{2} t^2$$

$$\dot{\lambda}_3(t) = 0 \qquad\qquad\qquad \lambda_3(t) = \lambda_{30}$$

and, by recalling the form of the H-minimizing control, the state motion is

$$x_1(t) = \beta + \gamma t - \frac{\lambda_{20}}{2} t^2 + \frac{\lambda_{10}}{6} t^3 - \frac{\lambda_{30}}{24} t^4$$

$$x_2(t) = \gamma - \lambda_{20} t + \frac{\lambda_{10}}{2} t^2 - \frac{\lambda_{30}}{6} t^3$$

$$x_3(t) = \beta t + \frac{\gamma}{2} t^2 - \frac{\lambda_{20}}{6} t^3 + \frac{\lambda_{10}}{24} t^4 - \frac{\lambda_{30}}{120} t^5$$

If we enforce feasibility (namely $x_1(t_f) = x_2(t_f) = x_3(t_f) = 0$), we obtain three linear equations which allow computing the three (unknown) parameters $\lambda_{10}, \lambda_{20}, \lambda_{30}$ as functions of β and γ. As an example, when $\beta = 1$, $\gamma = 0$, we obtain $\lambda_{10} = 192$, $\lambda_{20} = 36$, $\lambda_{30} = 360$. The corresponding state motion of the enlarged system is plotted in Fig. 8.1.

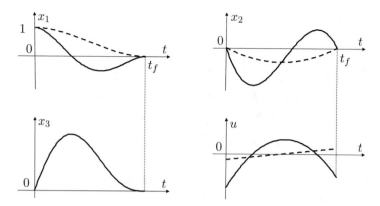

Fig. 8.1 Problem 8.1. State motion and control when $\beta = 1$, $\gamma = 0$ corresponding to the presence (*solid line*) and absence (*dashed line*) of the integral equality constraint

It is fairly interesting to compare this solution with the one resulting from the removal of the integral equality constraint. In this case the hamiltonian function and the H-minimizing control are

$$H = \frac{u^2}{2} + \nu_1 x_2 + \nu_2 u, \quad u_h = -\nu_2$$

and Eq. (2.5a) and its general solution over an interval beginning at time 0 are

$$\dot{\nu}_1(t) = 0 \qquad\qquad \nu_1(t) = \nu_{10}$$
$$\dot{\nu}_2(t) = -\nu_1(t) \qquad\qquad \nu_2(t) = \nu_{20} - \nu_{10}t$$

so that the state motion is

$$x_1(t) = \beta + \gamma t - \frac{\nu_{20}}{2}t^2 + \frac{\nu_{10}}{6}t^3$$
$$x_2(t) = \gamma - \nu_{20}t + \frac{\nu_{10}}{2}t^2$$

By enforcing feasibility ($x_1(t_f) = x_2(t_f) = 0$), we obtain two linear equations which allow computing the two (unknown) parameters ν_{10}, ν_{20} as functions of β e γ. As an example, if we give to β, γ the same values as above ($\beta = 1$, $\gamma = 0$), we get $\nu_{10} = 12$, $\nu_{20} = 6$.

Figure 8.1 clearly shows that both $x(\cdot)$ and $u(\cdot)$ are much smoother functions when the constraint is not present. This fact affects the value of the performance index in a somehow more impressive way. Indeed we have $J = J_v = 96$ when the constraint is present and $J = J_l = 6$ in the opposite case. ∎

Problem 8.2

Let

$$x(0) = \begin{bmatrix} \beta \\ 0 \end{bmatrix}, \quad J = \int_0^{t_f} l(x(t), u(t), t)\, dt, \quad l(x, u, t) = \frac{u^2}{2}, \quad u(t) \in R, \ \forall t, \quad t_f = 1$$

$$x(t_f) \in \widehat{S}_f = \{x | x \in S_f, \ x_1 \le b_1\}, \ b_1 = 1$$

$$S_f = \overline{S}_f = \{x | \alpha_f(x) = 0\}, \ \alpha_f(x) = x_2$$

$$\int_0^{t_f} w_e(x(t), u(t), t)\, dt = \overline{w}_e, \ w_e(x, u, t) = x_1, \ \overline{w}_e = 0$$

where the parameter β is given. We take into account the integral constraint by adding a new state variable x_3 defined by the equations

$$\dot{x}_3(t) = w_e(x(t), u(t), t) = x_1(t), \quad x_3(0) = 0, \quad x_3(t_f) = \overline{w}_e$$

In so doing, the system is *enlarged* because its state $x_a := \begin{bmatrix} x_1 & x_2 & x_3 \end{bmatrix}'$ is now a three-dimensional vector. Consistently, the set of the admissible final states is

$$x_a(t_f) \in \widehat{S}_{af} = \{x_a | x_a \in S_{af}, \ x_1 \le b_1\}$$

$$S_{af} = \overline{S}_{af} = \{x_a | \alpha_{fi}(x_a) = 0, \ i = 1, 2\}, \ \alpha_{f1}(x_a) = x_2, \ \alpha_{f2}(x_a) = x_3 - \overline{w}_e$$

The set $\overline{S}_{af}$ is a regular variety because

$$\Sigma_{af}(x_a) = \frac{d \begin{bmatrix} \alpha_{f1}(x_a) \\ \alpha_{f2}(x_a) \end{bmatrix}}{dx_a} = \begin{bmatrix} 0 & 1 & 0 \\ 0 & 0 & 1 \end{bmatrix}$$

and rank$(\Sigma_{af}(x_a)) = 2, \ \forall x_a \in \overline{S}_{af}$.

The hamiltonian function and the H-minimizing control are

$$H = \frac{u^2}{2} + \lambda_1 x_2 + \lambda_2 u + \lambda_3 x_1, \quad u_h = -\lambda_2$$

and Eq. (2.5a) and its general solution over an interval which begins at time 0 are

$$\dot{\lambda}_1(t) = -\lambda_3(t) \qquad\qquad \lambda_1(t) = \lambda_{10} - \lambda_{30} t$$

$$\dot{\lambda}_2(t) = -\lambda_1(t) \qquad\qquad \lambda_2(t) = \lambda_{20} - \lambda_{10} t + \frac{\lambda_{30}}{2} t^2$$

$$\dot{\lambda}_3(t) = 0 \qquad\qquad \lambda_3(t) = \lambda_{30}$$

We first consider the possibility that the inequality constraint on $x_1(t_f)$ is not binding, so that $x_a(t_f) \in \overline{S}_{af}$. In this case, the orthogonality condition at the final time

$$\begin{bmatrix} \lambda_1(t_f) \\ \lambda_2(t_f) \\ \lambda_3(t_f) \end{bmatrix} = \begin{bmatrix} \lambda_{10} - \lambda_{30} t_f \\ \lambda_{20} - \lambda_{10} t_f + \dfrac{\lambda_{30}}{2} t_f^2 \\ \lambda_{30} \end{bmatrix} = \sum_{i=1}^{2} \vartheta_{fi} \left. \frac{d\alpha_{fi}(x_a)}{dx_a} \right|'_{x_a = x_a(t_f)} =$$

$$= \vartheta_{f1} \begin{bmatrix} 0 \\ 1 \\ 0 \end{bmatrix} + \vartheta_{f2} \begin{bmatrix} 0 \\ 0 \\ 1 \end{bmatrix}$$

implies $\lambda_1(t_f) = \lambda_{10} - \lambda_{30} = 0$, namely $\lambda_{10} = \lambda_{30}$. By recalling the expression for the H-minimizing control, we conclude that the state motion of the enlarged system is

$$x_1(t) = \beta - \frac{\lambda_{20}}{2} t^2 + \frac{\lambda_{10}}{6} t^3 - \frac{\lambda_{30}}{24} t^4$$

$$x_2(t) = -\lambda_{20} t + \frac{\lambda_{10}}{2} t^2 - \frac{\lambda_{30}}{6} t^3$$

$$x_3(t) = \beta t - \frac{\lambda_{20}}{6} t^3 + \frac{\lambda_{10}}{24} t^4 - \frac{\lambda_{30}}{120} t^5$$

If we enforce feasibility ($x_2(t_f) = x_3(t_f) = 0$), we obtain

$$\lambda_{10} = 45\beta, \quad \lambda_{20} = 15\beta, \quad x_1(t_f) = -\frac{7\beta}{8}$$

because $\lambda_{10} = \lambda_{30}$. As a consequence, the constraint on $x_1(t_f)$ is not binding if $\beta \geq -8/7$. On the contrary the constraint is binding whenever $\beta < -8/7$ and it must be

$$x_a(t_f) \in \overline{S}_{afM} := \left\{ \begin{bmatrix} 1 \\ 0 \\ 0 \end{bmatrix} \right\}$$

The above given state motion, together with control feasibility ($x_a(t_f) \in \overline{S}_{afM}$), lead to

$$\lambda_{10} = 24(7 + 8\beta), \quad \lambda_{20} = 12(2 + 3\beta), \quad \lambda_{30} = 360(1 + \beta)$$

and the value J_v of the performance index is

$$J_v = \frac{1}{120} \left(3\lambda_{30}^2 + 20\lambda_{10}^2 + 60\lambda_{20}^2 - 15\lambda_{10}\lambda_{30} + 20\lambda_{20}\lambda_{30} - 60\lambda_{10}\lambda_{20} \right)$$

When we assume that the integral equality constraint is not present, we obtain a related and interesting scenario. We need not to introduce a further state variable, so that the hamiltonian function and the H-minimizing control are

$$H = \frac{1}{2} u^2 + \nu_1 x_2 + \nu_2 u, \quad u_h = -\nu_2$$

and Eq. (2.5a) and its general solution over an interval beginning at time 0 are

$$\dot{\nu}_1(t) = 0 \qquad\qquad \nu_1(t) = \nu_{10}$$
$$\dot{\nu}_2(t) = -\nu_1(t) \qquad\qquad \nu_2(t) = \nu_{20} - \nu_{10}t$$

so that

$$x_1(t) = \beta - \frac{\nu_{20}}{2}t^2 + \frac{\nu_{10}}{6}t^3$$
$$x_2(t) = -\nu_{20}t + \frac{\nu_{10}}{2}t^2$$

If we suppose that the inequality constraint on $x_1(t_f)$ is not binding, the orthogonality condition at the final time

$$\begin{bmatrix} \nu_1(t_f) \\ \nu_2(t_f) \end{bmatrix} = \begin{bmatrix} \nu_{10} \\ \nu_{20} - \nu_{10}t_f \end{bmatrix} = \vartheta_f \left.\frac{d\alpha_f(x)}{dx}\right|'_{x=x(t_f)} = \vartheta_f \begin{bmatrix} 0 \\ 1 \end{bmatrix}$$

implies $\nu_{10} = 0$. Thus, by enforcing feasibility ($x_2(t_f) = 0$), we obtain

$$\nu_{20} = 0, \quad x_1(t_f) = \beta$$

We conclude that the constraint on x_1 ($x_1(t_f) \leq 1$) can be ignored only if $\beta \leq 1$. In this case the value of the performance index is 0. On the contrary, if $\beta > 1$ we have

$$x(t_f) \in \overline{S}_{fM} := \left\{ \begin{bmatrix} 1 \\ 0 \end{bmatrix} \right\}$$

and, by enforcing feasibility ($x(t_f) \in \overline{S}_{fM}$), we get

$$\nu_{10} = 12(\beta - 1), \quad \nu_{20} = 6(\beta - 1)$$

The value of the performance index is

$$J_l = \frac{(\nu_{10} - \nu_{20})^3 + \nu_{20}^3}{6\nu_{10}}$$

A meaningful comparison between the two scenarios (presence or absence of the integral constraint) can be performed if we select the parameter β in such a way that the constraint $x_1(t_f) \leq 1$ is not binding in the first scenario and binding in the second one. A suitable choice is $\beta = 2$. The values J_v and J_f of the performance index relevant to the presence or absence of the constraint are

$$J_v = 90, \quad J_l = 6$$

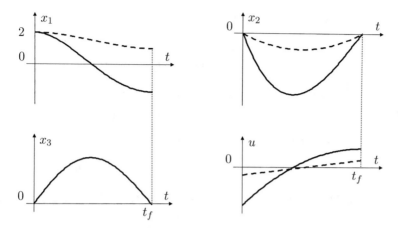

Fig. 8.2 Problem 8.2. State motion and control when $\beta = 2$ and the integral constraint is present (*solid line*) or absent (*dashed line*)

respectively. Thus we can state that, corresponding to the adopted β, the integral constraint is much more severe than the one on the final state. This conclusion can also be drawn from Fig. 8.2 where it is apparent that the absolute value of the control variable is much greater in the presence of the integral constraint. ■

Problem 8.3

Let

$$x(0) = \begin{bmatrix} 0 \\ 0 \end{bmatrix}, \quad x\left(t_f\right) \in \overline{S}_f = \left\{x | \alpha_f(x) = 0\right\}, \quad \alpha_f(x) = x_1 - 1$$

$$J = \int_0^{t_f} l(x(t), u(t), t) \, dt, \quad l(x, u, t) = 1, \quad u(t) \in \overline{U}, \ \forall t, \ \overline{U} = \{u | -1 \le u \le 1\}$$

$$\int_0^{t_f} w_e(x(t), u(t), t) \, dt = \overline{w}_e, \quad w_e(x, u, t) = x_1, \quad \overline{w}_e = 0$$

where the final time t_f is, obviously, free. We take into account the integral constraint by adding a new state variable x_3 defined by the equations

$$\dot{x}_3(t) = w_e(x(t), u(t), t) = x_1(t), \quad x_3(0) = 0, \quad x_3\left(t_f\right) = \overline{w}_e$$

In so doing the system is *enlarged* because its state $x_a := \begin{bmatrix} x_1 & x_2 & x_3 \end{bmatrix}'$ is a three-dimensional vector.

The set of the admissible final states is now

$$\overline{S}_{af} = \left\{x_a | \alpha_{afi}(x_a) = 0, \ i = 1, 2\right\}, \quad \alpha_{af1}(x_a) = x_1 - 1, \quad \alpha_{af2}(x_a) = x_3 - \overline{w}_e$$

where $\overline{S}_{af}$ is a regular variety because

$$\Sigma_{af}(x_a) := \frac{d\left(\begin{bmatrix} \alpha_{af1}(x_a) \\ \alpha_{af2}(x_a) \end{bmatrix}\right)}{dx_a} = \begin{bmatrix} 1 & 0 & 0 \\ 0 & 0 & 1 \end{bmatrix}$$

and $\text{rank}(\Sigma_{af}(x_a)) = 2, \forall x_a \in \overline{S}_{af}$.

The hamiltonian function and the H-minimizing control are

$$H = 1 + \lambda_1 x_2 + \lambda_2 u + \lambda_3 x_1, \quad u_h = -\text{sign}(\lambda_2)$$

Therefore Eq. (2.5a) and its general solution over an interval beginning at time 0 are

$$\dot{\lambda}_1(t) = -\lambda_3(t) \qquad\qquad \lambda_1(t) = \lambda_{10} - \lambda_{30}t$$
$$\dot{\lambda}_2(t) = -\lambda_1(t) \qquad\qquad \lambda_2(t) = \lambda_{20} - \lambda_{10}t + \frac{\lambda_{30}}{2}t^2$$
$$\dot{\lambda}_3(t) = 0 \qquad\qquad \lambda_3(t) = \lambda_{30}$$

The orthogonality condition at the final time

$$\begin{bmatrix} \lambda_1(t_f) \\ \lambda_2(t_f) \\ \lambda_3(t_f) \end{bmatrix} = \begin{bmatrix} \lambda_{10} - \lambda_{30}t_f \\ \lambda_{20} - \lambda_{10}t_f + \frac{\lambda_{30}}{2}t_f^2 \\ \lambda_{30} \end{bmatrix} = \sum_{i=1}^{2} \vartheta_{fi} \left. \frac{d\alpha_{afi}(x_a)}{dx_a} \right|'_{x_a=x_a(t_f)} =$$

$$= \vartheta_{f1} \begin{bmatrix} 1 \\ 0 \\ 0 \end{bmatrix} + \vartheta_{f2} \begin{bmatrix} 0 \\ 0 \\ 1 \end{bmatrix}$$

implies $\lambda_2(t_f) = 0$ and, with reference to the function $\lambda_2(\cdot)$, we must consider the following four cases.

Case 1. $\lambda_2(t) > 0, 0 \le t < t_f, \lambda_2(t_f) = 0$

Case 2. $\lambda_2(t) < 0, 0 \le t < t_f, \lambda_2(t_f) = 0$

Case 3. $\lambda_2(t) < 0, 0 \le t < \tau, \lambda_2(t) > 0, \tau < t < t_f, \lambda_2(\tau) = \lambda_2(t_f) = 0$

Case 4. $\lambda_2(t) > 0, 0 \le t < \tau, \lambda_2(t) < 0, \tau < t < t_f, \lambda_2(\tau) = \lambda_2(t_f) = 0$

The parameter τ in *Case 3* and *Case 4* is a suitable time to be determined.
We now discuss these cases in detail.

Case 1. $\lambda_2(t) > 0, 0 \le t < t_f, \lambda_2(t_f) = 0$

This case should not be considered as it violates the integral constraint. In fact

$$\left(x_1(t) = \frac{t^2}{2}, \ t \ge 0 \right) \Rightarrow \int_0^{t_f} x_1(t)\,dt > 0$$

because $t_f > 0$, necessarily.

Fig. 8.3 Problem 8.3.
Enlarged state motion

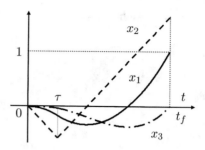

Case 2. $\lambda_2(t) < 0, 0 \le t < t_f, \lambda_2(t_f) = 0$

Also this case should not be considered as it violates the constraint on $x_1\,(t_f)$. In fact

$$x_1(t_f) = -\frac{t_f^2}{2}$$

Case 3. $\lambda_2(t) < 0, 0 \le t < \tau, \lambda_2(t) > 0, \tau < t < t_f, \lambda_2(\tau) = \lambda_2(t_f) = 0, \tau$ to be found

Also this case should not be considered. In fact we have

$$
\begin{array}{ll}
0 \le t < \tau & \tau < t \le t_f \\
u(t) = 1 & u(t) = -1 \\
0 \le t \le \tau & \tau \le t \le t_f \\
x_1(t) = \dfrac{1}{2}t^2 & x_1(t) = -\tau^2 + 2\tau t - \dfrac{1}{2}t^2 \\
x_2(t) = t & x_2(t) = 2\tau - t \\
x_3(t) = \dfrac{1}{6}t^3 & x_3(t) = \dfrac{1}{3}\tau^3 - \tau^2 t + \tau t^2 - \dfrac{1}{6}t^3
\end{array}
$$

so that, by enforcing feasibility, namely $x_1\,(t_f) = 1$ and $x_3\,(t_f) = \overline{w}_e$, we get the system of two equations

$$-\tau^2 + 2\tau t_f - \frac{1}{2}t_f^2 = 1$$

$$\frac{1}{3}\tau^3 - \tau^2 t_f + \tau t_f^2 - \frac{1}{6}t_f^3 = 0$$

for the two unknowns t_f and τ. No solution exists.

Case 4. $\lambda_2(t) > 0, 0 \le t < \tau, \lambda_2(t) < 0, \tau < t < t_f, \lambda_2(\tau) = \lambda_2(t_f) = 0, \tau$ to be found

A solution which satisfies the NC can be computed in this case.

In fact we have

$$
\begin{array}{ll}
0 \le t < \tau & \tau < t \le t_f \\
u(t) = -1 & u(t) = 1 \\
0 \le t \le \tau & \tau \le t \le t_f \\
x_1(t) = -\dfrac{1}{2}t^2 & x_1(t) = \tau^2 - 2\tau t + \dfrac{1}{2}t^2 \\
x_2(t) = -t & x_2(t) = -2\tau + t \\
x_3(t) = -\dfrac{1}{6}t^3 & x_3(t) = -\dfrac{1}{3}\tau^3 + \tau^2 t - \tau t^2 + \dfrac{1}{6}t^3
\end{array}
$$

and, if we enforce feasibility ($x_1(t_f) = 1$ and $x_3(t_f) = \overline{w}_e$), we obtain

$$t_f = 2.77, \quad \tau = 0.57$$

For the sake of completeness we compute the values of $\lambda_{i0}, i = 1, 2, 3$. The transversality condition at the final time, written at the initial time because the problem is time-invariant,

$$H(0) = 1 + \lambda_1(0)x_2(0) + \lambda_2(0)u(0) + \lambda_3(0)x_1(0) = 1 - \lambda_{20} = 0$$

yields $\lambda_{20} = 1$, whereas the two requests

$$\lambda_2(\tau) = 0, \quad \lambda_2(t_f) = 0$$

lead to the equations

$$\lambda_{20} - \lambda_{10}\tau + \frac{\lambda_{30}}{2}\tau^2 = 0$$

$$\lambda_{20} - \lambda_{10}t_f + \frac{\lambda_{30}}{2}t_f^2 = 0$$

which give $\lambda_{10} = 2.12$ and $\lambda_{30} = 1.27$.

The state motion of the enlarged system is plotted in Fig. 8.3.

Notice that $\dot{x}_2(\cdot)$ is discontinuous at time τ: this fact is a consequence of the discontinuity of $u(\cdot)$ at that time. ∎

Problem 8.4
Let

$$x(0) = \begin{bmatrix} 0 \\ 0 \end{bmatrix}, \quad x(t_f) = \text{free}, \quad t_f = 1,$$

$$J = m(x(t_f), t_f) + \int_0^{t_f} l(x(t), u(t), t)\, dt, \quad m(x, t) = x_1, \quad l(x, u, t) = \frac{u^2}{2}, \quad u(t) \in R, \ \forall t$$

$$\int_0^{t_f} w_e(x(t), u(t), t)\, dt = \overline{w}_e, \quad w(x, u, t) = x_1, \quad \overline{w}_e = 0$$

We take into account the integral constraint by adding a new state variable x_3 defined by the equations

$$\dot{x}_3(t) = w_e(x(t), u(t), t) = x_1(t), \quad x_3(0) = 0, \quad x_3(t_f) = \overline{w}_e$$

In so doing, the system is *enlarged* because its state $x_a := \begin{bmatrix} x_1 & x_2 & x_3 \end{bmatrix}'$ is now a three-dimensional vector. The set of the admissible final states is

$$\overline{S}_{af} = \{x_a | \alpha_{af}(x_a) = 0\}, \quad \alpha_{af}(x_a) = x_3 - \overline{w}_e$$

where $\overline{S}_{af}$ is a regular variety because

$$\Sigma_{af}(x_a) = \frac{d\alpha_{af}(x_a)}{dx_a} = \begin{bmatrix} 0 & 0 & 1 \end{bmatrix}$$

and $\text{rank}(\Sigma_{af}(x_a)) = 1, \forall x_a \in \overline{S}_{af}$.

The hamiltonian function and the H-minimizing control are

$$H = \frac{1}{2}u^2 + \lambda_1 x_2 + \lambda_2 u + \lambda_3 x_1, \quad u_h = -\lambda_2$$

Therefore Eq. (2.5a) and its general solution over an interval beginning at time 0 are

$$\dot{\lambda}_1(t) = -\lambda_3(t) \qquad\qquad \lambda_1(t) = \lambda_{10} - \lambda_{30} t$$
$$\dot{\lambda}_2(t) = -\lambda_1(t) \qquad\qquad \lambda_2(t) = \lambda_{20} - \lambda_{10} t + \frac{\lambda_{30}}{2} t^2$$
$$\dot{\lambda}_3(t) = 0 \qquad\qquad\quad \lambda_3(t) = \lambda_{30}$$

If we let $m_a(x_a, t) := x_1$, the orthogonality condition at the final time

$$\begin{bmatrix} \lambda_1(t_f) \\ \lambda_2(t_f) \\ \lambda_3(t_f) \end{bmatrix} - \frac{\partial m_a(x_a, t_f)}{\partial x_a} \bigg|'_{x_a = x_a(t_f)} = \begin{bmatrix} \lambda_{10} - \lambda_{30} t_f \\ \lambda_{20} - \lambda_{10} t_f + \frac{\lambda_{30}}{2} t_f^2 \\ \lambda_{30} \end{bmatrix} - \begin{bmatrix} 1 \\ 0 \\ 0 \end{bmatrix} =$$

$$= \vartheta_f \frac{d\alpha_{af}(x_a)}{dx_a} \bigg|'_{x = x(t_f)} = \vartheta_f \begin{bmatrix} 0 \\ 0 \\ 1 \end{bmatrix}$$

yields $\lambda_2(t_f) = 0$ and $\lambda_1(t_f) = 1$. By taking into account the solution of Eq. (2.5a) we obtain

$$\lambda_{30} = \lambda_{10} - 1, \quad \lambda_{20} = \frac{\lambda_{10} + 1}{2}$$

On the other hand, when $u(t) = -\lambda_2(t)$, we have

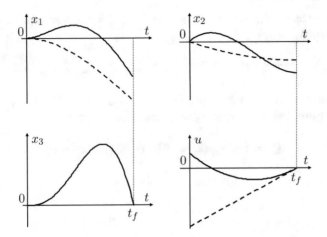

Fig. 8.4 Problem 8.4. State motion and control. *Solid lines* refer to the presence of the integral constraint, whereas *dashed lines* refer to the absence of the constraint. When the constraint is present the plotted variables are $10x_1(t)$ and $10x_2(t)$

$$x_1(t) = -\frac{\lambda_{20}}{2}t^2 + \frac{\lambda_{10}}{6}t^3 - \frac{\lambda_{30}}{24}t^4$$

$$x_2(t) = -\lambda_{20}t + \frac{\lambda_{10}}{2}t^2 - \frac{\lambda_{30}}{6}t^3$$

$$x_3(t) = -\frac{\lambda_{20}}{6}t^3 + \frac{\lambda_{10}}{24}t^4 - \frac{\lambda_{30}}{120}t^5$$

Thus, by enforcing feasibility ($x_3(t_f) = 0$), it follows that $\lambda_{10} = -1.5$, $\lambda_{20} = -0.25$, $\lambda_{30} = -2.5$. The state motion and the control are shown in Fig. 8.4 (solid lines).

Also for this problem it is meaningful to compare the solution above with the one relevant to the case where the integral equality constraint is not present. In this new framework the hamiltonian function and the H-minimizing control are

$$H = \frac{1}{2}u^2 + \nu_1 x_2 + \nu_2 u, \quad u_h = -\nu_2$$

and, consistently, Eq. (2.5a) and its general solution over an interval beginning at time 0 are

$$\dot\nu_1(t) = 0 \qquad\qquad \nu_1(t) = \nu_{10}$$
$$\dot\nu_2(t) = -\nu_1(t) \qquad\qquad \nu_2(t) = \nu_{20} - \nu_{10}t$$

The orthogonality condition at the final time

$$\begin{bmatrix} \nu_1(t_f) \\ \nu_2(t_f) \end{bmatrix} - \frac{\partial m(x, t_f)}{\partial x}\bigg|'_{x=x(t_f)} = \begin{bmatrix} \nu_{10} \\ \nu_{20} - \nu_{10}t_f \end{bmatrix} - \begin{bmatrix} 1 \\ 0 \end{bmatrix} = \begin{bmatrix} 0 \\ 0 \end{bmatrix}$$

implies $\nu_{10} = 1$ and $\nu_{20} = \nu_{10}$. The state motion and the control are shown in Fig. 8.4 with dashed lines. Once more it has to be noted that the solution is substantially different in the two considered cases. Also the value of the performance index undergoes substantial changes as it moves from J_v (the constraint is present) to J_l (the constraint is not present) because

$$J_v = -1.04 \times 10^{-2}, \quad J_l = -16.67 \times 10^{-2}$$

The amount of the ratio ($\simeq 16$) of these two values points out that the constraint greatly deteriorates the system performance. ∎

We now discuss three problems which constitutes significant variations of Problem 8.1. The integral constraint appearing in each one of them is either a function of the state (Problem 8.5-1) or of the state and control (Problem 8.5-2), or else of the state and time (Problem 8.5-3).

Problem 8.5

For the three problems we have

$$x(0) = \begin{bmatrix} \beta \\ \gamma \end{bmatrix}, \quad x(t_f) = \begin{bmatrix} 0 \\ 0 \end{bmatrix}, \quad t_f = 1$$

$$J = \int_0^{t_f} l(x(t), u(t), t)\, dt, \quad l(x, u, t) = \frac{u^2}{2}, \quad u(t) \in R, \ \forall t$$

where the parameters β and γ are given and not both zero. As above mentioned, the integral constraint has a different form in each problem.

Problem 8.5-1 $\int_0^{t_f} w_e(x(t), u(t), t)\, dt = \overline{w}_e, \quad w_e(x, u, t) = x_1 + x_2, \quad \overline{w}_e = 0$

We take into account the integral constraint by adding a new state variable x_3 defined by the equations

$$\dot{x}_3(t) = w_e(x(t), u(t), t) = x_1(t) + x_2(t), \quad x_3(0) = 0, \quad x_3(t_f) = \overline{w}_e$$

In so doing, the system is *enlarged* because its state $x_a := \begin{bmatrix} x_1 & x_2 & x_3 \end{bmatrix}'$ is now a three-dimensional vector which is completely specified at the final time. The hamiltonian function and the H-minimizing control are

$$H = \frac{1}{2}u^2 + \lambda_1 x_2 + \lambda_2 u + \lambda_3(x_1 + x_2), \quad u_h = -\lambda_2$$

Consequently, Eq. (2.5a) and its general solution over an interval beginning at time 0 are

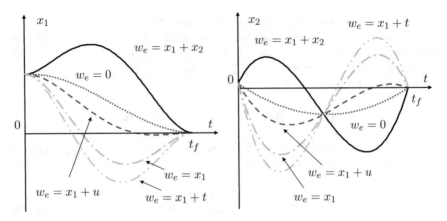

Fig. 8.5 Problem 8.5. State motion when $\beta = 3$ and $\gamma = 1$. For an useful comparison, the state motion relevant to Problem 8.1 (*dash-single-dotted line*) and that resulting from the absence of the constraint (*dotted line*) are reported as well

$$\dot{\lambda}_1(t) = -\lambda_3(t) \qquad\qquad\qquad \lambda_1(t) = \lambda_{10} - \lambda_{30}t$$

$$\dot{\lambda}_2(t) = -\lambda_1(t) - \lambda_3(t) \qquad\quad \lambda_2(t) = \lambda_{20} - (\lambda_{10} + \lambda_{30})t + \frac{\lambda_{30}}{2}t^2$$

$$\dot{\lambda}_3(t) = 0 \qquad\qquad\qquad\qquad \lambda_3(t) = \lambda_{30}$$

and the state motion of the enlarged system is

$$x_1(t) = \beta + \gamma t - \frac{\lambda_{20}}{2}t^2 + \frac{\lambda_{10} + \lambda_{30}}{6}t^3 - \frac{\lambda_{30}}{24}t^4$$

$$x_2(t) = \gamma - \lambda_{20}t + \frac{\lambda_{10} + \lambda_{30}}{2}t^2 - \frac{\lambda_{30}}{6}t^3$$

$$x_3(t) = (\beta + \gamma)t + \frac{\gamma - \lambda_{20}}{2}t^2 + \frac{\lambda_{10} - \lambda_{20} + \lambda_{30}}{6}t^3 + \frac{\lambda_{10}}{24}t^4 - \frac{\lambda_{30}}{120}t^5$$

By enforcing feasibility ($x_1(t_f) = x_2(t_f) = 0$, $x_3(t_f) = \overline{w}_e$), we obtain a system of three linear equations which supply the three (unknown) parameters λ_{10}, λ_{20}, λ_{30} as functions of β and γ. As an example, if $\beta = 3$ and $\gamma = 1$, we get $\lambda_{10} = -1488$, $\lambda_{20} = -63$, $\lambda_{30} = -1020$. The corresponding state motion is plotted in Fig. 8.5 with a solid line.

Problem 8.5-2 $\displaystyle\int_0^{t_f} w(x(t), u(t), t)\, dt = \overline{w}_e, \quad w_e(x, u, t) = x_1 + u, \quad \overline{w}_e = 0$

We take into account the integral constraint by adding a new state variable x_3 defined by the equations

$$\dot{x}_3(t) = w_e(x(t), u(t), t) = x_1(t) + u(t), \quad x_3(0) = 0, \quad x_3(t_f) = \overline{w}_e$$

In so doing, the system is *enlarged* because its state $x_a := \begin{bmatrix} x_1 & x_2 & x_3 \end{bmatrix}'$ is now a three-dimensional vector which is completely specified at the final time. The hamiltonian function and the H-minimizing control are

$$H = \frac{1}{2}u^2 + \lambda_1 x_2 + \lambda_2 u + \lambda_3(x_1 + u), \quad u_h = -(\lambda_2 + \lambda_3)$$

Consequently, Eq. (2.5a) and its general solution over an interval beginning at time 0 are

$$\dot{\lambda}_1(t) = -\lambda_3(t) \qquad\qquad \lambda_1(t) = \lambda_{10} - \lambda_{30}t$$

$$\dot{\lambda}_2(t) = -\lambda_1(t) \qquad\qquad \lambda_2(t) = \lambda_{20} - \lambda_{10}t + \frac{\lambda_{30}}{2}t^2$$

$$\dot{\lambda}_3(t) = 0 \qquad\qquad \lambda_3(t) = \lambda_{30}$$

and the state motion of the enlarged system is

$$x_1(t) = \beta + \gamma t - \frac{\lambda_{20} + \lambda_{30}}{2}t^2 + \frac{\lambda_{10}}{6}t^3 - \frac{\lambda_{30}}{24}t^4$$

$$x_2(t) = \gamma - (\lambda_{20} - \lambda_{30})t + \frac{\lambda_{10}}{2}t^2 - \frac{\lambda_{30}}{6}t^3$$

$$x_3(t) = (\beta - \lambda_{20} - \lambda_{30})t + \frac{\gamma + \lambda_{10}}{2}t^2 - \frac{\lambda_{20} + 2\lambda_{30}}{6}t^3 + \frac{\lambda_{10}}{24}t^4 - \frac{\lambda_{30}}{120}t^5$$

By enforcing feasibility $(x_1(t_f) = x_2(t_f) = 0, x_3(t_f) = \overline{w}_e)$, we obtain a system of three linear equations which supply the three parameters $\lambda_{10}, \lambda_{20}, \lambda_{30}$ as functions of β and γ. As an example, if we choose β and γ as in Problem 8.5-1, that is $\beta = 3$ and $\gamma = 1$, we get $\lambda_{10} = 252$, $\lambda_{20} = -363$, $\lambda_{30} = 420$. The corresponding state motion is plotted in Fig. 8.5 with a dashed line.

Problem 8.5-3 $\displaystyle\int_0^{t_f} w(x(t), u(t), t)\, dt = \overline{w}_e, \quad w_e(x, u, t) = x_1 + t, \quad \overline{w}_e = 0$

We take into account the integral constraint by adding a new state variable x_3 defined by the equations

$$\dot{x}_3(t) = w_e(x(t), u(t), t) = x_1(t) + t, \quad x_3(0) = 0, \quad x_3(t_f) = \overline{w}_e$$

In so doing, the system is *enlarged* because its state $x_a := \begin{bmatrix} x_1 & x_2 & x_3 \end{bmatrix}'$ is now a three-dimensional vector which is completely specified at the final time.

The hamiltonian function and the H-minimizing control are

$$H = \frac{1}{2}u^2 + \lambda_1 x_2 + \lambda_2 u + \lambda_3(x_1 + t), \quad u_h = -\lambda_2$$

Table 8.1 Problems 8.1 and 8.5. The values of the performance index when $\beta = 3$ and $\gamma = 1$.

Problem	$w_e(x, u, t)$	J
8.5-1	$x_1 + x_2$	796.5
8.5-2	$x_1 + u$	196.5
8.5-3	$x_1 + t$	1636.5
8.1	0	74.0
8.1	x_1	976.5

Consequently, Eq. (2.5a) and its general solution over an interval beginning at time 0 are

$$\dot{\lambda}_1(t) = -\lambda_3(t) \qquad\qquad \lambda_1(t) = \lambda_{10} - \lambda_{30}t$$

$$\dot{\lambda}_2(t) = -\lambda_1(t) \qquad\qquad \lambda_2(t) = \lambda_{20} - \lambda_{10}t + \frac{\lambda_{30}}{2}t^2$$

$$\dot{\lambda}_3(t) = 0 \qquad\qquad \lambda_3(t) = \lambda_{30}$$

and the state motion of the enlarged system is

$$x_1(t) = \beta + \gamma t - \frac{\lambda_{20}}{2}t^2 + \frac{\lambda_{10}}{6}t^3 - \frac{\lambda_{30}}{24}t^4$$

$$x_2(t) = \gamma - \lambda_{20}t + \frac{\lambda_{10}}{2}t^2 - \frac{\lambda_{30}}{6}t^3$$

$$x_3(t) = \beta t + \frac{\gamma + 1}{2}t^2 - \frac{\lambda_{20}}{6}t^3 + \frac{\lambda_{10}}{24}t^4 - \frac{\lambda_{30}}{120}t^5$$

By enforcing feasibility ($x_1(t_f) = x_2(t_f) = 0$, $x_3(t_f) = \overline{w}_e$), we obtain a system of three linear equations which supply the three parameters $\lambda_{10}, \lambda_{20}, \lambda_{30}$ as functions of β and γ. As an example, if we choose β and γ as in Problems 8.5-1 and 8.5-2, that is $\beta = 3$ and $\gamma = 1$, we get $\lambda_{10} = 792$, $\lambda_{20} = 147$, $\lambda_{30} = 1500$. The corresponding state motion is plotted in Fig. 8.5 with a dash-double-dotted line. Note how different are the functions $x(\cdot)$.

For a significant comparison, we compute the values of the parameters which define the solution of Problem 8.1 when $\beta = 3$ and $\gamma = 1$. If $w(x, u, t) = x_1$, we obtain $\lambda_{10} = 612, \lambda_{20} = 117, \lambda_{30} = 1140$, whereas, if $w(x, u, t) = 0$, we have $\lambda_{10} = 42$, $\lambda_{20} = 22$. The corresponding state motions are plotted in Fig. 8.5 with a dash-single-dotted line and dotted line, respectively. Finally, it is worth considering the values of the performance index relevant to the five considered situations. They are given in Table 8.1. As it should be expected, the smallest value corresponds to the absence of the integral constraint. Furthermore it is possible to verify that, when the constraint refers to the first state variable only, we obtain a result worse than the one which results from also including in the constraint either the second state variable or the control. The second of these two situations is the most favorable. The case when the constraint is an explicit function of time is the worst one. ∎

Fig. 8.6 Problem 8.6. State trajectories corresponding to various values of β

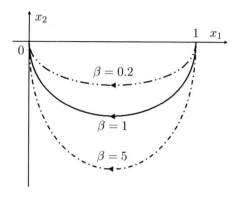

Problem 8.6

Let

$$x(0) := \begin{bmatrix} 1 \\ 0 \end{bmatrix}, \quad x(t_f) = \begin{bmatrix} 0 \\ 0 \end{bmatrix}, \quad J = \int_0^{t_f} l(x(t), u(t), t)\, dt, \ l(x, u, t) = 1$$

$$\int_0^{t_f} w_e(x(t), u(t), t)\, dt = \overline{w}_e, \ w_e(x, u, t) = \frac{u^2}{2}, \ \overline{w}_e = \beta, \ u(t) \in R, \ \forall t$$

where the positive parameter β is given and the final time t_f is free. We take into account the integral constraint by adding a new state variable x_3 defined by the equations

$$\dot{x}_3(t) = w_e(x(t), u(t), t) = \frac{u^2(t)}{2}, \quad x_3(0) = 0, \quad x_3(t_f) = \overline{w}_e$$

In so doing, the system is *enlarged* because its state $x_a := \begin{bmatrix} x_1 & x_2 & x_3 \end{bmatrix}'$ is now a three-dimensional vector which is given at the final time. Consequently, the hamiltonian function is

$$H = 1 + \lambda_1 x_2 + \lambda_2 u + \lambda_3 \frac{u^2}{2}$$

and the H-minimizing control can be found only if $\lambda_3 > 0$. If so, we have

$$u_h = -\frac{\lambda_2}{\lambda_3}$$

Thus Eq. (2.5a) and its general solution over an interval beginning at time 0 are

$$\dot{\lambda}_1(t) = 0, \qquad\qquad \lambda_1(t) = \lambda_{10}$$
$$\dot{\lambda}_2(t) = -\lambda_1(t), \qquad \lambda_2(t) = \lambda_{20} - \lambda_{10}t$$
$$\dot{\lambda}_3(t) = 0, \qquad\qquad \lambda_3(t) = \lambda_{30}$$

By recalling the expression of the H-minimizing control, the transversality condition at the final time, written at the initial time because the problem is time-invariant, is

$$H(0) = 1 + \lambda_1(0)x_2(0) + \lambda_2(0)u(0) + \lambda_3(0)\frac{u^2(0)}{2} = 1 - \frac{\lambda_{20}^2}{2\lambda_{30}} = 0$$

and implies $\lambda_{30} = \dfrac{\lambda_{20}^2}{2}$. Therefore it follows

$$u(t) = -2\frac{\lambda_{20} - \lambda_{10}t}{\lambda_{20}^2}$$

$$x_1(t) = 1 - \frac{t^2}{\lambda_{20}} + \frac{\lambda_{10}t^3}{3\lambda_{20}^2}$$

$$x_2(t) = -\frac{2t}{\lambda_{20}} + \frac{\lambda_{10}t^2}{\lambda_{20}^2}$$

$$x_3(t) = \frac{2}{\lambda_{20}^2}\left(\frac{\lambda_{10}^2}{3\lambda_{20}^2}t^3 - \frac{\lambda_{10}}{\lambda_{20}}t^2 + t\right)$$

and feasibility $(x(t_f) = 0, \lambda_3(t_f) = \overline{w}_e)$ together with the relation between λ_{20} and λ_{30}, lead to

$$\lambda_{10} = \sqrt[3]{\frac{16}{9\beta}}, \quad t_f = \frac{3}{2}\lambda_{10}, \quad \lambda_{20} = \frac{3}{4}\lambda_{10}^2, \quad \lambda_{30} = \frac{9}{32}\lambda_{10}^4$$

First, we ascertain that $\lambda_{30} > 0$. Secondly, we notice that, as it was to be foreseen, the length of the control interval decreases as β increases. Thus we also expect that the maximal deviation from zero of the second state variable becomes greater and greater. Some state trajectories corresponding to various values of β are plotted in Fig. 8.6. Their shapes confirm that the conclusions above are correct. ∎

8.2 Integral Inequality Constraints

The performance indexes of the forthcoming Problems 8.7, 8.8 and 8.10 are simply the length of the control interval. The integral constraint only concerns the control variable in Problem 8.7, the second state variable in Problems 8.8 and 8.9 (the final time is given in Problem 8.9), the control variable and time in Problem 8.10.

The discussion makes reference to the material presented in Sect. 2.2.3 and 2.3.2 and to the examples considered in Sect. 8.1.

Problem 8.7

Let

$$x(0) = \begin{bmatrix} 0 \\ 0 \end{bmatrix}, \quad x\left(t_f\right) \in \overline{S}_f = \left\{ x | \alpha_f(x) = 0 \right\}, \quad \alpha_f(x) = x_1 - 1$$

$$J = \int_0^{t_f} l(x(t), u(t), t)\, dt, \quad l(x, u, t) = 1$$

$$\int_0^{t_f} w_d(x(t), u(t), t)\, dt \leq \overline{w}_d, \quad w_d(x, u, t) = \frac{u^2}{2}, \quad \overline{w}_d = \beta, \quad u(t) \in R, \quad \forall t$$

where $\beta > 0$ is a given parameter. Obviously, the final time t_f is free.

We take into account the integral inequality constraint by adding a new state variable x_3 defined by the equations

$$\dot{x}_3(t) = w_d(x(t), u(t), t) = \frac{u^2(t)}{2}, \quad x_3(0) = 0, \quad x_3\left(t_f\right) \leq \overline{w}_d$$

In so doing, the system is *enlarged*, because its state $x_a := \begin{bmatrix} x_1 & x_2 & x_3 \end{bmatrix}'$ is now a three-dimensional vector. Thus the set $\widehat{S}_{af}$ of the admissible final states is

$$\widehat{S}_{af} = \left\{ x_a | x_a \in S_{af}, \; x_3 \leq b_3 \right\}, \; b_3 = \overline{w}_d$$

$$S_{af} = \overline{S}_{af} = \left\{ x_a | \alpha_{af}(x_a) = 0 \right\}, \; \alpha_{af}(x_a) = x_1 - 1$$

where $\overline{S}_{af}$ is a regular variety because

$$\Sigma_{af}(x_a) = \frac{d\alpha_{af}(x_a)}{dx_a} = \begin{bmatrix} 1 & 0 & 0 \end{bmatrix}$$

and $\text{rank}(\Sigma_{af}(x_a)) = 1, \forall x_a \in \overline{S}_{af}$.

The hamiltonian function and the H-minimizing control are

$$H = 1 + \lambda_1 x_2 + \lambda_2 u + \frac{1}{2}\lambda_3 u^2, \quad u_h = -\frac{\lambda_2}{\lambda_3}$$

provided that $\lambda_3 > 0$. The form of the hamiltonian function implies that Eq. (2.5a) and its general solution over an interval beginning at time 0 are

$$\begin{aligned} \dot{\lambda}_1(t) &= 0 \\ \dot{\lambda}_2(t) &= -\lambda_1(t) \\ \dot{\lambda}_3(t) &= 0 \end{aligned} \qquad \begin{aligned} \lambda_1(t) &= \lambda_{10} \\ \lambda_2(t) &= \lambda_{20} - \lambda_{10}t \\ \lambda_3(t) &= \lambda_{30} \end{aligned}$$

Therefore we state that the integral constraint is binding because, otherwise, the set of the admissible final states is $\overline{S}_{af}$ and the orthogonality condition at the final time

$$\begin{bmatrix} \lambda_1(t_f) \\ \lambda_2(t_f) \\ \lambda_3(t_f) \end{bmatrix} = \begin{bmatrix} \lambda_{10} \\ \lambda_{20} - \lambda_{10}t_f \\ \lambda_{30} \end{bmatrix} = \vartheta_{af} \left. \frac{d\alpha_{af}(x_a)}{dx_a} \right|'_{x_a=x_a(t_f)} = \vartheta_{af} \begin{bmatrix} 1 \\ 0 \\ 0 \end{bmatrix}$$

implies $\lambda_{30} = 0$ so that no H-minimizing control exists. We conclude that

$$x_3(t_f) = \overline{w}_d, \quad x_a(t_f) \in \overline{S}_{afM}$$

where

$$\overline{S}_{afM} := \left\{ x_a | \alpha_{afM_i}(x_a) = 0, \ i = 1, 2 \right\}, \quad \alpha_{afM_1}(x_a) = x_1 - 1, \quad \alpha_{afM_2}(x_a) = x_3 - \overline{w}_d$$

and

$$u(t) = -\frac{\lambda_2(t)}{\lambda_{30}}$$

provided that $\lambda_{30} > 0$. The set $\overline{S}_{afM}$ is a regular variety because

$$\Sigma_{afM}(x_a) = \frac{d\left(\begin{bmatrix} \alpha_{afM_1}(x_a) \\ \alpha_{afM_2}(x_a) \end{bmatrix} \right)}{dx_a} = \begin{bmatrix} 1 & 0 & 0 \\ 0 & 0 & 1 \end{bmatrix}$$

and $\text{rank}(\Sigma_{afM}(x_a)) = 2, \forall x_a \in \overline{S}_{afM}$. The new orthogonality condition at the final time

$$\begin{bmatrix} \lambda_1(t_f) \\ \lambda_2(t_f) \\ \lambda_3(t_f) \end{bmatrix} = \begin{bmatrix} \lambda_{10} \\ \lambda_{20} - \lambda_{10}t_f \\ \lambda_{30} \end{bmatrix} = \sum_{i=1}^{2} \vartheta_{afM_i} \left. \frac{d\alpha_{afM_i}(x_a)}{dx_a} \right|'_{x_a=x_a(t_f)} =$$

$$= \vartheta_{afM_1} \begin{bmatrix} 1 \\ 0 \\ 0 \end{bmatrix} + \vartheta_{afM_2} \begin{bmatrix} 0 \\ 0 \\ 1 \end{bmatrix}$$

implies $\lambda_{20} - \lambda_{10}t_f = 0$. By recalling the expression of the H-minimizing control, the transversality condition at the final time, written at the initial time because the problem is time-invariant, is

$$H(0) = 1 + \lambda_1(0)x_2(0) + \lambda_2(0)u(0) + \frac{1}{2}\lambda_3(0)u^2(0) = 1 - \frac{\lambda_{20}^2}{2\lambda_{30}} = 0$$

From this equations and the previous one ($\lambda_{20} = \lambda_{10}t_f$) we obtain

$$\lambda_2(t) = \lambda_{10}(t_f - t), \quad \lambda_{30} = \frac{\lambda_{10}^2 t_f^2}{2}$$

so that

$$u(t) = \frac{2}{\lambda_{10}t_f^2}(t - t_f)$$

Therefore the state motion of the enlarged system is

$$x_1(t) = \frac{t^3}{3\lambda_{10}t_f^2} - \frac{t^2}{\lambda_{10}t_f}$$

$$x_2(t) = \frac{t^2}{\lambda_{10}t_f^2} - 2\frac{t}{\lambda_{10}t_f}$$

$$x_3(t) = 2\frac{(t - t_f)^3 + t_f^3}{3\lambda_{10}^2 t_f^4}$$

By enforcing feasibility ($x_3((t_f) = \overline{w}_e$ and $x_1(t_f) = 1$), we get

$$t_f = \sqrt[3]{\frac{3}{2\beta}}, \quad \lambda_{10} = -\frac{2t_f}{3}$$

As it was easily predictable, the value of the final time, that is the value of the performance index, decreases as the value of the parameter β increases, consistently with the fact that β can be interpreted as the allowed amount of the control effort. ∎

Problem 8.8

Let

$$x(0) = \begin{bmatrix} 0 \\ 0 \end{bmatrix}, \quad x\left(t_f\right) \in \overline{S}_f = \left\{ x | \alpha_f(x) = 0 \right\}, \quad \alpha_f(x) = x_1 - 1$$

$$J = \int_0^{t_f} l(x(t), u(t), t)\, dt, \quad l(x, u, t) = 1, \quad u(t) \in \overline{U}, \ \forall t, \ \overline{U} = \{u | -1 \le u \le 1\}$$

$$\int_0^{t_f} w_d(x(t), u(t), t)\, dt \le \overline{w}_d, \quad w_d(x, u, t) = \frac{x_2^2}{2}, \quad \overline{w}_d = \beta$$

where $\beta > 0$ is a given parameter. The final time t_f is free.

We take into account the integral inequality constraint by adding a new state variable x_3 defined by the equations

$$\dot{x}_3(t) = w_d(x(t), u(t), t) = \frac{x_2^2(t)}{2}, \quad x_3(0) = 0, \quad x_3\left(t_f\right) \le \overline{w}_d = \beta$$

In so doing, the system is *enlarged* because its state $x_a := \begin{bmatrix} x_1 & x_2 & x_3 \end{bmatrix}'$ is now a three-dimensional vector. The set $\widehat{S}_{af}$ of the admissible final states of the enlarged system is

$$\widehat{S}_{af} = \left\{ x_a | x_a \in S_{af}, \ x_3 \leq b_3 \right\}, \ b_3 = \overline{w}_d$$
$$S_{af} = \overline{S}_{af} = \left\{ x_a | \alpha_{af}(x_a) = 0 \right\}, \ \alpha_{af}(x_a) = x_1 - 1$$

where $\overline{S}_{af}$ is a regular variety because

$$\Sigma_{af}(x_a) = \frac{d\alpha_{af}(x_a)}{dx_a} = \begin{bmatrix} 1 & 0 & 0 \end{bmatrix}$$

and $\mathrm{rank}(\Sigma_{af}(x_a)) = 1, \forall x_a \in \overline{S}_{af}$.

The hamiltonian function and the H-minimizing control are

$$H = 1 + \lambda_1 x_2 + \lambda_2 u + \frac{1}{2} \lambda_3 x_2^2, \quad u_h = -\mathrm{sign}\,(\lambda_2)$$

from which Eq. (2.5a) are easily derived together with the general solution for $\lambda_1(\cdot)$ and $\lambda_3(\cdot)$ over an interval beginning at time 0 yielding

$$\begin{aligned} \dot{\lambda}_1(t) &= 0 & \lambda_1(t) &= \lambda_{10} \\ \dot{\lambda}_2(t) &= -\lambda_1(t) - \lambda_3(t)x_2(t) & & \\ \dot{\lambda}_3(t) &= 0 & \lambda_3(t) &= \lambda_{30} \end{aligned}$$

We first discuss the possibility that the constraint on $x_3(t_f)$ is not binding. In this case, the set of the admissible final states $x_a(t_f)$ is $\overline{S}_{af}$ and, consequently, the orthogonality condition at the final time

$$\begin{bmatrix} \lambda_1(t_f) \\ \lambda_2(t_f) \\ \lambda_3(t_f) \end{bmatrix} = \begin{bmatrix} \lambda_{10} \\ \lambda_2(t_f) \\ \lambda_{30} \end{bmatrix} = \vartheta_{af} \frac{d\alpha_{af}(x_a)}{dx_a} \Bigg|_{x_a = x_a(t_f)}' = \vartheta_{af} \begin{bmatrix} 1 \\ 0 \\ 0 \end{bmatrix}$$

implies $\lambda_{30} = \lambda_2(t_f) = 0$. Thus the equation for $\lambda_2(\cdot)$ gives $\lambda_2(t) = \lambda_{20} - \lambda_{10}t$ so that $\lambda_2(t) = \lambda_{10}(t_f - t)$. Therefore the sign of $\lambda_2(\cdot)$ does not change and the control $u(\cdot)$ is constant and equal to ± 1. The transversality condition at the final time, written at the initial time because the problem is time-invariant,

$$H(0) = 1 + \lambda_1(0)x_2(0) + \lambda_2(0)u(0) + \frac{1}{2}\lambda_3(0)x_2^2(0) = 1 + \lambda_{20}u(0) = 0$$

entails $\lambda_{20} = \mp 1$. Feasibility, namely $\alpha_{af}(x_a) = 0$, requires $u(\cdot) = 1$. Thus the relations above give

$$\lambda_2(0) = -1, \quad \lambda_{10} = -\frac{1}{t_f}$$

and the state motion of the enlarged system is

$$x_1(t) = \frac{t^2}{2}, \quad x_2(t) = t, \quad x_3(t) = \frac{t^3}{6}$$

The fulfillment of the condition $\alpha_{af}(x_a) = 0$ requires $t_f = \sqrt{2}$ and, consequently, $x_3(t_f) = \sqrt{2}/3$. Therefore the integral constraint is not binding if $\beta \geq \bar{\beta} := \sqrt{2}/3$.

On the contrary, if $\beta < \bar{\beta}$ the integral constraint is binding, so that $x_3(t_f) = \beta$. Thus the set of the admissible final states $x_a(t_f)$ is now

$$\bar{S}_{afM} := \{x_a | \alpha_{afM_i}(x_a) = 0, \ i = 1, 2\}, \ \alpha_{afM_1}(x_a) = x_1 - 1, \ \alpha_{afM_2}(x_a) = x_3 - \beta$$

which is a regular variety because

$$\Sigma_{afM}(x_a) := \frac{d\left(\begin{bmatrix} \alpha_{afM_1}(x_a) \\ \alpha_{afM_2}(x_a) \end{bmatrix}\right)}{dx_a} = \begin{bmatrix} 1 & 0 & 0 \\ 0 & 0 & 1 \end{bmatrix}$$

and $\text{rank}(\Sigma_{afM}(x_a)) = 2, \forall x_a \in \bar{S}_{afM}$. The new orthogonality condition at the final time is

$$\begin{bmatrix} \lambda_1(t_f) \\ \lambda_2(t_f) \\ \lambda_3(t_f) \end{bmatrix} = \sum_{i=1}^{2} \vartheta_{afM_i} \frac{d\alpha_{afM_i}(x_a)}{dx_a}\bigg|'_{x_a=x_a(t_f)} = \vartheta_{afM_1} \begin{bmatrix} 1 \\ 0 \\ 0 \end{bmatrix} + \vartheta_{afM_2} \begin{bmatrix} 0 \\ 0 \\ 1 \end{bmatrix}$$

and implies $\lambda_2(t_f) = 0$. Observe that the sign of $\lambda_2(\cdot)$ must change because we have already ascertained that, corresponding to the considered values of β, a constant control is not feasible. Actually this is made possible by $\lambda_3(0)$ being not necessarily zero. Thus the control switches at a suitable time τ, namely $u(t) = 1$, $t < \tau$ and $u(t) = -1$, $t > \tau$. Thanks to the nature of the problem, the switching from the value -1 to the value 1 must plainly be discarded. In fact the value of the first state variable must increase and, being the performance index the length of the control interval, it is obviously not convenient to first apply a negative control. The transversality condition at the final time is the one previously shown and again supplies $\lambda_{20} = -1$. As a consequence, the state motion of the enlarged system and the solution of Eq. (2.5a) is

$0 \le t \le \tau$

$x_1(t) = \frac{1}{2}t^2$

$x_2(t) = t$

$x_3(t) = \frac{1}{6}t^3$

$\lambda_1(t) = \lambda_{10}$

$\lambda_2(t) = -1 - \lambda_{10}t - \frac{\lambda_{30}}{2}t^2$

$\lambda_3(t) = \lambda_{30}$

$\tau \le t \le t_f$

$x_1(t) = x_1(\tau) + x_2(\tau)(t - \tau) - \frac{1}{2}(t - \tau)^2$

$x_2(t) = x_2(\tau) - (t - \tau)$

$x_3(t) = x_3(\tau) + \frac{x_2^2(\tau)}{2}(t - \tau) +$

$\qquad - \frac{x_2(\tau)}{2}(t - \tau)^2 + \frac{1}{6}(t - \tau)^3$

$\lambda_1(t) = \lambda_{10}$

$\lambda_2(t) = -\lambda_{10}(t - \tau) - \lambda_{30}x_2(\tau)(t - \tau) +$

$\qquad + \frac{1}{2}\lambda_{30}(t - \tau)^2$

$\lambda_3(t) = \lambda_{30}$

because $u(t) = 1, 0 \le t < \tau$ and $u(t) = -1, \tau < t < t_f$. In writing down the equation for $\lambda_2(t), \tau \le t \le t_f$ we have taken into account that the control switches at time τ only if the sign of $\lambda_2(\cdot)$ changes at that time, namely only if $\lambda_2(\tau) = 0$. The four parameters to be determined are $\lambda_{10}, \lambda_{30}, \tau, t_f$ and the available equations are

$$x_3(t_f) = \beta, \quad x_1(t_f) = 1, \quad \lambda_2(\tau) = 0, \quad \lambda_2(t_f) = 0$$

The first two of them account for feasibility $(\alpha_{afM_i}(x_a) = 0, i = 1, 2)$, whereas the last one originates from the orthogonality condition at the final time.

The above parameters can be computed in the following way. First, assume that t_f is given. Observe that it has to be greater than $\sqrt{2}$ because the absolute value of $x_2(t)$ must be lower than the value resulting from having ignored the integral constraint. Corresponding to any given t_f, we compute τ from the equation

$$x_1(t_f) = -\tau^2 + 2\tau t_f - \frac{t_f^2}{2} = 1 \Rightarrow \tau = t_f - \sqrt{\frac{t_f^2}{2} - 1}$$

where we select, out of the two roots, the one that gives $\tau \le t_f$. Subsequently, we find

$$x_3(t_f) = \frac{1}{6}\tau^3 + \frac{\tau^2}{2}(t_f - \tau) - \frac{\tau}{2}(t_f - \tau)^2 + \frac{1}{6}(t_f - \tau)^3$$

if the expressions for the state motion are exploited. The plots of $\tau(\cdot)$ (dashed line) and $x_3(\cdot)$ (solid line) as functions of t_f are reported in Fig. 8.7. We ascertain that $\tau \le t_f$ (see the straight line ϱ_2 defined by $\tau(t_f) := t_f$) and $x_3(t_f)$ attains its minimum value, $1/3$, at $t_f = 2$. Therefore a solution does not exist if $\beta < \beta_{min} := 1/3$. When β belongs to the interval $[\beta_{min}, \bar{\beta})$ the final time t_f corresponds to the first intersection of the curve $x_3(t_f)$ with the straight line ϱ_1 defined by $x_3(t_f) := \beta$ (recall that the performance index is t_f). Once t_f is obtained, we compute τ. For the sake of completeness we can also find $\lambda_{10}, \lambda_{30}$, which, however, are not expedient to characterize the solution. The state motions are reported in Fig. 8.8 if $\beta \ge \bar{\beta}$ (solid

Fig. 8.7 Problem 8.8. The functions $\tau(\cdot)$ (*dashed line*) e $x_3(\cdot)$ (*solid line*)

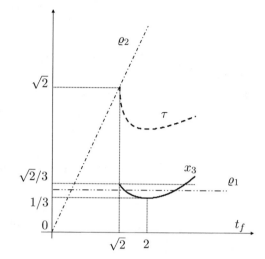

Fig. 8.8 Problem 8.8. State motion when $\beta \geq \bar{\beta}$ (*solid line*), when $\beta = \beta_{min}$ (*dashed line*) and when $\beta = 0.4 > \beta_{min}$ (*dash-single-dotted line*)

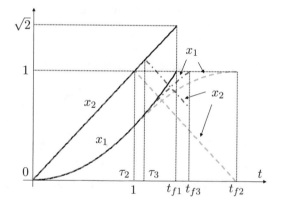

lines), $\beta = \beta_{min}$ (dashed lines) and $\beta = 0.4$ which is an interior point of the interval $[\beta_{min}, \bar{\beta})$ (dash-single-dotted lines). When $t \leq 1$ the state motions are the same for all the three values of β. The time where the switching takes place is denoted with τ_2 and τ_3 in the second and third case, whereas the final time is t_{f1}, t_{f2}, t_{f3} in the three cases. When the integral constraint becomes binding we observe that the maximum absolute value of x_2 decreases as β decreases.

As a conclusion, if $\beta \geq \bar{\beta}$, the control is $u(\cdot) = 1$, if $\beta_{min} \leq \beta < \bar{\beta}$, the control switches from 1 to -1, if $\beta < \beta_{min}$, no solution exists. ∎

Problem 8.9

Let

$$x(0) = \begin{bmatrix} 1 \\ 0 \end{bmatrix}, \quad x(t_f) = \begin{bmatrix} 0 \\ 0 \end{bmatrix}, \quad t_f = 1$$

$$J = \int_0^{t_f} l(x(t), u(t), t)\, dt, \quad l(x, u, t) = \frac{u^2(t)}{2}, \quad u(t) \in R, \; \forall t$$

$$\int_0^{t_f} w_d(x(t), u(t), t)\, dt \le \overline{w}_d, \quad w_d(x, u, t) = \frac{x_2^2}{2}, \quad \overline{w}_d = \beta$$

where $\beta > 0$ is a given parameter.

We take into account the integral inequality constraint by adding a new state variable x_3 defined by the equations

$$\dot{x}_3(t) = w_d(x(t), u(t), t) = \frac{x_2^2(t)}{2}, \quad x_3(0) = 0, \quad x_3(t_f) \le \overline{w}_d$$

In so doing, the system is *enlarged* because its state $x_a := \begin{bmatrix} x_1 & x_2 & x_3 \end{bmatrix}'$ is now a three-dimensional vector. The set $\widehat{S}_{af}$ of the admissible final states of the enlarged system is

$$\widehat{S}_{af} := \{ x_a | x_a \in S_{af}, \; x_3 \le b_3 \}, \quad b_3 = \overline{w}_d$$
$$S_{af} = \overline{S}_{af} := \{ x_a | \alpha_{af_i}(x_a) = 0, \; i = 1, 2 \}, \quad \alpha_{af_1}(x_a) = x_1, \; \alpha_{af_2}(x_a) = x_2$$

where $\overline{S}_{af}$ is a regular variety because

$$\Sigma_{af}(x_a) = \frac{d\left(\begin{bmatrix} \alpha_{af_1}(x_a) \\ \alpha_{af_2}(x_a) \end{bmatrix} \right)}{dx_a} = \begin{bmatrix} 1 & 0 & 0 \\ 0 & 1 & 0 \end{bmatrix}$$

and $\mathrm{rank}(\Sigma_{af}(x_a)) = 2, \; \forall x_a \in \overline{S}_{af}$.

The hamiltonian function and the H-minimizing control are

$$H = \frac{u^2}{2} + \lambda_1 x_2 + \lambda_2 u + \frac{1}{2}\lambda_3 x_2^2, \quad u_h = -\lambda_2$$

Therefore Eq. (2.5a) and $\lambda_1(\cdot)$ and $\lambda_3(\cdot)$ are

$$\dot{\lambda}_1(t) = 0 \qquad\qquad\qquad\qquad\qquad \lambda_1(t) = \lambda_{10}$$
$$\dot{\lambda}_2(t) = -\lambda_1(t) - \lambda_3(t)x_2(t)$$
$$\dot{\lambda}_3(t) = 0 \qquad\qquad\qquad\qquad\qquad \lambda_3(t) = \lambda_{30}$$

First, we discuss the case when the integral constraint is not binding. If so, the set of the admissible final states is $\overline{S}_{af}$.

The orthogonality condition at the final time

$$\begin{bmatrix} \lambda_1(t_f) \\ \lambda_2(t_f) \\ \lambda_3(t_f) \end{bmatrix} = \begin{bmatrix} \lambda_{10} \\ \lambda_2(t_f) \\ \lambda_{30} \end{bmatrix} = \sum_{i=1}^{2} \vartheta_{af_i} \left. \frac{d\alpha_{af_i}(x_a)}{dx_a} \right|'_{x_a = x_a(t_f)} = \begin{bmatrix} \vartheta_{af_1} \\ \vartheta_{af_2} \\ 0 \end{bmatrix}$$

implies $\lambda_3(\cdot) = \lambda_{30} = 0$ and also $\lambda_2(t) = \lambda_{20} - \lambda_{10}t$ (recall that the initial time is 0). By enforcing feasibility, (namely $x_1(t_f) = x_2(t_f) = 0$), we obtain

$$\lambda_{10} = 12, \quad \lambda_{20} = 6, \quad x_{3f} := x_3(t_f) = 0.6$$

Therefore the integral constraint is not binding if $\beta \geq 0.6$.

We now discuss the case $\beta < 0.6$. We have $x_3(t_f) = \beta$ because the constraint is binding. Thus the final state $x_a(t_f)$ is completely specified. The equations of the hamiltonian system (2.6), (2.5a) to be considered are

$$\dot{\lambda}_2(t) = -\lambda_{10} - \lambda_{30}x_2(t)$$
$$\dot{x}_2(t) = -\lambda_2(t)$$

If we let $k := \sqrt{|\lambda_{30}|}$, this system of differential equations leads to

$\lambda_{30} > 0$	$\lambda_{30} < 0$
$x_2(t) = ae^{kt} + be^{-kt} + c$	$x_2(t) = a\sin(kt + b) + c$

For a given k, the constants a, b, c can be determined by enforcing feasibility ($x_2(0) = x_2(t_f) = x_1(t_f) = 0$). According to the sign of λ_{30}, these conditions imply

$$a + b + c = 0$$
$$ae^{kt_f} + be^{-kt_f} + c = 0$$
$$1 + ct_f + \frac{1}{k}\left(a\left(e^{kt_f} - 1\right) - b\left(e^{-kt_f} - 1\right)\right) = 0$$

if $\lambda_{30} > 0$, whereas, if $\lambda_{30} < 0$,

$$a\sin(b) + c = 0$$
$$a\sin(kt_f + b) + c = 0$$
$$1 + ct_f - \frac{a}{k}\left(\cos(kt_f + b) - \cos(b)\right) = 0$$

When $\lambda_{30} > 0$, we obtain

$$c = -(a + b), \quad b = -a\frac{e^{kt_f} - 1}{e^{-kt_f} - 1}, \quad a = \left(1 - \frac{e^{kt_f} - 1}{e^{-kt_f} - 1} - \frac{2\left(e^{kt_f} - 1\right)}{k}\right)^{-1}$$

It is obvious that $x_3(t_f)$ depends on a, b, c, that is on k. Therefore we can refer to he function $x_{3f}(\cdot)$ which, for a given k, gives the value of $x_3(t_f)$. Recall that we have already established that at the final time $x_3 = \beta$: thus the equation $x_{3f}(k) = \beta$ gives k as a function of β. It is easy to verify that the plot of $x_{3f}(\cdot)$ is the one shown in Fig. 8.9 and that $x_{3f}(k)$ tends to 0.5 as k tends to infinity. Therefore when $\lambda_{30} > 0$,

Fig. 8.9 Problem 8.9. The plot of x_{3f} as a function of k when $\lambda_{30} = k^2$

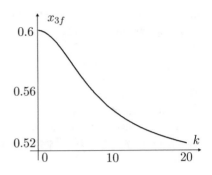

we can find a value of k corresponding to which $x_{3f} = \beta$ only if $0.6 > \beta > 0.5$, whereas, when $\beta \leq 0.5$, this is not possible.

When $\lambda_{30} < 0$ we notice that the solution of the equations deriving from the request $x_2(0) = x_2(t_f) = 0$ is not $a = c = 0$ (not consistent with $x_1(t_f) = 0$) only if

$$\tan(b) = \frac{\sin(kt_f)}{1 - \cos(kt_f)}$$

Therefore if $0 < kt_f < 2\pi$, we obtain, besides the above relation,

$$c = -a\sin(b), \quad a = \frac{k}{k\sin(b) + \cos(kt_f + b) - \cos(b)}$$

It is easy to verify that $x_{3f}(\cdot) \geq 0.6$, so that no solution exists.

We can conclude that, if $\beta \leq 0.5$, no solution exists, if $0.5 < \beta < 0.6$, the control is

$$u(t) = -\lambda_2(t) = \dot{x}_2(t) = ake^t - bke^{-t}$$

where the parameters a and b can be computed from the expressions relevant to the case $\lambda_{30} > 0$ and k is such that $x_{3f}(k) = \beta$. Finally, if $\beta \geq 0.6$ the control is $u(t) = 12t - 6$. ∎

Problem 8.10

Let

$$x(0) = \begin{bmatrix} 0 \\ 0 \end{bmatrix}, \quad x(t_f) \in \bar{S}_f = \{x | \alpha_f(x) = 0\}, \quad \alpha_f(x) = x_1 - 1$$

$$J = \int_0^{t_f} l(x(t), u(t), t)\, dt, \quad l(x, u, t) = 1$$

$$\int_0^{t_f} w_d(x(t), u(t), t)\, dt \leq \bar{w}_d, \quad w_d(x, u, t) = \frac{u^2}{2} + \gamma t, \quad \bar{w}_d = \beta, \quad u(t) \in R, \quad \forall t$$

where the parameters β and γ are positive and given. Obviously, the final time t_f is free.

We take into account the integral inequality constraint by adding a new state variable x_3 defined by the equations

$$\dot{x}_3(t) = w_d(x(t), u(t), t) = \frac{u^2(t)}{2} + \gamma t, \quad x_3(0) = 0, \quad x_3(t_f) \le \overline{w}_d$$

In so doing, the system is *enlarged* because its state $x_a := \begin{bmatrix} x_1 & x_2 & x_3 \end{bmatrix}'$ is now a three-dimensional vector. The set $\widehat{S}_{af}$ of the admissible final states is

$$\widehat{S}_{af} := \{x_a | x_a \in S_{af}, \ x_3 \le b_3\}, \ b_3 := \overline{w}_d$$
$$S_{af} = \overline{S}_{af} = \{x_a | \alpha_{af}(x_a) = 0\}, \ \alpha_{af}(x_a) = x_1 - 1$$

where $\overline{S}_{af}$ is a regular variety because

$$\Sigma_{af}(x_a) = \frac{d\alpha_{af}(x_a)}{dx_a} = \begin{bmatrix} 1 & 0 & 0 \end{bmatrix}$$

and $\text{rank}(\Sigma_{af}(x_a)) = 1, \ \forall x_a \in \overline{S}_{af}$.

The hamiltonian function and the H-minimizing control are

$$H = 1 + \lambda_1 x_2 + \lambda_2 u + \lambda_3(\frac{u^2}{2} + \gamma t), \quad u_h = -\frac{\lambda_2}{\lambda_3}$$

provided that $\lambda_3 > 0$. Equations (2.5a) and its general solution over an interval beginning at time 0 are

$$\dot{\lambda}_1(t) = 0 \qquad\qquad \lambda_1(t) = \lambda_{10}$$
$$\dot{\lambda}_2(t) = -\lambda_1(t) \qquad\qquad \lambda_2(t) = \lambda_{20} - \lambda_{10}t$$
$$\dot{\lambda}_3(t) = 0 \qquad\qquad \lambda_3(t) = \lambda_{30}$$

If the integral constraint is not binding, namely if the only constraint is $x_a(t_f) \in \overline{S}_{af}$, the orthogonality condition at the final time

$$\begin{bmatrix} \lambda_1(t_f) \\ \lambda_2(t_f) \\ \lambda_3(t_f) \end{bmatrix} = \begin{bmatrix} \lambda_{10} \\ \lambda_{20} - \lambda_{10}t_f \\ \lambda_{30} \end{bmatrix} = \vartheta_{af} \frac{d\alpha_{af}(x_a)}{dx_a}' \Bigg|_{x_a = x_a(t_f)} = \vartheta_{af} \begin{bmatrix} 1 \\ 0 \\ 0 \end{bmatrix}$$

implies $\lambda_3(\cdot) = \lambda_{30} = 0$ and the H-minimizing control does not exist. Therefore $x_3(t_f) = \beta$ and the set of the admissible final states is

$$\overline{S}_{afM} := \{x_a | \alpha_{afM_i}(x_a) = 0, \ i = 1, 2\}, \ \alpha_{afM_1}(x_a) = x_1 - 1, \ \alpha_{afM_2}(x_a) = x_3 - \beta$$

which is a regular variety because

$$\Sigma_{afM}(x_a) = \frac{d\begin{bmatrix} \alpha_{afM_1}(x_a) \\ \alpha_{afM_2}(x_a) \end{bmatrix}}{dx_a} = \begin{bmatrix} 1 & 0 & 0 \\ 0 & 0 & 1 \end{bmatrix}$$

and $\mathrm{rank}(\Sigma_{afM}(x_a)) = 2, \forall x_a \in \overline{S}_{afM}$.

The new orthogonality condition at the final time

$$\begin{bmatrix} \lambda_1(t_f) \\ \lambda_2(t_f) \\ \lambda_3(t_f) \end{bmatrix} = \begin{bmatrix} \lambda_{10} \\ \lambda_{20} - \lambda_{10}t_f \\ \lambda_{30} \end{bmatrix} = \sum_{i=1}^{2} \vartheta_{afM_i} \left. \frac{d\alpha_{afM_i}(x_a)}{dx_a} \right|'_{x_a = x_a(t_f)} =$$

$$= \vartheta_{afM_1} \begin{bmatrix} 1 \\ 0 \\ 0 \end{bmatrix} + \vartheta_{afM_2} \begin{bmatrix} 0 \\ 0 \\ 1 \end{bmatrix}$$

implies $\lambda_2(t_f) = \lambda_{20} - \lambda_{10}t_f = 0$. In view of this fact and the expression of u_h (under the assumption $\lambda_{30} > 0$), the transversality condition at the final time, to be necessarily written at this time because the problem is not time-invariant (see the equation for $\dot{x}_3(t)$), is

$$H(t_f) = 1 + \lambda_1(t_f)x_2(t_f) + \lambda_2(t_f)u(t_f) + \lambda_3(t_f)(\frac{u^2(t_f)}{2} + \gamma t_f) =$$

$$= 1 + \lambda_{10}x_2(t_f) + \lambda_{30}\gamma t_f = 0$$

The state motion is easily found as a function of the three parameters $\lambda_{i0}, i = 1, 2, 3$, namely

$$x_1(t) = -\frac{\lambda_{20}}{2\lambda_{30}}t^2 + \frac{\lambda_{10}}{6\lambda_{30}}t^3$$

$$x_2(t) = -\frac{\lambda_{20}}{\lambda_{30}}t + \frac{\lambda_{10}}{2\lambda_{30}}t^2$$

$$x_3(t) = \frac{(\lambda_{10}t - \lambda_{20})^3 + \lambda_{20}^3}{6\lambda_{10}\lambda_{30}^2} + \frac{\gamma}{2}t^2$$

The four unknowns $t_f, \lambda_{10}, \lambda_{20}, \lambda_{30}$ which completely specify the solution can be computed by solving the four equations relevant to the orthogonality, transversality and feasibility conditions. The latter is $x_a(t_f) \in \overline{S}_{af}$, that is $x_1(t_f) = 1, x_3(t_f) = \beta$, where $x_1(t_f)$ and $x_3(t_f)$ are given by the equations of the state motion. We get

$$\lambda_{10} = \frac{6t_f}{2\gamma t_f^5 - 9}, \quad \lambda_{20} = \lambda_{10}t_f \quad \lambda_{30} = -\frac{\lambda_{10}t_f^3}{3}$$

Table 8.2 Problem 8.10. The parameters t_f, λ_{10}, λ_{20}, λ_{30} corresponding to various choices of β and γ.

		$\beta = 5$	$\beta = 10$	$\beta = 50$
$\gamma = 1$	t_f	0.68	0.53	0.31
	λ_{10}	-0.47	-0.36	-0.21
	λ_{20}	-0.32	-0.19	-0.06
	λ_{30}	0.05	0.02	0.002
$\gamma = 10$	t_f	$\nexists$ solution	0.56	0.31
	λ_{10}	$\nexists$ solution	-0.43	-0.21
	λ_{20}	$\nexists$ solution	-0.24	-0.07
	λ_{30}	$\nexists$ solution	0.03	0.002
$\gamma = 50$	t_f	$\nexists$ solution	$\nexists$ solution	0.32
	λ_{10}	$\nexists$ solution	$\nexists$ solution	-0.22
	λ_{20}	$\nexists$ solution	$\nexists$ solution	-0.07
	λ_{30}	$\nexists$ solution	$\nexists$ solution	0.002

t_f being a root of the polynomial $q(\cdot)$, where

$$q(z) = \gamma z^5 - 2\beta z^3 + 3$$

Obviously, various cases can occur. First, it is possible that positive roots do not exist: then the problem does not admit a solution. Secondly, it may happen that we have one or more positive roots: then we select the smallest of them (the performance index is the length of the control interval), provided that it entails $\lambda_{30} > 0$. The values of the four parameters of interest are shown in Table 8.2 corresponding to various choices of β and γ. We observe that the solution does not exist if, for a given β, γ is sufficiently large. We also notice that, for a given γ, t_f decreases as β increases. On the contrary, for a given β, t_f increases as γ increases. Finally, the solution is more sensitive to variations of β than to variations γ. ∎

Chapter 9
Punctual and Isolated Constrains

Abstract In this chapter we consider punctual and isolated equality constraints which concern the values taken on by the state variables at isolated time instants inside the control interval. These instants may or may not be specified. First we present seven problems where the initial and final state are given. Furthermore: (1) the constraints are either functions of the state only or also functions of time; (2) the final time is given or free; (3) different scenarios are discussed according to whether the time of constraints satisfaction is given or free; (4) occasionally the comparison with the constraint-free situation is presented. We end this chapter by considering a problem which leads to a seemingly contradictory conclusion because the necessary conditions can not be satisfied, even if an optimal solution is shown to exist.

In this chapter we consider punctual and isolated equality constraints. We recall that their denomination refers to the fact that they concern the values taken on by the state variables at isolated time instants inside the control interval. These instants may or may not be specified.

We have formally described this kind of constraints in Sect. 2.3.1 (point (b)), whereas we have outlined the relevant way of handling them in Sect. 2.3.3.

We first present seven problems which, as customary, concern a double integrator and are characterized by the following main features: (1) the initial and final states are given in all of them; (2) the constraints are functions of the state only in Problems 9.1–9.3 and 9.7, whereas they are also functions of time in Problems 9.4–9.6; (3) the final time is given in Problems 9.1 and 9.2, whereas it is free in Problems 9.3–9.7; (4) Problems 9.1–9.3 and 9.7 present different scenarios according to whether the time of constraints satisfaction is given or free. Finally, the comparison with the constraint-free situation is presented in Problems 9.1 and 9.3.

We end this chapter by considering Problem 9.8 which leads to a seemingly contradictory conclusion because the NC can not be satisfied, even if we show that an optimal solution exists.

In dealing with the forthcoming problems it is expedient resorting to the material in Sect. 2.3.3. Furthermore we adopt the following notation which applies to any time function $\psi(\cdot)$. Suppose that both the left and right limits of $\psi(\cdot)$ are defined at time τ. Then we set

© Springer International Publishing Switzerland 2017

A. Locatelli, *Optimal Control of a Double Integrator*, Studies in Systems, Decision and Control 68, DOI 10.1007/978-3-319-42126-1_9

$$\psi^-(\tau) := \lim_{t \to \tau^-} \psi(t) \tag{9.1}$$

$$\psi^+(\tau) := \lim_{t \to \tau^+} \psi(t) \tag{9.2}$$

Problem 9.1

Let

$$x(0) = \begin{bmatrix} 0 \\ 0 \end{bmatrix}, \quad x(t_f) = \begin{bmatrix} 1 \\ 0 \end{bmatrix}, \quad t_f = 1$$

$$J = \int_0^{t_f} l(x(t), u(t), t) \, dt, \quad l(x, u, t) = \frac{u^2}{2}, \quad u(t) \in R, \ \forall t$$

Within this framework we discuss the three following cases

Problem 9.1-1. No further constraints are present.

Problem 9.1-2. $w(x(t_1), t_1) = 0$, $w(x, t) = x_1 + 1$, t_1 free.

Problem 9.1-3. $w(x(t_1), t_1) = 0$, $w(x, t) = x_1 + 1$, t_1 given.

In all problems the hamiltonian function and the H-minimizing control are

$$H = \frac{u^2}{2} + \lambda_1 x_2 + \lambda_2 u, \quad u_h = -\lambda_2$$

so that the hamiltonian system (2.6) and (2.5a) is

$$\begin{aligned} \dot{\lambda}_1(t) &= 0 & \dot{x}_1(t) &= x_2(t) \\ \dot{\lambda}_2(t) &= -\lambda_1(t) & \dot{x}_2(t) &= -\lambda_2(t) \end{aligned}$$

and its general solution over an interval beginning at time t_0 is

$$\lambda_1(t) = \lambda_1(t_0) \qquad\qquad x_1(t) = x_1(t_0) + x_2(t_0)(t - t_0) +$$
$$-\frac{\lambda_2(t_0)}{2}(t - t_0)^2 + \frac{\lambda_1(t_0)}{6}(t - t_0)^3$$

$$\lambda_2(t) = \lambda_2(t_0) - \lambda_1(t_0)(t - t_0) \qquad x_2(t) = x_2(t_0) - \lambda_2(t_0)(t - t_0) +$$
$$+\frac{\lambda_1(t_0)}{2}(t - t_0)^2$$

Problem 9.1-1 No further constraints are present.

By enforcing feasibility $(x_1(t_f) = 1, x_2(t_f) = 0)$, we obtain

$$x_1(t) = -\frac{\lambda_2(0)}{2}t^2 + \frac{\lambda_1(0)}{6}t^3, \quad x_2(t) = -\lambda_2(0)t + \frac{\lambda_1(0)}{2}t^2$$

where

$$\lambda_1(0) = -12, \quad \lambda_2(0) = -6$$

The state motion and the control are shown in Fig. 9.1 (dashed lines). The value of the performance index is $J = J_1 := 6$.

Problem 9.1-2 $w(x(t_1), t_1) = 0$, $w(x, t) = x_1 + 1$, t_1 free.

Equation (2.10) entails that the function $\lambda_1(\cdot)$ can be discontinuous at time t_1 because $w(\cdot, \cdot)$ explicitly depends on x_1. On the contrary, Eqs. (2.10) and (2.11) imply that both $\lambda_2(\cdot)$ and the hamiltonian function are continuous at t_1 because t_1 is free and $w(\cdot, \cdot)$ does not explicitly depend on x_2 and t. In view of Eq. (9.1) we conclude

$$\lambda_1^-(t_1) = \lambda_1^+(t_1) + \left.\frac{\partial w(x, t_1)}{\partial x}\right|_{x=x(t_1)} \mu = \lambda_1^+(t_1) + \mu$$

$$\lambda_2^-(t_1) = \lambda_2^+(t_1)$$

$$H^-(t_1) = -\frac{(\lambda_2^-(t_1))^2}{2} + \lambda_1^-(t_1)x_2(t_1) =$$

$$= H^+(t_1) - \left.\frac{\partial w(x(t_1), t)}{\partial t}\right|_{t=t_1} \mu = -\frac{(\lambda_2^+(t_1))^2}{2} + \lambda_1^+(t_1)x_2(t_1)$$

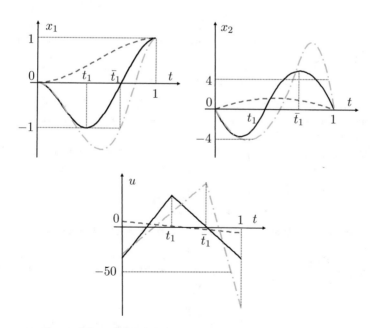

Fig. 9.1 Problem 9.1. State motion and control for Problem 9.1-1 (*dashed line*), for Problem 9.1-2 (*solid line*), for Problem 9.1-3 (*dash-single-dotted line*)

These equations imply $\lambda_1^-(t_1) = \lambda_1^+(t_1)$ and/or $x_2(t_1) = 0$. The first occurrence (continuity of $\lambda_1(\cdot)$) must be disregarded as it will lead to Problem 9.1-1 and the constraint $x_1(t_1) + 1 = 0$ is violated. Therefore if we enforce feasibility, namely

$$x_1(t_1) = -1, \quad x_2(t_1) = 0, \quad x_1(t_f) = 1, \quad x_2(t_f) = 0$$

we can compute the values of the four unknowns $\lambda_1(0), \lambda_2(0), \lambda_1^+(t_1), t_1$. The parameter μ needs not be determined as it is plainly unimportant. If we consider the functions $x_1(\cdot)$ and $x_2(\cdot)$

$0 \le t < t_1$

$u(t) = -\lambda_2(0) + \lambda_1(0)t$

$0 \le t \le t_1$

$x_1(t) = -\dfrac{\lambda_2(0)}{2}t^2 + \dfrac{\lambda_1(0)}{6}t^3$

$x_2(t) = -\lambda_2(0)t + \dfrac{\lambda_1(0)}{2}t^2$

$t_1 < t \le t_f$

$u(t) = -\lambda_2(t_1) + \lambda_1^+(t_1)(t - t_1)$

$t_1 \le t \le t_f$

$x_1(t) = -1 - \dfrac{\lambda_2(t_1)}{2}(t - t_1)^2 + \dfrac{\lambda_1^+(t_1)}{6}(t - t_1)^3$

$x_2(t) = -\lambda_2(t_1)(t - t_1) + \dfrac{\lambda_1^+(t_1)}{2}(t - t_1)^2$

where $\lambda_2(t_1) = \lambda_2(0) - \lambda_1(0)t_1$, we obtain

$$t_1 = \sqrt{2} - 1, \quad \lambda_1(0) = \frac{12}{t_1^3}, \quad \lambda_2(0) = \frac{6}{t_1^2}, \quad \lambda_1^+(t_1) = 2\frac{\lambda_2(t_1)}{(1 - t_1)}$$

The state motion and the control are shown in Fig. 9.1 (solid lines). The value of the performance index is $J = J_2 := 203.82$.

Problem 9.1-3 $w(x(t_1), t_1) = 0, w(x, t) = x_1 + 1, t_1$ given.

We proceed as in Problem 9.1-2, a part from imposing the continuity of the hamiltonian function at $t = t_1$ because t_1 is now given. Therefore it is no longer necessary that $x_2(t_1) = 0$.

By enforcing feasibility ($x_1(t_1) = -1, x_1(t_f) = 1, x_2(t_f) = 0$) and considering the state motion, we compute the values of the three unknowns $\lambda_1(0), \lambda_2(0), \lambda_1^+(t_1)$. More in detail, we have

$0 \le t < t_1$

$u(t) = -\lambda_2(0) + \lambda_1(0)t$

$0 \le t \le t_1$

$x_1(t) = -\dfrac{\lambda_2(0)}{2}t^2 + \dfrac{\lambda_1(0)}{6}t^3$

$x_2(t) = -\lambda_2(0)t + \dfrac{\lambda_1(0)}{2}t^2$

$t_1 < t \le t_f$

$u(t) = -\lambda_2(t_1) + \lambda_1^+(t_1)(t - t_1)$

$t_1 \le t \le t_f$

$x_1(t) = -1 + x_2(t_1)(t - t_1) - \dfrac{\lambda_2(t_1)}{2}(t - t_1)^2 +$
$\quad + \dfrac{\lambda_1^+(t_1)}{6}(t - t_1)^3$

$x_2(t) = x_2(t_1) - \lambda_2(t_1)(t - t_1) + \dfrac{\lambda_1^+(t_1)}{2}(t - t_1)^2$

where

$$\lambda_2(t_1) = \lambda_2(0) - \lambda_1(0)t_1$$

$$x_2(t_1) = -\lambda_2(0)t_1 + \frac{\lambda_1(0)}{2}t_1^2$$

and we have exploited the notation (9.1).

If, as an example, we choose $t_1 = \bar{t}_1 := 0.7$, it results

$$\lambda_1(0) = 112.83, \quad \lambda_2(0) = 30.41, \quad \lambda_1^+(\bar{t}_1) = -465.08$$

The state motion and the control are shown in Fig. 9.1 (dash-single-dotted lines). The value of the performance index is $J = J_3 := 521.50$.

We notice that another time $\hat{t}_1$ exists where $x_1(\hat{t}_1) = -1$. This fact should not be surprising.

Furthermore we observe that the control actions becomes more and more intense when passing from Problems 9.1-1 to 9.1-3. This outcome was to be expected. Indeed the successive introduction of new constraints (first the constraint on x_1, then also the constraint on t_1) implies *moving away* from the unconstrained framework more and more. The consequence of this is clearly enhanced by comparing the values J_1, J_2, J_3, of the performance index. ∎

Problem 9.2
Let

$$x(0) = \begin{bmatrix} 0 \\ 0 \end{bmatrix}, \quad x(t_f) = \begin{bmatrix} 0 \\ 0 \end{bmatrix}, \quad t_f = 3$$

$$J = \int_0^{t_f} l(x(t), u(t), t)\, dt, \ l(x, u, t) = \frac{u^2}{2}, \ u(t) \in R, \ \forall t$$

Two punctual and isolated equality constraints are present. The constraints take on three different forms with reference to the two times where the state must have a given value which is the same in all cases. Therefore we always have

$$w^{(1)}(x(t_1), t_1) = 0, \ w^{(1)}(x, t) = \begin{bmatrix} x_1 - 1 \\ x_2 \end{bmatrix}$$

$$w^{(2)}(x(t_2), t_2) = 0, \ w^{(2)}(x, t) = \begin{bmatrix} x_1 \\ x_2 - 1 \end{bmatrix}$$

whereas t_1 and t_2 comply with three different requirements. Both times are given in Problem 9.2-1, only one is given in Problem 9.2-2 and, finally, they are free in Problem 9.2-3. Therefore the constraints are less and less demanding when passing from the first to the third scenario. More specifically we consider the following problems.

Problem 9.2-1. $t_1 = 1, \ t_2 = 2$.

Problem 9.2-2. t_1 free, $t_1 < t_2, \ t_2 = 2$.

Problem 9.2-3. t_1 free, t_2 free, $t_1 < t_2$.

Before discussing them in detail, it is important to notice that the hamiltonian function, the H-minimizing control, Eq. (2.5a) and its general solution over an interval beginning at t_0 are the same for all problems. More precisely, we always have

$$H = \frac{u^2}{2} + \lambda_1 x_2 + \lambda_2 u \qquad\qquad u_h = -\lambda_2$$
$$\dot{\lambda}_1(t) = 0 \qquad\qquad\qquad \lambda_1(t) = \lambda_1(t_0)$$
$$\dot{\lambda}_2(t) = -\lambda_1(t) \qquad\qquad \lambda_2(t) = \lambda_2(t_0) - \lambda_1(t_0)(t - t_0)$$

In view of Eqs. (2.10) we know that $\lambda(\cdot)$ can be discontinuous at t_1 and t_2: in fact the functions $w^{(i)}(\cdot, \cdot)$, $i = 1, 2$, explicitly depend on both the state variables. Furthermore we notice that the times t_1 and t_2 partition the control interval into three subintervals where the initial values of $\lambda_1(\cdot)$ and $\lambda_2(\cdot)$ are denoted with $\lambda_1(0)$, $\lambda_1^+(t_1)$, $\lambda_1^+(t_2)$, $\lambda_2(0)$, $\lambda_2^+(t_1)$, $\lambda_2^+(t_2)$, according to the notation (9.1).

Problem 9.2-1 $x_1(t_1) = 1, x_2(t_1) = 0, x_1(t_2) = 0, x_2(t_2) = 1, t_1 = 1, t_2 = 2$.

Observe that the problem is indeed constituted by three independent subproblems defined over the given control intervals $0 \leq t \leq t_1, t_1 \leq t \leq t_2, t_2 \leq t \leq t_f$. In fact the state is completely specified at the beginning and at the end of each interval.

The six unknown parameters are the values of the functions $\lambda_1(\cdot)$ and $\lambda_2(\cdot)$ at the beginning of the above intervals: no relation has to be imposed between them because the functions $\lambda_1(\cdot)$ and $\lambda_2(\cdot)$ can be discontinuous at $t = t_1$ and $t = t_2$, as already claimed.

The six equations for the state motion are

$$0 \leq t < t_1, \quad \lambda_2(t) = \lambda_2(0) - \lambda_1(0)t, \quad u(t) = -\lambda_2(t)$$
$$0 \leq t \leq t_1, \quad x(t) = \begin{bmatrix} -\dfrac{\lambda_2(0)}{2}t^2 + \dfrac{\lambda_1(0)}{6}t^3 \\[2ex] -\lambda_2(0)t + \dfrac{\lambda_1(0)}{2}t^2 \end{bmatrix}$$

$$t_1 < t < t_2, \quad \lambda_2(t) = \lambda_2^+(t_1) - \lambda_1^+(t_1)(t - t_1), \quad u(t) = -\lambda_2(t)$$
$$t_1 \leq t \leq t_2, \quad x(t) = \begin{bmatrix} 1 - \dfrac{\lambda_2^+(t_1)}{2}(t - t_1)^2 + \dfrac{\lambda_1^+(t_1)}{6}(t - t_1)^3 \\[2ex] -\lambda_2^+(t_1)(t - t_1) + \dfrac{\lambda_1^+(t_1)}{2}(t - t_1)^2 \end{bmatrix}$$

$$t_2 < t \le t_f, \quad \lambda_2(t) = \lambda_2^+(t_2) - \lambda_1^+(t_2)(t - t_2), \quad u(t) = -\lambda_2(t)$$

$$t_2 \le t \le t_f, \quad x(t) = \begin{bmatrix} (t - t_2) - \dfrac{\lambda_2^+(t_2)}{2}(t - t_2)^2 + \dfrac{\lambda_1^+(t_2)}{6}(t - t_2)^3 \\[2mm] 1 - \lambda_2^+(t_2)(t - t_2) + \dfrac{\lambda_1^+(t_2)}{2}(t - t_2)^2 \end{bmatrix}$$

If we exploit these equations and enforce feasibility (the satisfaction of the constraints on $x_1(t)$ and $x_2(t)$ when $t = t_1, t = t_2, t = t_f$), we obtain a system of six equations for the six unknown parameters. It leads to

$$\lambda_1(0) = -12, \quad \lambda_2(0) = -6, \quad \lambda_1^+(t_1) = 18, \quad \lambda_2^+(t_1) = 8, \quad \lambda_1^+(t_2) = 6, \quad \lambda_2^+(t_2) = 4$$

The state motion and the control are shown in Fig. 9.2 (solid lines).

Problem 9.2-2 $x_1(t_1) = 1, x_2(t_1) = 0, x_1(t_2) = 0, x_2(t_2) = 1, t_1$ free, $t_1 < t_2, t_2 = 2$.

As before, we know that the functions $\lambda_1(\cdot)$ and $\lambda_2(\cdot)$ can be discontinuous at t_1 and t_2. On the contrary, the hamiltonian function must be continuous at $t = t_1$ thanks to Eq. (2.11) and the independence on time of $w^{(1)}(\cdot, \cdot)$ (the constraint relevant to t_1).

The seven unknown parameters are t_1 together with the six values of the functions $\lambda_1(\cdot)$ and $\lambda_2(\cdot)$ at the beginning of the three time intervals.

The six equations for the state motion are

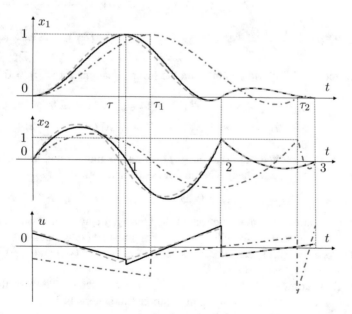

Fig. 9.2 Problem 9.2. State motions and control for Problem 9.2-1 (*solid lines*), for Problem 9.2-2 (*dashed lines*), for Problem 9.2-3 (*dash-single-dotted lines*)

$$0 \le t < t_1, \quad \lambda_2(t) = \lambda_2(0) - \lambda_1(0)t, \quad u(t) = -\lambda_2(t)$$

$$0 \le t \le t_1, \quad x(t) = \begin{bmatrix} -\dfrac{\lambda_2(0)}{2}t^2 + \dfrac{\lambda_1(0)}{6}t^3 \\[2mm] -\lambda_2(0)t + \dfrac{\lambda_1(0)}{2}t^2 \end{bmatrix}$$

$$t_1 < t < t_2, \quad \lambda_2(t) = \lambda_2^+(t_1) - \lambda_1^+(t_1)(t - t_1), \quad u(t) = -\lambda_2(t)$$

$$t_1 \le t \le t_2, \quad x(t) = \begin{bmatrix} 1 - \dfrac{\lambda_2^+(t_1)}{2}(t - t_1)^2 + \dfrac{\lambda_1^+(t_1)}{6}(t - t_1)^3 \\[2mm] -\lambda_2^+(t_1)(t - t_1) + \dfrac{\lambda_1^+(t_1)}{2}(t - t_1)^2 \end{bmatrix}$$

$$t_2 < t \le t_f, \quad \lambda_2(t) = \lambda_2^+(t_2) - \lambda_1^+(t_2)(t - t_2), \quad u(t) = -\lambda_2(t)$$

$$t_2 \le t \le t_f, \quad x(t) = \begin{bmatrix} (t - t_2) - \dfrac{\lambda_2^+(t_2)}{2}(t - t_2)^2 + \dfrac{\lambda_1^+(t_2)}{6}(t - t_2)^3 \\[2mm] 1 - \lambda_2^+(t_2)(t - t_2) + \dfrac{\lambda_1^+(t_2)}{2}(t - t_2)^2 \end{bmatrix}$$

We add the equation for the continuity of the hamiltonian function at t_1

$$H^-(t_1) = \frac{(u^-(t_1))^2}{2} + \lambda_1^-(t_1)x_2(t_1) + \lambda_2^-(t_1)u^-(t_1) = -\frac{(\lambda_{20} - \lambda_{10}t_1)^2}{2} =$$

$$= H^+(t_1) = \frac{(u^+(t_1))^2}{2} + \lambda_1^+(t_1)x_2(t_1) + \lambda_2^+(t_1)u^+(t_1) = -\frac{(\lambda_2^+(t_1))^2}{2}$$

If we enforce feasibility, we obtain, from the seven equations above, $t_1 = 0.92$ and

$$\lambda_1(0) = -15.24, \quad \lambda_1^+(t_1) = 14.80, \quad \lambda_1^+(t_2) = 6$$
$$\lambda_2(0) = -7.04, \quad \lambda_2^+(t_1) = 7.04, \quad \lambda_2^+(t_2) = 4$$

where we have exploited the notation (9.1). The state motion and the control are shown in Fig. 9.2 (dashed lines). The time t_1 is denoted with τ.

Problem 9.2-3 $x_1(t_1) = 1$, $x_2(t_1) = 0$, $x_1(t_2) = 0$, $x_2(t_2) = 1$, t_1, t_2 free, $t_1 < t_2$.

As before, we know that the functions $\lambda_1(\cdot)$ and $\lambda_2(\cdot)$ can be discontinuous at t_1 and t_2. On the contrary, the hamiltonian function must be continuous at $t = t_1$ and $t = t_2$ thanks to Eq. (2.11) and independence on time of $w^{(1)}(\cdot, \cdot)$ and $w^{(2)}(\cdot, \cdot)$ (the constraints relevant to t_1 and t_2).

The eight unknown parameters are t_1, t_2 together with the six values of the functions $\lambda_1(\cdot)$ and $\lambda_2(\cdot)$ at the beginning of the three time intervals.

The six equations for the state motion are

$$0 \leq t < t_1, \quad \lambda_2(t) = \lambda_2(0) - \lambda_1(0)t, \quad u(t) = -\lambda_2(t)$$

$$0 \leq t \leq t_1, \quad x(t) = \begin{bmatrix} -\dfrac{\lambda_2(0)}{2}t^2 + \dfrac{\lambda_1(0)}{6}t^3 \\[2mm] -\lambda_2(0)t + \dfrac{\lambda_1(0)}{2}t^2 \end{bmatrix}$$

$$t_1 < t < t_2, \quad \lambda_2(t) = \lambda_2^+(t_1) - \lambda_1^+(t_1)(t - t_1), \quad u(t) = -\lambda_2(t)$$

$$t_1 \leq t \leq t_2, \quad x(t) = \begin{bmatrix} 1 - \dfrac{\lambda_2^+(t_1)}{2}(t - t_1)^2 + \dfrac{\lambda_1^+(t_1)}{6}(t - t_1)^3 \\[2mm] -\lambda_2^+(t_1)(t - t_1) + \dfrac{\lambda_1^+(t_1)}{2}(t - t_1)^2 \end{bmatrix}$$

$$t_2 < t \leq t_f, \quad \lambda_2(t) = \lambda_2^+(t_2) - \lambda_1^+(t_2)(t - t_2), \quad u(t) = -\lambda_2(t)$$

$$t_2 \leq t \leq t_f, \quad x(t) = \begin{bmatrix} (t - t_2) - \dfrac{\lambda_2^+(t_2)}{2}(t - t_2)^2 + \dfrac{\lambda_1^+(t_2)}{6}(t - t_2)^3 \\[2mm] 1 - \lambda_2^+(t_2)(t - t_2) + \dfrac{\lambda_1^+(t_2)}{2}(t - t_2)^2 \end{bmatrix}$$

where we have used the notation (9.1). The equations for the continuity at t_1 and t_2 of the hamiltonian function are

$$H^-(t_1) = \frac{(u^-(t_1))^2}{2} + \lambda_1^-(t_1)x_2(t_1) + \lambda_2^-(t_1)u^-(t_1) = -\frac{(\lambda_{20} - \lambda_{10}t_1)^2}{2} =$$

$$= H^+(t_1) = \frac{(u^+(t_1))^2}{2} + \lambda_1^+(t_1)x_2(t_1) + \lambda_2^+(t_1)u^+(t_1) = -\frac{(\lambda_2^+(t_1))^2}{2}$$

and

$$H^-(t_2) = \frac{(u^-(t_2))^2}{2} + \lambda_1^-(t_2)x_2(t_2) + \lambda_2^-(t_2)u^-(t_2) =$$

$$-\frac{(\lambda_2^+(t_1) - \lambda_1^+(t_1)(t_2 - t_1))^2}{2} + \lambda_1^+(t_1) = H^+(t_2) =$$

$$= \frac{(u^+(t_2))^2}{2} + \lambda_1^+(t_2)x_2(t_2) + \lambda_2^+(t_2)u^+(t_2) = -\frac{(\lambda_2^+(t_2))^2}{2} + \lambda_1^+(t_2)$$

respectively. If we enforce feasibility, we obtain, from the eight equations above, $t_1 = 1.16$, $t_2 = 2.56$ and

$$\lambda_1(0) = -7.78, \quad \lambda_1^+(t_1) = 7.44, \quad \lambda_1^+(t_2) = 30.30$$
$$\lambda_2(0) = -4.49, \quad \lambda_2^+(t_1) = 4.49, \quad \lambda_2^+(t_2) = 8.99$$

The state motion and the control are shown in Fig. 9.2 (dash-single-dotted lines). The times t_1 and t_2 are denoted with τ_1 and τ_2, respectively.

It is worth comparing the values J_1, J_2, J_3 of the performance index of the three considered problems. We have $J_1 = 22.00$, $J_2 = 21.50$, $J_3 = 15.06$, so that $J_1 > J_2 > J_3$. This result is completely consistent with the fact that the minimum value of the performance index must decrease when, in sequence, we eliminate more and more constraints.

As a conclusion of the above discussion, it is somehow interesting to describe a possible way of solving Problem 9.2-3 which is the most intriguing one from an algebraic point of view.

First, we assume that the time t_1 is given and then compute $\lambda_i(0)$, $\lambda_i^+(t_j)$, $i = 1, 2$, $j = 1, 2$, t_2 as functions of it. The equation relevant to the continuity of the hamiltonian function at time t_2 can be exploited, if not satisfied, to modify the initial guess of t_1.

More in detail we can make the following statements each one corresponding to a step of a computational procedure.

(1) Guess a value for t_1.
(2) Compute $\lambda_1(0)$ and $\lambda_2(0)$ from the two feasibility equations at t_1.
(3) Compute $\lambda_2^+(t_1) = \lambda_2(0) - \lambda_1(0)t_1$ from the continuity equation of the hamiltonian function at t_1. Note that the other possible outcome, namely $\lambda_2^+(t_1) = -(\lambda_2(0) - \lambda_1(0)t_1)$, must be discarded because the constraints on the state variables at times t_1 and t_2 surely require that $\lambda_2(\cdot)$ has the same sign as t goes to t_1 from the left and from the right. Therefore this function is continuous at t_1.
(4) Compute $\lambda_1^+(t_1)$ and t_2 from the two feasibility equations at t_2.
(5) Compute $\lambda_1^+(t_2)$ and $\lambda_2^+(t_2)$ from the two feasibility equations at the final time.
(6) If the continuity equation of the hamiltonian function at t_2 is not verified, change the value of t_1 and iterate the procedure.   ∎

Problem 9.3

Let

$$x(0) = \begin{bmatrix} 0 \\ 0 \end{bmatrix}, \quad x\left(t_f\right) = \begin{bmatrix} 1 \\ 0 \end{bmatrix}, \quad J = \int_0^{t_f} l(x(t), u(t), t)\, dt, \quad l(x, u, t) = 1$$

$$u(t) \in \overline{U}, \ \forall t, \ \overline{U} = \{u | -1 \le u \le 1\}$$

Obviously, the final time t_f is free.

We discuss three problems which differ because of the (possible) presence of other constraints.

Problem 9.3-1. No other constraints are imposed.

Problem 9.3-2. $w^{(1)}(x(\tau), \tau) = 0$, $w^{(1)}(x, \tau) = x_1 + 1$, τ free.

Problem 9.3-3.

$$w^{(1)}(x(t_1), t_1) = 0, \quad w^{(1)}(x, t) = x_1 + 1, \quad t_1 \text{ free}$$
$$w^{(2)}(x(t_2), t_2) = 0, \quad w^{(2)}(x, t) = x_2 + 2, \quad t_2 \text{ free}, \quad t_2 > t_1$$

The hamiltonian function and the H-minimizing control of all these problems are

$$H = 1 + \lambda_1 x_2 + \lambda_2 u, \quad u_h = -\text{sign}(\lambda_2)$$

so that also Eq. (2.5a) and its general solution over an interval beginning at time t_0 are the same, namely

$$\dot{\lambda}_1(t) = 0 \qquad\qquad \lambda_1(t) = \lambda_1(t_0)$$
$$\dot{\lambda}_2(t) = -\lambda_1(t) \qquad\qquad \lambda_2(t) = \lambda_2(t_0) - \lambda_1(t_0)(t - t_0)$$

We now consider the three problems.

Problem 9.3-1 No other constraints are imposed.

The sign of $\lambda_2(\cdot)$ can at most change once at a time denoted with τ_c. When $u(\cdot) = \pm 1$ we have

$$x_2(t) = \pm t, \quad x_1(t) = \pm \frac{t^2}{2}$$

Therefore the equations

$$x_2(t_f) = 0 = \pm t_f, \quad x_1(t_f) = 1 = \pm \frac{t_f^2}{2}$$

do not have a solution and feasibility does not result. Consequently, it is mandatory that $\lambda_2(\cdot)$ switches or, in other words, $\lambda_2(\tau_c) = 0$ for some suitable τ_c. It is apparent that $u(0)$ should be positive, so that the transversality condition at the final time, written at the initial time because the problem is time-invariant,

$$H(0) = 1 + \lambda_1(0)x_2(0) + \lambda_2(0)u(0) = 1 + \lambda_2(0)u(0) = 0$$

supplies $\lambda_2(0) = -1$. Thus we have

$0 \le t < \tau_c$	$\tau_c < t \le t_f$
$u(t) = 1$	$u(t) = -1$
$0 \le t \le \tau_c$	$\tau_c \le t \le t_f$
$x_1(t) = \dfrac{t^2}{2}$	$x_1(t) = \dfrac{t^2}{2} + \tau_c(t - \tau_c) - \dfrac{(t - \tau_c)^2}{2}$
$x_2(t) = t$	$x_2(t) = \tau_c - (t - \tau_c)$

Fig. 9.3 Problem 9.3. State
trajectories relevant to:
Problem 9.3-1 (0–a–b),
Problem 9.3-2 (0–c–d–a–b),
Problem 9.3-3 when the sign
of $\lambda_2(\cdot)$ changes three times
(0–c–d–g–e–f–b), Problem
9.3-3 when the sign of $\lambda_2(\cdot)$
changes four times
(0–c–d–g–h–i–d–a–b)

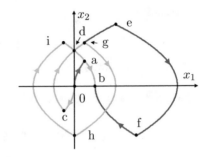

By requiring $x_1(t_f) = 1$, $x_2(t_f) = 0$ (feasibility) and $\lambda_2(\tau_c) = 0$ (control switching), we obtain

$$t_f := t_{f_1} = 2, \quad \tau_c := \tau_{c1} = 1, \quad \lambda_1(0) = -1$$

The corresponding state trajectory is the curve (0-a-b) plotted in Fig. 9.3.

Problem 9.3-2 $w^{(1)}(x(\tau), \tau) = 0$, $w^{(1)}(x, \tau) = x_1 + 1$, τ free.

From Eq. (2.10) it follows that the function $\lambda_1(\cdot)$ can be discontinuous at time τ because $w^{(1)}(\cdot, \cdot)$ (the constraint relevant to this time) explicitly depends on x_1. On the contrary, Eqs. (2.10) and (2.11) imply that both the hamiltonian function and $\lambda_2(\cdot)$ are continuous at τ. In fact $w^{(1)}(\cdot, \cdot)$ explicitly depends neither on x_2 nor on t and τ is free.

Thanks to the continuity of the hamiltonian function, the transversality condition at the final time can be written at any instant (recall that the problem is time-invariant). This fact, together with the continuity of $\lambda_2(\cdot)$, implies that

$$H^-(\tau) = 1 + \lambda_1^-(\tau)x_2(\tau) + \lambda_2(\tau)u^-(\tau) = 1 + \lambda_1^+(\tau)x_2(\tau) + \lambda_2(\tau)u^+(\tau) = H^+(\tau)$$

where we have used the notation (9.1). We observe that $\lambda_2(\tau)u^-(\tau) = \lambda_2(\tau)u^+(\tau)$: indeed this is obvious if $\lambda_2(\tau) = 0$, whereas, if $\lambda_2(\tau) \neq 0$, we have $u^-(\tau) = u^+(\tau)$. From the above equation concerning the hamiltonian function we conclude that either $x_2(\tau) = 0$ or, in the opposite case, $\lambda_1^-(\tau) = \lambda_1^+(\tau)$. The second alternative must be discarded. In fact if $\lambda_1(\cdot)$ is continuous, then $\lambda_2(\cdot)$ is a linear function of t over the whole interval $[0, t_f]$ and its sign either does not change or changes once. In both cases it is not possible that $x_1(\tau) = -1$, $x_1(t_f) = 1$ and $x_2(t_f) = 0$. This conclusion easily follows if reference is made to the shape of the state trajectories when $u(\cdot) = \pm 1$. They are plotted in Fig. 9.4 and defined by the equations.

$$x_1 = \frac{x_2^2}{2} + k, \text{ if } u(\cdot) = 1, \quad x_1 = -\frac{x_2^2}{2} + k, \text{ if } u(\cdot) = -1$$

where k is a constant.

Fig. 9.4 Problem 9.3-2.
State trajectories
corresponding to $u(\cdot) = 1$
(*solid lines*) and $u(\cdot) = -1$
(*dashed lines*)

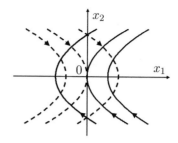

Therefore $x_2(\tau) = 0$.

Thanks to the discussion above, the transversality condition at the final time can be written at the initial time yielding

$$H(0) = 1 + \lambda_1(0)x_2(0) + \lambda_2(0)u(0) = 1 \pm \lambda_2(0) = 0$$

and implies $\lambda_2(0) = \pm 1$. The nature of the problem suggests that $u(0) = -1$ and, in turn, that $\lambda_2(0) = 1$. Furthermore $\lambda_2(\cdot)$ is anyhow a linear function of time both when $t \leq \tau$ and $t \geq \tau$. Consequently, two times $\tau_{c_{21}} < \tau$ and $\tau_{c_{22}} > \tau$ can exist where the sign of $\lambda_2(\cdot)$ changes, thus causing the control to switch from -1 to 1 and viceversa, respectively. We notice that only one change of sign might exist, but, as remarked above, feasibility would not be achieved. Within the interval $[0, \tau_{c_{21}}]$ the state motion and the function $\lambda_2(\cdot)$ are

$0 \leq t < \tau_{c_{21}}$	$\tau_{c_{21}} < t \leq \tau$
$u(t) = -1$	$u(t) = 1$
$0 \leq t \leq \tau_{c_{21}}$	$\tau_{c_{21}} \leq t \leq \tau$
$\lambda_2(t) = 1 - \lambda_1(0)t$	$\lambda_2(t) = 1 - \lambda_1(0)t$
$x_1(t) = -\dfrac{t^2}{2}$	$x_1(t) = -\dfrac{\tau_{c_{21}}^2}{2} - \tau_{c_{21}}(t - \tau_{c_{21}}) + \dfrac{(t - \tau_{c_{21}})^2}{2}$
$x_2(t) = -t$	$x_2(t) = -\tau_{c_{21}} + (t - \tau_{c_{21}})$

where we have taken into account that $\lambda_2(0) = 1$. We obtain

$$\tau_{c_{21}} = 1, \quad \tau = 2, \quad \lambda_1(0) = 1, \quad \lambda_2(\tau) = -1$$

because $x_1(\tau) = -1$, $x_2(\tau) = 0$, $\lambda_2(\tau_{c_{21}}) = 0$.

Subsequently, we consider the time interval beginning at τ and ending at the final time t_f. Within the interval $[\tau, t_f]$ the state motion and the function $\lambda_2(\cdot)$ are

$$\tau < t < \tau_{c_{22}}$$
$$u(t) = 1$$
$$\tau \leq t \leq \tau_{c_{22}}$$
$$\lambda_2(t) = -1 - \lambda_1^+(\tau)(t - \tau)$$
$$x_1(t) = -1 + \frac{(t - \tau)^2}{2}$$

$$x_2(t) = t - \tau$$

$$\tau_{c_{22}} < t \leq t_f$$
$$u(t) = -1$$
$$\tau_{c_{22}} \leq t \leq t_f$$
$$\lambda_2(t) = -1 - \lambda_1^+(\tau)(t - \tau)$$
$$x_1(t) = -1 + \frac{(\tau_{c_{22}} - \tau)^2}{2} +$$
$$+ (\tau_{c_{22}} - \tau)(t - \tau_{c_{22}}) - \frac{(t - \tau_{c_{22}})^2}{2}$$

$$x_2(t) = \tau_{c_{22}} - \tau - (t - \tau_{c_{22}})$$

where we have taken into account the previously found values for $x_1(\tau), x_2(\tau), \lambda_2(\tau)$. Since $x_1(t_f) = 1, x_2(t_f) = 0, \lambda_2(\tau_{c_{22}}) = 0$ we obtain

$$\tau_{c_{22}} = 2 + \sqrt{2}, \quad t_f := t_{f_2} = 2 + 2\sqrt{2}, \quad \lambda_1^+(\tau) = -\frac{1}{\sqrt{2}}$$

The corresponding state trajectory (0-c-d-a-b) is shown in Fig. 9.3.

Problem 9.3-3

$$w^{(1)}(x(t_1), t_1) = 0, \quad w^{(1)}(x, t) = x_1 + 1, \quad t_1 \text{ free}$$
$$w^{(2)}(x(t_2), t_2) = 0, \quad w^{(2)}(x, t) = x_2 + 2, \quad t_2 \text{ free}, t_2 > t_1$$

In view of Eq. (2.10) the presence of constraints and their nature imply that $\lambda_1(\cdot)$ and $\lambda_2(\cdot)$ can be discontinuous at t_1 and t_2, respectively. On the contrary, Eq. (2.11), together with the fact that t_1 and t_2 are free and $w^{(1)}(\cdot, \cdot)$ and $w^{(2)}(\cdot, \cdot)$ do not explicitly depend on time, entails the continuity of the hamiltonian function. Therefore the transversality condition at the final time t_f, written at the initial time (recall that the problem is time-invariant),

$$H(0) = 1 + \lambda_1(0)x_2(0) + \lambda_2(0)u(0) = 1 \pm \lambda_2(0) = 0$$

requires $\lambda_2(0) = \pm 1$. The nature of the problem suggests $u(0) = -1$, that is $\lambda_2(0) = 1$. By making use of the notation (9.1), the equation for the continuity of the hamiltonian function at t_1 is

$$H^-(t_1) = 1 + \lambda_1^-(t_1)x_2(t_1) + \lambda_2(t_1)u(t_1) = 1 + \lambda_1^+(t_1)x_2(t_1) + \lambda_2(t_1)u(t_1) = H^+(t_1)$$

As in Problem 9.3-2, we notice that $\lambda_2(t_1)u^-(t_1) = \lambda_2(t_1)u^+(t_1)$: indeed this is obvious if $\lambda_2(t_1) = 0$, whereas, if $\lambda_2(t_1) \neq 0$, we have $u^-(t_1) = u^+(t_1)$. From the above equation we conclude that either $x_2(t_1) = 0$ or $\lambda_1^-(t_1) = \lambda_1^+(t_1)$. If we again make use of the notation (9.1), the equation for the continuity of the hamiltonian function at t_2 is

$$H^-(t_2) = 1 + \lambda_1(t_2)x_2(t_2) + \lambda_2^-(t_2)u^-(t_2) = 1 + \lambda_1(t_2)x_2(t_2) + \lambda_2^-(t_2)u^-(t_2) = H^+(t_2)$$

We conclude that either $\lambda_2^-(t_2)u^-(t_2) = \lambda_2^+(t_2)u^+(t_2)$, that is $\lambda_2^-(t_2) = \lambda_2^+(t_2)$ or $\lambda_2^-(t_2) = -\lambda_2^+(t_2)$. Consequently, the following eleven *alternatives* must be discussed.

Alternative 1. $\lambda_1^-(t_1) = \lambda_1^+(t_1)$, $\lambda_2^-(t_2) = \lambda_2^+(t_2)$.

Alternative 2. $\lambda_1^-(t_1) = \lambda_1^+(t_1)$, $\lambda_2^-(t_2) = \lambda_2^+(t_2)$, $\lambda_2(t_1) = 0, 0 < t_1 < t_f$.

Alternative 3. $\lambda_1^-(t_1) = \lambda_1^+(t_1)$, $\lambda_2^-(t_2) = -\lambda_2^+(t_2)$, $\lambda_2(\tau_{c_{31}}) = 0$, $\lambda_2(\tau_{c_{32}}) = 0$
$0 < \tau_{c_{31}} < t_2, t_2 < \tau_{c_{32}} < t_f$.

Alternative 4. $\lambda_1^-(t_1) = \lambda_1^+(t_1)$, $\lambda_2^-(t_2) = -\lambda_2^+(t_2)$, $\lambda_2(\tau_{c_{31}}) = 0, 0 < \tau_{c_{31}} < t_2$

Alternative 5. $\lambda_1^-(t_1) = \lambda_1^+(t_1)$, $\lambda_2^-(t_2) = -\lambda_2^+(t_2)$, $\lambda_2(\tau_{c_{32}}) = 0, t_2 < \tau_{c_{32}} < t_f$

Alternative 6. $\lambda_1^-(t_1) = \lambda_1^+(t_1)$, $\lambda_2^-(t_2) = -\lambda_2^+(t_2)$.

Alternative 7. $x_2(t_1) = 0$, $\lambda_2^-(t_2) = \lambda_2^+(t_2)$, $\lambda_2(\tau_{c_{31}}) = 0$, $\lambda_2(\tau_{c_{32}}) = 0$.
$0 < \tau_{c_{31}} < t_2, t_2 < \tau_{c_{32}} < t_f$.

Alternative 8. $x_2(t_1) = 0$, $\lambda_2^-(t_2) = \lambda_2^+(t_2)$, $\lambda_2(\tau_{c_{31}}) = 0, 0 < \tau_{c_{31}} < t_1$.

Alternative 9. $x_2(t_1) = 0$, $\lambda_2^-(t_2) = -\lambda_2^+(t_2)$, $\lambda_2(\tau_{c_{31}}) = 0, 0 < \tau_{c_{31}} < t_1$.

Alternative 10. $x_2(t_1) = 0$, $\lambda_2^-(t_2) = -\lambda_2^+(t_2)$, $\lambda_2(\tau_{c_{31}}) = 0$, $\lambda_2(\tau_{c_{32}}) = 0$, $\lambda_2(\tau_{c_{33}}) = 0$.
$0 < \tau_{c_{31}} < t_1, t_1 < \tau_{c_{32}} < t_2, t_2 < \tau_{c_{33}} < t_f$.

Alternative 11. $x_2(t_1) = 0$, $\lambda_2^-(t_2) = -\lambda_2^+(t_2)$, $\lambda_2(\tau_{c_{31}}) = 0$, $\lambda_2(\tau_{c_{32}}) = 0$.
$0 < \tau_{c_{31}} < t_1, t_1 < \tau_{c_{32}} < t_2$.

We recall that $\lambda_2(\cdot)$ is a piecewise linear function of time. Thus the possible discontinuities (pointed out above) allow the existence of at most four times where its sign can change. The shapes of $\lambda_2(\cdot)$ consistent with the eleven alternatives are shown in Fig. 9.5: many of them do not satisfy the NC. This is made clear by the analysis below, where it is expedient to recall the shapes of the state trajectories resulting from $u(\cdot) = \pm 1$ (see Fig. 9.4). It is also helpful looking at Fig. 9.6: there we have plotted a possible trajectory consistent with each one of the alternatives above. Obviously, trajectories like (0–a–b), (0–a–c–d), (0–a–c–e–f) do not allow complying with the four requests $x_1(t_1) = -1, x_2(t_2) = -2, x_1(t_f) = 1, x_2(t_f) = 0$ which, on the contrary, can be satisfied by trajectories like (0–a–c–e–g–h) or (0–a–c–e–g–k). Thus we focus the attention on the last two trajectories which are consistent with *Alternative 11* or *Alternative 10* and verify whether they satisfy the NC.

Alternative 11. Trajectories of the kind (0–a–c–e–g–h) in Fig. 9.6
The sign of $\lambda_2(\cdot)$ changes three times at $\tau_{c_{31}}, \tau_{c_{32}}$ and t_2 where $\tau_{c_{31}} < t_1 < \tau_{c_{32}} < t_2 \leq t_f$. Thus

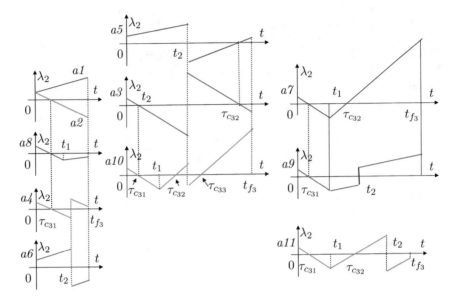

Fig. 9.5 Problem 9.3-3. The significantly different forms of $\lambda_2(\cdot)$

Fig. 9.6 Problem 9.3-3.
State trajectories consistent
with *Alternative 1* (0–a–b);
Alternative 2 or *Alternative 6*
or *Alternative 8* (0–a–c–d);
Alternative 4 or *Alternative 5*
or *Alternative 7* or
Alternative 9 (0–a–c–e–f);
Alternative 3 or *Alternative*
11 (0–a–c–e–g–h);
Alternative 10
(0–a–c–e–g–k)

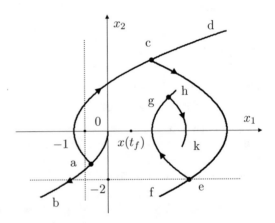

$$\lambda_2(\tau_{c_{31}}) = \lambda_2(0) - \lambda_1(0)\tau_{c_{31}} = 0, \quad \lambda_2(\tau_{c_{32}}) = \lambda_2(t_1) - \lambda_1^+(t_1)(\tau_{c_{32}} - t_1) = 0$$

$$\lambda_2(t_1) = \lambda_2(0) - \lambda_1(0)t_1, \quad \lambda_2^-(t_2) = -\lambda_1^+(t_1)(t_2 - \tau_{c_{32}})$$

while at t_2 the discontinuity

$$\lambda_2^+(t_2) = -\lambda_2^-(t_2)$$

occurs. Consistently, the control $u(\cdot)$ is

$$u(t) = \begin{cases} -1, & 0 \leq t < \tau_{c_{31}} \\ 1, & \tau_{c_{31}} < t < \tau_{c_{32}} \\ -1, & \tau_{c_{32}} < t < t_2 \\ 1, & t_2 < t \leq t_f \end{cases}$$

and entails

$$x_1(\tau_{c_{31}}) = -\frac{\tau_{c_{31}}^2}{2}, \quad x_2(\tau_{c_{31}}) = -\tau_{c_{31}}$$

$$x_1(t_1) = x_1(\tau_{c_{31}}) + x_2(\tau_{c_{31}})(t_1 - \tau_{c_{31}}) + \frac{(t_1 - \tau_{c_{31}})^2}{2}$$

$$x_2(t_1) = x_2(\tau_{c_{31}}) + (t_1 - \tau_{c_{31}})$$

$$x_1(\tau_{c_{32}}) = x_1(t_1) + x_2(t_1)(\tau_{c_{32}} - t_1) + \frac{(\tau_{c_{32}2} - t_1)^2}{2}$$

$$x_2(\tau_{c_{32}}) = x_2(t_1) + (\tau_{c_{32}} - t_1)$$

$$x_1(t_2) = x_1(\tau_{c_{32}}) + x_2(\tau_{c_{32}})(t_2 - \tau_{c_{32}}) - \frac{(t_2 - \tau_{c_{32}})^2}{2}$$

$$x_2(t_2) = x_2(\tau_{c_{32}}) - (t_2 - \tau_{c_{32}})$$

$$x_1(t_f) = x_1(t_2) + x_2(t_2)(t_f - t_2) + \frac{(t_f - t_2)^2}{2}$$

$$x_2(t_f) = x_2(t_2) + (t_f - t_2)$$

These ten relations, where we have taken into account that $x_1(0) = x_2(0) = 0$, together with the five ones concerning $\lambda_2(\cdot)$ and the (feasibility) constraints

$$x_1(t_1) = -1, \quad x_2(t_1) = 0, \quad x_2(t_2) = -2, \quad x_1(t_f) = 1, \quad x_2(t_f) = 0$$

constitute a system of equations which supplies the parameters of the solution (recall that we have already established that $\lambda_2(0) = 1$). More in detail we have

$$\lambda_1(0) = 1, \quad \lambda_1^+(t_1) = -\frac{1}{\sqrt{6}}, \quad \lambda_2^-(t_2) = \frac{2 + \sqrt{6}}{\sqrt{6}}, \quad \lambda_2^+(t_2) = -\frac{2 + \sqrt{6}}{\sqrt{6}}$$

$$\tau_{c_{31}} = 1, \quad \tau_{c_{32}} = 2 + \sqrt{6}, \quad t_1 = 2, \quad t_2 = 4 + 2\sqrt{6}, \quad t_f := t_{f_3} = 6 + 2\sqrt{6}$$

The corresponding state trajectory (0–c–d–g–e–f–b) is plotted in Fig. 9.3.

Alternative 10. Trajectories of the kind (0–a–c–e–g–k) in Fig. 9.6.

The sign of $\lambda_2(\cdot)$ changes four times at $\tau_{c_{31}}, \tau_{c_{32}}, t_2$ and $\tau_{c_{33}}$ where $\tau_{c_{31}} < t_1 < \tau_{c_{32}} < t_2 < \tau_{c_{33}} \leq t_{f_3}$.

   In the present discussion, we denote the various parameters of interest with the same symbols used in dealing with *Alternative 11*: however, in order to avoid possible misunderstandings, we put a bar over them. Thus we have

$$\bar{\lambda}_2(\bar{\tau}_{c_{31}}) = \bar{\lambda}_2(0) - \bar{\lambda}_1(0)\bar{\tau}_{c_{31}} = 0, \quad \bar{\lambda}_2(\bar{t}_1) = \bar{\lambda}_2(0) - \bar{\lambda}_1(0)\bar{t}_1$$

$$\bar{\lambda}_2(\bar{\tau}_{c_{32}}) = \bar{\lambda}_2(\bar{t}_1) - \bar{\lambda}_1^+(\bar{t}_1)(\bar{\tau}_{c_{32}} - \bar{t}_1) = 0, \quad \bar{\lambda}_2^-(\bar{t}_2) = -\bar{\lambda}_1^+(\bar{t}_1)(\bar{t}_2 - \bar{\tau}_{c_{32}})$$

$$\bar{\lambda}_2(\bar{\tau}_{c_{33}}) = \bar{\lambda}_2^+(\bar{t}_2) - \bar{\lambda}_1^+(\bar{t}_1)(\bar{\tau}_{c_{33}} - \bar{t}_2) = 0$$

whereas the discontinuity

$$\bar{\lambda}_2^+(\bar{t}_2) = -\bar{\lambda}_2^-(\bar{t}_2)$$

occurs at $\bar{t}_2$. Consistently, the control $u(\cdot)$ is

$$u(t) = \begin{cases} -1, & 0 \le t < \bar{\tau}_{c_{31}} \\ 1, & \bar{\tau}_{c_{31}} < t < \bar{\tau}_{c_{32}} \\ -1, & \bar{\tau}_{c_{32}} < t < \bar{t}_2 \\ 1, & \bar{t}_2 < t < \bar{\tau}_{c_{33}} \\ -1, & \bar{\tau}_{c_{33}} < t \le \bar{t}_f \end{cases}$$

and entails

$$x_1(\bar{\tau}_{c_{31}}) = -\frac{\bar{\tau}_{c_{31}}^2}{2}, \quad x_2(\bar{\tau}_{c_{31}}) = -\bar{\tau}_{c_{31}}$$

$$x_1(\bar{t}_1) = x_1(\bar{\tau}_{c_{31}}) + x_2(\bar{\tau}_{c_{31}}) + \frac{(\bar{t}_1 - \bar{\tau}_{c_{31}})^2}{2}$$

$$x_2(\bar{t}_1) = x_2(\bar{\tau}_{c_{31}}) + (\bar{t}_1 - \bar{\tau}_{c_{31}})$$

$$x_1(\bar{\tau}_{c_{32}}) = x_1(\bar{t}_1) + x_2(\bar{t}_1)(\bar{\tau}_{c_{32}} - \bar{t}_1) + \frac{(\bar{\tau}_{c_{32}} - \bar{t}_1)^2}{2}$$

$$x_2(\bar{\tau}_{c_{32}}) = x_2(\bar{t}_1) + (\bar{\tau}_{c_{32}} - \bar{t}_1)$$

$$x_1(\bar{t}_2) = x_1(\bar{\tau}_{c_{32}}) + x_2(\bar{\tau}_{c_{32}})(\bar{t}_2 - \bar{\tau}_{c_{32}}) - \frac{(\bar{t}_2 - \bar{\tau}_{c_{32}})^2}{2}$$

$$x_2(\bar{t}_2) = x_2(\bar{\tau}_{c_{32}}) - (\bar{t}_2 - \bar{\tau}_{c_{32}})$$

$$x_1(\bar{\tau}_{c_{33}}) = x_1(\bar{t}_2) + x_2(\bar{t}_2)(\bar{\tau}_{c_{33}} - \bar{t}_2) + \frac{(\bar{\tau}_{c_{33}} - \bar{t}_2)^2}{2}$$

$$x_2(\bar{\tau}_{c_{33}}) = x_2(\bar{t}_2) + (\bar{\tau}_{c_{33}} - \bar{t}_2)$$

$$x_1(\bar{t}_f) = x_1(\bar{\tau}_{c_{33}}) + x_2(\bar{\tau}_{c_{33}})(\bar{t}_f - \bar{\tau}_{c_{33}}) - \frac{(\bar{\tau}_{c_{33}} - \bar{t}_2)^2}{2}$$

$$x_2(\bar{t}_f) = x_2(\bar{\tau}_{c_{33}}) - (\bar{t}_f - \bar{\tau}_{c_{33}})$$

These twelve relations, together with those concerning $\lambda_2(\cdot)$ and the (feasibility) constraints

$$x_1(\bar{t}_1) = -1, \quad x_2(\bar{t}_1) = 0 \quad x_2(\bar{t}_2) = -2, \quad x_1(\bar{t}_f) = 1, \quad x_2(\bar{t}_f) = 0$$

constitute a system of equations for the parameters which specify the solution (recall that we have already established that $\bar{\lambda}_2(0) = 1$). More precisely, we obtain

$$\bar{\lambda}_1(0) = 1, \quad \bar{\lambda}_1^+(\bar{t}_1) = -0.58, \quad \bar{\lambda}_2^-(\bar{t}_2) = 2.15, \quad \bar{\lambda}_2^+(\bar{t}_2) = -2.15$$
$$\bar{\tau}_{c_{31}} = 1, \quad \bar{\tau}_{c_{32}} = 3.73, \quad \bar{\tau}_{c_{33}} = 11.20, \quad \bar{t}_1 = 2, \ \bar{t}_2 = 7.46, \quad \bar{t}_f := \bar{t}_{f_3} = 12.93$$

The resulting state trajectory (0–c–d–g–h–i–d–a–b) is plotted in Fig. 9.3.

The solution where the sign of $\lambda_2(\cdot)$ changes three times is preferred because $\bar{t}_{f_3} > t_{f_3}$.

As it should be expected, the value of the performance index, namely the length of the control interval, increases when passing from Problems 9.3-1 to 9.3-2 and 9.3-3. Indeed each problem requires the satisfaction of a further constraint, not present in the preceding one. Furthermore we have found $x_2(\tau) = 0$ in Problem 9.3-2 and $x_2(t_1) = 0$ in Problem 9.3-3: thus $\tau = t_1$ and the state motion is the same over the interval $[0, \tau]$ for the two problems: this is consistent with their nature. ∎

Problem 9.4

Let

$$x(0) = \begin{bmatrix} 0 \\ 0 \end{bmatrix}, \quad x(t_f) = \begin{bmatrix} 1 \\ 0 \end{bmatrix}, \quad w(x(t_1), t_1) = 0, \ w(x, t) = x_1 + 1 + t$$

$$J = \int_0^{t_f} l(x(t), u(t), t)\, dt, \ l(x, u, t) = 1 + \frac{u^2}{2}, \ u(t) \in R, \ \forall t$$

The intermediate time t_1 and the final one t_f are free.

The hamiltonian function and the H-minimizing control are

$$H = 1 + \frac{u^2}{2} + \lambda_1 x_2 + \lambda_2 u, \quad u_h = -\lambda_2$$

In view of Eq. (2.10) $\lambda_1(\cdot)$ can be discontinuous at time t_1 because $w(\cdot, \cdot)$ explicitly depends on da x_1. On the contrary, $\lambda_2(\cdot)$ is continuous at this time because $w(\cdot, \cdot)$ does not explicitly depend on x_2. Thanks to Eq. (2.11), the hamiltonian function can be discontinuous at t_1: in fact $w(\cdot, \cdot)$ explicitly depends on time and t_1 is free.

Equation (2.5a) and its general solution over an interval beginning at t_0 are

$$\dot{\lambda}_1(t) = 0 \qquad\qquad \lambda_1(t) = \lambda_1(t_0)$$
$$\dot{\lambda}_2(t) = -\lambda_1(t) \qquad\qquad \lambda_2(t) = \lambda_2(t_0) - \lambda_1(t_0)(t - t_0)$$

Therefore $\lambda_1(\cdot)$ and $\lambda_2(\cdot)$ are functions of the time t which are piecewise constant and linear, respectively. More specifically, by exploiting, as customary, the notation (9.1), we have

$$\lambda_1^-(t_1) = \lambda_1(0) = \lambda_1^+(t_1) + \left.\frac{\partial w(x, t_1)}{\partial x}\right|_{x=x(t_1)} \mu = \lambda_1^+(t_1) + \mu$$

$$H^-(t_1) = 1 - \frac{(\lambda_2^-(t_1))^2}{2} + \lambda_1^-(t_1)x_2(t_1) =$$

$$= 1 - \frac{(\lambda_2^+(t_1))^2}{2} + \lambda_1^+(t_1)x_2(t_1) - \frac{\partial w(x(t_1), t)}{\partial t}\bigg|_{t=t_1} \mu =$$

$$= 1 - \frac{(\lambda_2^+(t_1))^2}{2} + \lambda_1^+(t_1)x_2(t_1) - \mu = H^+(t_1) - \mu$$

We recall that $\lambda_2(\cdot)$ is continuous: thus from these equations it follows that $x_2(t_1) = -1$, because the constraint is violated if $\lambda_1^-(t_1) = \lambda_1^+(t_1)$.

The values of the five unknowns $\lambda_1(0), \lambda_2(0), \lambda_1^+(t_1), t_1, t_f$ can be computed by enforcing feasibility ($x_1(t_1) = -1 - t_1, x_2(t_1) = -1, x_1(t_f) = 1, x_2(t_f) = 0$) and requiring the satisfaction of the transversality condition at the final time

$$H(t_f) = 1 + \frac{u^2(t_f)}{2} + \lambda_1(t_f)x_2(t_f) + \lambda_2(t_f)u(t_f) = 1 - \frac{(\lambda_2(t_f))^2}{2} = 0$$

Thus the last equation implies

$$\lambda_2(t_f) = \lambda_2(t_1) - \lambda_1^+(t_1)(t_f - t_1) = \lambda_2(0) - \lambda_1(0)t_1 - \lambda_1^+(t_1)(t_f - t_1) = k$$

where $k := \pm\sqrt{2}$. We can ascertain that the above equations have a meaningful solution only if $k = -\sqrt{2}$. In this case we have

$$\lambda_1(0) = 1.80, \quad \lambda_2(0) = 2.62, \quad \lambda_1^+(t_1) = -0.63, \quad t_1 = 2.46, \quad t_f = 7.57$$

■

Problem 9.5
Let

$$x(0) = \begin{bmatrix} 0 \\ 0 \end{bmatrix}, \quad x(t_f) = \begin{bmatrix} 1 \\ 0 \end{bmatrix}, \quad J = \int_0^{t_f} l(x(t), u(t), t)\, dt, \quad l(x, u, t) = 1$$

$$u(t) \in \overline{U}, \ \forall t, \ \overline{U} = \{u| -1 \le u \le 1\}$$

$$w(x(t_1), t_1) = 0, \quad w(x, t) = x_1 + 1 + t$$

The intermediate time t_1 and the final one t_f are free.

This problem constitutes a small variation of Problem 9.3-2. In fact the punctual and isolated equality constraint is now specified by $w(x, t) = x_1 + 1 + t$ rather than $w(x, t) = x_1 + 1$.

The hamiltonian function and the H-minimizing control are

$$H = 1 + \lambda_1 x_2 + \lambda_2 u, \quad u_h = -\text{sign}(\lambda_2)$$

In view of Eq. (2.10) $\lambda_1(\cdot)$ can be discontinuous at time t_1 because $w(\cdot, \cdot)$ explicitly depends on x_1. On the contrary, $\lambda_2(\cdot)$ is continuous at this time because $w(\cdot, \cdot)$ does not explicitly depend on x_2. Finally, thanks to Eq. (2.11), the hamiltonian function can be discontinuous at t_1: in fact $w(\cdot, \cdot)$ explicitly depends on time and t_1 is free.

Equation (2.5a) and its general solution over an interval beginning at t_0 are

$$\dot{\lambda}_1(t) = 0 \qquad\qquad \lambda_1(t) = \lambda_1(t_0)$$
$$\dot{\lambda}_2(t) = -\lambda_1(t) \qquad\qquad \lambda_2(t) = \lambda_2(t_0) - \lambda_1(t_0)(t - t_0)$$

Therefore $\lambda_1(\cdot)$ and $\lambda_2(\cdot)$ are piecewise constant and linear functions of time, respectively. By exploiting the notation (9.1), the satisfaction of the NC requires

$$\lambda_1^-(t_1) = \lambda_1(0) = \lambda_1^+(t_1) + \left.\frac{\partial w(x, t_1)}{\partial x}\right|_{x=x(t_1)} \mu = \lambda_1^+(t_1) + \mu$$

$$H^-(t_1) = 1 + \lambda_1^-(t_1)x_2(t_1) - \lambda_2^-(t_1)\text{sign}(\lambda_2^-(t_1)) =$$
$$= 1 + \lambda_1^+(t_1)x_2(t_1) - \lambda_2^+(t_1)\text{sign}(\lambda_2^+(t_1)) - \left.\frac{\partial w(x(t_1), t)}{\partial t}\right|_{t=t_1} \mu =$$
$$= 1 + \lambda_1^+(t_1)x_2(t_1) - \lambda_2^+(t_1)\text{sign}(\lambda_2^+(t_1)) - \mu = H^+(t_1) - \mu$$

If the continuity of $\lambda_2(\cdot)$ is taken into account, we deduce from these equations that either $x_2(t_1) = -1$ or $\mu = 0$. The last possibility must be discarded since it implies $\lambda_1^-(t_1) = \lambda_1^+(t_1)$, so that the sign of $\lambda_2(\cdot)$ can at most change once and it is not possible to satisfy the constraints on system state (see Fig. 9.4 where we have drawn the shapes of the state trajectories corresponding to $u(\cdot) = \pm 1$). Therefore $x_2(t_1) = -1$. The nature of the problem suggests that (i) the control switches from -1 to 1 at time t_{c_1} and from 1 to -1 at time t_{c_2}, $t_{c_1} < t_{c_2}$, (ii) the time t_1 is an interior point of the interval (t_{c_1}, t_{c_2}) (again make reference to Fig. 9.4). Consistently, we have

$$0 \le t \le t_{c_1}$$
$$x_1(t) = -\frac{t^2}{2}, \quad x_2(t) = -t$$
$$t_{c_1} \le t \le t_1$$
$$x_1(t) = x_1(t_{c_1}) + x_2(t_{c_1})(t - t_{c_1}) + \frac{(t - t_{c_1})^2}{2}, \quad x_2(t) = x_2(t_{c_1}) + (t - t_{c_1})$$
$$t_1 \le t \le t_{c_2}$$
$$x_1(t) = x_1(t_1) + x_2(t_1)(t - t_1) + \frac{(t - t_1)^2}{2}, \quad x_2(t) = x_2(t_1) + (t - t_1)$$
$$t_{c_2} \le t \le t_f$$
$$x_1(t) = x_1(t_{c_2}) + x_2(t_{c_2})(t - t_{c_2}) - \frac{(t - t_{c_2})^2}{2}, \quad x_2(t) = x_2(t_{c_2}) - (t - t_{c_2})$$

By requiring that

$$x_1(t_1) + 1 + t_1 = 0, \quad x_2(t_1) = -1, \quad x_1(t_f) = 1, \quad x_2(t_f) = 0$$

we obtain

$$t_1 = 3.45, \quad t_{c_1} = 2.22, \quad t_f = 9.33, \quad t_{c_2} = 6.89$$

The equations

$$\lambda_2(t_{c_1}) = \lambda_2(0) - \lambda_1(0)t_{c_1} = 0$$
$$\lambda_2(t_{c_2}) = \lambda_2(t_1) - \lambda_1^+(t_1)(t_{c_2} - t_1) = 0$$
$$\lambda_2(t_1) = \lambda_2(0) - \lambda_1(0)t_1$$

together with the transversality condition at the final time

$$H(t_f) = 1 + \lambda_1(t_f)x_2(t_f) + \lambda_2(t_f)u(t_f) = 1 + \lambda_1^+(t_f - t_{c_2}) = 0$$

supply the values of $\lambda_1(0)$, $\lambda_2(0)$, $\lambda_1^+(t_1)$.

The state trajectory corresponding to the solution above is plotted in Fig. 9.7 (line (0–a–d–e–b–c)) together with the one relevant to Problem 9.3-2 (line (0–a–b–c)). We notice that the solutions of the two problems are substantially different in spite of the fact that one seems to constitute a small variation of the other. Indeed when passing from Problems 9.3-2 to 9.5, the control interval becomes remarkably longer and both the state variables take on much larger absolute values.   ∎

Fig. 9.7 Problem 9.5. State trajectories when the constraint is $x_1(\tau) = -1$ (*line* (0–a–b–c)) or $x_1(\tau) = -1 - \tau$ (*line* (0–a–d–e–b–c))

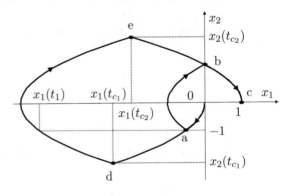

Problem 9.6

Let

$$x(0) = x(t_f) = \begin{bmatrix} 0 \\ 0 \end{bmatrix}, \quad J = \int_0^{t_f} l(x(t), u(t), t)\,dt, \quad l(x, u, t) = 1 + \frac{u^2}{2}, \quad u(t) \in R, \ \forall t$$

$$w(x(t_1), t_1) = 0, \quad w(x, t) = \begin{bmatrix} x_1 - 1 - t - \dfrac{1}{6} t^3 \\ x_2 - 1 - \dfrac{1}{2} t^2 \end{bmatrix}$$

The intermediate time t_1 and the final one t_f are free.

The hamiltonian function and the H-minimizing control are

$$H = 1 + \frac{u^2}{2} + \lambda_1 x_2 + \lambda_2 u, \quad u_h = -\lambda_2$$

In view of Eq. (2.10) $\lambda_1(\cdot)$ and $\lambda_2(\cdot)$ can be discontinuous at time t_1 because $w(\cdot, \cdot)$ explicitly depends on x_1 and x_2. Thanks to Eq. (2.11), also the hamiltonian function can be discontinuous at t_1: in fact $w(\cdot, \cdot)$ explicitly depends on time and t_1 is free.

Therefore Eq. (2.5a) and its general solution over the interval $[0, t_f]$ are

$$\dot\lambda_1(t) = 0 \qquad\qquad 0 \le t < t_1 \qquad\qquad \lambda_1(t) = \lambda_1(0)$$
$$\qquad\qquad\qquad\qquad t_1 < t \le t_f \qquad\qquad \lambda_1(t) = \lambda_1^+(t_1)$$
$$\dot\lambda_2(t) = -\lambda_1(t) \qquad 0 \le t < t_1 \qquad\qquad \lambda_2(t) = \lambda_2(0) - \lambda_1(0)t$$
$$\qquad\qquad\qquad\qquad t_1 < t \le t_f \qquad\qquad \lambda_2(t) = \lambda_2^+(t_1) - \lambda_1^+(t_1)(t - t_1)$$

where the notation (9.1) has been adopted. The NC lead to the equations

$$x_1(t_1) = \frac{\lambda_1(0)}{6} t_1^3 - \frac{\lambda_2(0)}{2} t_1^2 = 1 + t_1 + \frac{t_1^3}{6}$$
$$x_2(t_1) = \frac{\lambda_1(0)}{2} t_1^2 - \lambda_2(0) t_1 = 1 + \frac{t_1^2}{2}$$

$$\begin{bmatrix} \lambda_1^-(t_1) \\ \lambda_2^-(t_1) \end{bmatrix} = \begin{bmatrix} \lambda_1(0) \\ \lambda_2(0) - \lambda_1(0)t_1 \end{bmatrix} = \begin{bmatrix} \lambda_1^+(t_1) \\ \lambda_2^+(t_1) \end{bmatrix} + \frac{\partial w(x, t_1)}{\partial x}\bigg|'_{x = x(t_1)} \begin{bmatrix} \mu_1 \\ \mu_2 \end{bmatrix} =$$

$$= \begin{bmatrix} \lambda_1^+(t_1) + \mu_1 \\ \lambda_2^+(t_1) + \mu_2 \end{bmatrix}$$

$$H^-(t_1) = 1 + \frac{(u^-(t_1))^2}{2} + \lambda_1^-(t_1)x_2(t_1) + \lambda_2^-(t_1)u^-(t_1) =$$

$$= 1 - \frac{(\lambda_2(0) - \lambda_1(0)t_1)^2}{2} + \lambda_1(0)x_2(t_1) = H^+(t_1) + \left.\frac{\partial w(x(t_1))}{\partial t}\right|'_{t=t_1} \begin{bmatrix} \mu_1 \\ \mu_2 \end{bmatrix} =$$

$$= 1 - \frac{(\lambda_2^+(t_1))^2}{2} + \lambda_1^+(t_1)x_2(t_1) + \mu_1\left(1 + \frac{t_1^2}{2}\right) + \mu_2 t_1$$

$$x_1(t_f) = x_1(t_1) + x_2(t_1)(t_f - t_1) + \frac{\lambda_1^+(t_1)}{6}(t_f - t_1)^3 - \frac{\lambda_2^+(t_1)}{2}(t_f - t_1)^2 = 0$$

$$x_2(t_f) = x_2(t_1) + \frac{\lambda_1^+(t_1)}{2}(t_f - t_1)^2 - \lambda_2^+(t_1)(t_f - t_1) = 0$$

$$H(t_f) = 1 + \frac{u^2(t_f)}{2} + \lambda_1(t_f)x_2(t_f) + \lambda_2(t_f)u(t_f) = 1 - \frac{(\lambda_2(t_f))^2}{2} =$$

$$= 1 - \frac{(\lambda_2^+(t_1) - \lambda_1^+(t_1)(t_f - t_1))^2}{2} = 0$$

The solution of the problem can be deduced from this system of equations, the last one of which is the transversality condition at the final time. We obtain

$$\lambda_1(0) = -1.26, \quad \lambda_2(0) = -2.98, \quad \lambda_1^+(t_1) = 0.48, \quad \lambda_2^+(t_1) = 2.32$$
$$t_1 = 2.24, \quad t_f = 10.00$$

The plots of $x_1(\cdot)$ and $x_2(\cdot)$ are shown in Fig. 9.8 where $x_1(t_1) = f_1 := 1 + t_1 + \frac{1}{6}(t_1)^3$ and $x_2(t_1) = f_2 := (t_1) = 1 + \frac{t_1^2}{2}$.   ∎

Fig. 9.8 Problem 9.6. State motion

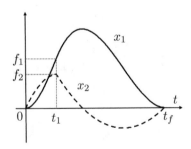

Problem 9.7

Let

$$x(0) = x(t_f) = \begin{bmatrix} 0 \\ 0 \end{bmatrix}, \quad J = \int_0^{t_f} l(x(t), u(t), t)\, dt, \quad l(x, u, t) = 1 + \frac{u^2}{2}, \quad u(t) \in R, \; \forall t$$

The final time t_f is free.

We consider six cases where different punctual and isolated equality constraints must be satisfied. Thus we face six problems where the hamiltonian function, the H-minimizing control, the transversality condition at the final time, Eq. (2.5a) and its general solution over an interval beginning at time t_0 are the same, precisely,

$$H = 1 + \frac{u^2}{2} + \lambda_1 x_2 + \lambda_2 u, \quad u_h = -\lambda_2$$

$$H(t_f) = 1 + \frac{u^2(t_f)}{2} + \lambda_1(t_f) x_2(t_f) + \lambda_2(t_f) u(t_f) = 1 - \frac{\lambda_2^2(t_f)}{2} = 0$$

$$\dot{\lambda}_1(t) = 0 \qquad\qquad \lambda_1(t) = \lambda_1(t_0)$$
$$\dot{\lambda}_2(t) = -\lambda_1(t) \qquad\qquad \lambda_2(t) = \lambda_2(t_0) - \lambda_1(t_0)(t - t_0)$$

The six functions which specify the constraints are listed below.

Problem 9.7-1. $w(x(t_1), t_1) = 0$, $w(x, t) = \begin{bmatrix} x_1 - 1 \\ x_2 - 1 \end{bmatrix}$, $t_1 = 3$.

Problem 9.7-2. $w(x(t_1), t_1) = 0$, $w(x, t) = \begin{bmatrix} x_1 - 1 \\ x_2 - 1 \end{bmatrix}$, t_1 free.

Problem 9.7-3. $w(x(t_1), t_1) = 0$, $w(x, t) = x_1 - 1$, $t_1 = 3$.

Problem 9.7-4. $w(x(t_1), t_1) = 0$, $w(x, t) = x_1 - 1$, t_1 free.

Problem 9.7-5. $w(x(t_1), t_1) = 0$, $w(x, t) = x_2 - 1$, $t_1 = 3$.

Problem 9.7-6. $w(x(t_1), t_1) = 0$, $w(x, t) = x_2 - 1$, t_1 free.

As customary, we use the notation (9.1).

Problem 9.7-1 $x_1(t_1) = 1$, $x_2(t_1) = 1$, $t_1 = 3$.

In view of Eq. (2.10) $\lambda_1(\cdot)$ and $\lambda_2(\cdot)$ can be discontinuous at time t_1 because $w(\cdot, \cdot)$ explicitly depends on x_1 and x_2. This fact implies that also the hamiltonian function can be discontinuous at t_1 because Eq. (2.11) needs not to be satisfied (the time t_1 is given). Therefore we have

$0 \le t < t_1$

$\lambda_1(t) = \lambda_1(0)$

$\lambda_2(t) = \lambda_2(0) - \lambda_1(0)t$

$0 \le t \le t_1$

$$x_1(t) = -\frac{\lambda_2(0)}{2}t^2 + \frac{\lambda_1(0)}{6}t^3$$

$$x_2(t) = -\lambda_2(0)t + \frac{\lambda_1(0)}{2}t^2$$

$t_1 < t \le t_f$

$\lambda_1(t) = \lambda_1^+(t_1)$

$\lambda_2(t) = \lambda_2^+(t_1) - \lambda_1^+(t_1)(t - t_1)$

$t_1 \le t \le t_f$

$$x_1(t) = x_1(t_1) + x_2(t_1)(t - t_1) + $$
$$-\frac{\lambda_2^+(t_1)}{2}(t - t_1)^2 + \frac{\lambda_1^+(t_1)}{6}(t - t_1)^3$$

$$x_2(t) = x_2(t_1) - \lambda_2^+(t_1)(t - t_1) + $$
$$+\frac{\lambda_1^+(t_1)}{2}(t - t_1)^2$$

By enforcing feasibility, namely

$$x(t_1) = \begin{bmatrix} 1 \\ 1 \end{bmatrix}, \quad x(t_f) = \begin{bmatrix} 0 \\ 0 \end{bmatrix}$$

together with the fulfilment of the transversality condition at the final time, we can write five equations for the five unknowns parameters $\lambda_1(0)$, $\lambda_2(0)$, $\lambda_1^+(t_1)$, $\lambda_2^+(t_1)$ and t_f, where the relations

$$\lambda_1^-(t_1) = \lambda_1(0), \quad \lambda_2^-(t_1) = \lambda_2(0) - \lambda_1(0)t_1, \quad \lambda_2(t_f) = \lambda_2^+(t_1) - \lambda_1^+(t_1)(t_f - t_1)$$

are taken into account. Their values are given in Table 9.1 and allow the computation of $x_1(\cdot)$ and $x_2(\cdot)$. These functions are plotted with solid lines in Figs. 9.9 and 9.10, the final time being denoted with t_{f_1}. In the quoted table the value J_1 of the performance index is given as well.

The same results are obtained in a different way by noticing that the problem at hand can be subdivided into two independent subproblems. In the first of them the initial and final times and states are given, whereas in the second subproblem the initial and final states, the initial time are given and the final time is free.

Table 9.1 The values of the parameters which specify the solution of Problems 9.7-1–9.7-6

Problem	$\lambda_1(0)$	$\lambda_2(0)$	$\lambda_1^+(t_1)$	$\lambda_2^+(t_1)$	t_1	t_f	J
9.7-1	0.22	0.00	1.22	2.11	3.00	5.88	7.77
9.7-2	−1.00	−1.41	1.22	2.11	1.47	4.36	6.49
9.7-3	−0.65	−0.87	1.35	1.08	3.00	4.85	5.88
9.7-4	−1.37	−1.41	1.37	1.41	2.06	4.12	5.49
9.7-5	0.79	0.86	0.79	1.89	3.00	3.60	5.31
9.7-6	2.00	1.41	2.00	2.45	1.93	2.45	4.90

Problem 9.7-2 $x_1(t_1) = 1$, $x_2(t_1) = 1$, t_1 free.

In view of Eq. (2.10) $\lambda_1(\cdot)$ and $\lambda_2(\cdot)$ can be discontinuous at time t_1 because $w(\cdot, \cdot)$ explicitly depends on x_1 and x_2. Thanks to Eq. (2.11) the hamiltonian function is continuous at t_1: in fact $w(\cdot, \cdot)$ does not explicitly depend on time and t_1 is free. Therefore we have

$0 \le t < t_1$

$\lambda_1(t) = \lambda_1(0)$

$\lambda_2(t) = \lambda_2(0) - \lambda_1(0)t$

$0 \le t \le t_1$

$x_1(t) = -\dfrac{\lambda_2(0)}{2}t^2 + \dfrac{\lambda_1(0)}{6}t^3$

$x_2(t) = -\lambda_2(0)t + \dfrac{\lambda_1(0)}{2}t^2$

$t_1 < t \le t_f$

$\lambda_1(t) = \lambda_1^+(t_1)$

$\lambda_2(t) = \lambda_2^+(t_1) - \lambda_1^+(t_1)(t - t_1)$

$t_1 \le t \le t_f$

$x_1(t) = x_1(t_1) + x_2(t_1)(t - t_1) +$
$\qquad - \dfrac{\lambda_2^+(t_1)}{2}(t - t_1)^2 + \dfrac{\lambda_1^+(t_1)}{6}(t - t_1)^3$

$x_2(t) = x_2(t_1) - \lambda_2^+(t_1)(t - t_1) +$
$\qquad + \dfrac{\lambda_1^+(t_1)}{2}(t - t_1)^2$

The values of the six parameters $\lambda_1(0)$, $\lambda_2(0)$, t_1, $\lambda_1^+(t_1)$, $\lambda_2^+(t_1)$ and t_f can be computed from the system of six equations resulting from feasibility (the four equations $x_1(t_1) = 1$, $x_2(t_1) = 1$, $x_1(t_f) = 0$, $x_2(t_f) = 0$), the transversality condition at the final time and the continuity at t_1 of the hamiltonian function, namely

$$H^-(t_1) = 1 + \frac{(u^-(t_1))^2}{2} + \lambda_1^-(t_1)x_2(t_1) = 1 - \frac{(\lambda_2^-(t_1))^2}{2} + \lambda_1^-(t_1) =$$

$$= H^+(t_1) = 1 + \frac{(u^+(t_1))^2}{2} + \lambda_1^+(t_1)x_2(t_1) = 1 - \frac{(\lambda_2^+(t_1))^2}{2} + \lambda_1^+(t_1)$$

We recall that in these equations $\lambda_2^-(t_1) = \lambda_2(0) - \lambda_1(0)t_1$ and $\lambda_1^-(t_1) = \lambda_1(0)$: then we have the results given in Table 9.1 where the value J_2 of the performance index is reported as well.

The plots of $x_1(\cdot)$ and $x_2(\cdot)$ are shown in Figs. 9.9 and 9.10 with long-dashed lines. The final time is denoted with t_{f_2}, whereas t_{1_2} denotes the intermediate time t_1.

Problem 9.7-3 $x_1(t_1) = 1$, $t_1 = 3$

In view of Eq. (2.10) $\lambda_1(\cdot)$ can be discontinuous at time t_1 because $w(\cdot, \cdot)$ explicitly depends on x_1. On the contrary, $\lambda_2(\cdot)$ is continuous at time t_1 because $w(\cdot, \cdot)$ does not explicitly depend on x_2. Finally, Eq. (2.11) must not be satisfied since t_1 is given: thus

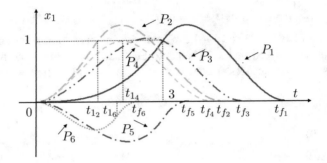

Fig. 9.9 Problem 9.7. The plots of $x_1(\cdot)$ relevant to Problems 9.7-1–9.7-6

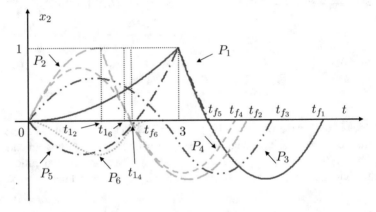

Fig. 9.10 Problem 9.7. The plots of $x_2(\cdot)$ relevant to Problems 9.7-1–9.7-6

the hamiltonian function can be discontinuous because of the possible discontinuity of $\lambda_1(\cdot)$. Therefore we have

$0 \le t < t_1$

$\lambda_1(t) = \lambda_1(0)$

$0 \le t \le t_1$

$\lambda_2(t) = \lambda_2(0) - \lambda_1(0)t$

$$x_1(t) = -\frac{\lambda_2(0)}{2}t^2 + \frac{\lambda_1(0)}{6}t^3$$

$$x_2(t) = -\lambda_2(0)t + \frac{\lambda_1(0)}{2}t^2$$

$t_1 < t \le t_f$

$\lambda_1(t) = \lambda_1^+(t_1)$

$t_1 \le t \le t_f$

$\lambda_2(t) = \lambda_2(t_1) - \lambda_1^+(t_1)(t - t_1)$

$x_1(t) = x_1(t_1) + x_2(t_1)(t - t_1) +$

$$-\frac{\lambda_2(t_1)}{2}(t - t_1)^2 + \frac{\lambda_1^+(t_1)}{6}(t - t_1)^3$$

$x_2(t) = x_2(t_1) - \lambda_2(t_1)(t - t_1) +$

$$+\frac{\lambda_1^+(t_1)}{2}(t - t_1)^2$$

where $\lambda_2(t_1) = \lambda_2(0) - \lambda_1(0)t_1$. The values of the four parameters $\lambda_1(0)$, $\lambda_2(0)$, $\lambda_1^+(t_1)$ and t_f can be computed from the system of four equations resulting from feasibility (the three equations $x_1(t_1) = 1$, $x_1(t_f) = 0$, $x_2(t_f) = 0$) and the transversality condition at the final time. They are given in Table 9.1 where the value J_3 of the performance index is reported as well. The plots of $x_1(\cdot)$ and $x_2(\cdot)$ are shown in Figs. 9.9 and 9.10 with dash-double-dotted lines. The final time is denoted with t_{f_3}.

Problem 9.7-4 $x_1(t_1) = 1$, t_1 free.

In view of Eq. (2.10) $\lambda_1(\cdot)$ can be discontinuous at time t_1 because $w(\cdot, \cdot)$ explicitly depends on x_1. On the contrary, $\lambda_2(\cdot)$ is continuous at time t_1 because $w(\cdot, \cdot)$ does not explicitly depend on x_2. Finally, Eq. (2.11) implies that also the hamiltonian function is continuous at t_1 because this time is free and $w(\cdot, \cdot)$ does not explicitly depend on t. Therefore we have

$$0 \leq t < t_1$$
$$\lambda_1(t) = \lambda_1(0)$$
$$0 \leq t \leq t_1$$
$$\lambda_2(t) = \lambda_2(0) - \lambda_1(0)t$$
$$x_1(t) = -\frac{\lambda_2(0)}{2}t^2 + \frac{\lambda_1(0)}{6}t^3$$

$$x_2(t) = -\lambda_2(0)t + \frac{\lambda_1(0)}{2}t^2$$

$$t_1 < t \leq t_f$$
$$\lambda_1(t) = \lambda_1^+(t_1)$$
$$t_1 \leq t \leq t_f$$
$$\lambda_2(t) = \lambda_2(t_1) - \lambda_1^+(t_1)(t - t_1)$$
$$x_1(t) = x_1(t_1) + x_2(t_1)(t - t_1) +$$
$$-\frac{\lambda_2(t_1)}{2}(t - t_1)^2 + \frac{\lambda_1^+(t_1)}{6}(t - t_1)^3$$
$$x_2(t) = x_2(t_1) - \lambda_2(t_1)(t - t_1) +$$
$$+\frac{\lambda_1^+(t_1)}{2}(t - t_1)^2$$

where $\lambda_2(t_1) = \lambda_2(0) - \lambda_1(0)t_1$. The values of the five parameters $\lambda_1(0)$, $\lambda_2(0)$, t_1, $\lambda_1^+(t_1)$ and t_f can be computed through the system of five equations resulting from feasibility (the three equations $x_1(t_1) = 1$, $x_1(t_f) = 0$, $x_2(t_f) = 0$), the transversality condition at the final time and the continuity at t_1 of the hamiltonian function, namely

$$H^-(t_1) = 1 + \frac{u^2(t_1)}{2} + \lambda_1^-(t_1)x_2(t_1) + \lambda_2(t_1)u(t_1) =$$

$$= 1 - \frac{\lambda_2^2(t_1)}{2} + \lambda_1^-(t_1)x_2(t_1) = H^+(t_1) =$$

$$= 1 + \frac{u^2(t_1)}{2} + \lambda_1^+(t_1)x_2(t_1) + \lambda_2(t_1)u(t_1) = 1 - \frac{\lambda_2^2(t_1)}{2} + \lambda_1^+(t_1)x_2(t_1)$$

The last equation implies $\lambda_1^-(t_1)x_2(t_1) = \lambda_1^+(t_1)x_2(t_1)$ which, in turn, requires that either $x_2(t_1) = 0$ or $\lambda_1^-(t_1) = \lambda_1^+(t_1)$. The second alternative must be discarded because it prevents the fulfillment of the constraints on the state at the final state. In fact we have $\lambda_2(t) = \lambda_2(0) - \lambda_1(0)t$, $0 \leq t \leq t_f$ and

$$x_1(t_f) = -\frac{\lambda_2(0)}{2}t_f^2 + \frac{\lambda_1(0)}{6}t_f^3, \quad x_2(t_f) = -\lambda_2(0)t_f + \frac{\lambda_1(0)}{2}t_f^2$$

Values for $\lambda_1(0)$ and $\lambda_2(0)$ such that $x_1(t_1) = 1, x_1(t_f) = 0, x_2(t_f) = 0$ do not exist. Therefore $x_2(t_1) = 0$. The resulting five parameters are shown in Table 9.1 where the value J_4 of the performance index is reported as well.

The plots of $x_1(\cdot)$ and $x_2(\cdot)$ are shown in Figs. 9.9 and 9.10 with short-dashed lines. The final time is denoted with t_{f_4} and the intermediate time t_1 with t_{1_4}.

Problem 9.7-5 $x_2(t_1) = 1, \ t_1 = 3$

In view of Eq. (2.10) $\lambda_2(\cdot)$ can be discontinuous at time t_1 because $w(\cdot, \cdot)$ explicitly depends on x_2. On the contrary, $\lambda_1(\cdot)$ is continuous at time t_1 because $w(\cdot, \cdot)$ does not explicitly depend on x_1. Finally, Eq. (2.11) must not be satisfied because t_1 is given: thus the hamiltonian function can be discontinuous because of the possible discontinuity of $\lambda_2(\cdot)$. Therefore we have

$0 \le t < t_1$

$\lambda_2(t) = \lambda_2(0) - \lambda_1(0)t$

$0 \le t \le t_1$

$\lambda_1(t) = \lambda_1(0)$

$x_1(t) = -\frac{\lambda_2(0)}{2}t^2 + \frac{\lambda_1(0)}{6}t^3$

$x_2(t) = -\lambda_2(0)t + \frac{\lambda_1(0)}{2}t^2$

$t_1 < t \le t_f$

$\lambda_2(t) = \lambda_2^+(t_1) - \lambda_{10}(t - t_1)$

$t_1 \le t \le t_f$

$\lambda_1(t) = \lambda_1(0)$

$x_1(t) = x_1(t_1) + x_2(t_1)(t - t_1) +$
$\qquad -\frac{\lambda_2^+(t_1)}{2}(t - t_1)^2 + \frac{\lambda_1(0)}{6}(t - t_1)^3$

$x_2(t) = x_2(t_1) - \lambda_2^+(t_1)(t - t_1) +$
$\qquad +\frac{\lambda_1(0)}{2}(t - t_1)^2$

The values of the four parameters $\lambda_1(0)$, $\lambda_2(0)$, $\lambda_2^+(t_1)$ and t_f can be computed through the system of four equations resulting from feasibility (the three equations $x_2(t_1) = 1, x_1(t_f) = 0, x_2(t_f) = 0$) and the transversality condition at the final time. They are given in Table 9.1 where the value J_5 of the performance index is also reported. The plots of $x_1(\cdot)$ and $x_2(\cdot)$ are shown in Figs. 9.9 and 9.10 with dash-single-dotted lines. The final time is denoted with t_{f_5}.

Problem 9.7-6 $x_2(t_1) = 1, \ t_1$ free.

In view of Eq. (2.10) $\lambda_2(\cdot)$ can be discontinuous at time t_1 because $w(\cdot, \cdot)$ explicitly depends on x_2. On the contrary, $\lambda_1(\cdot)$ is continuous at time t_1 because $w(\cdot, \cdot)$ does not explicitly depend on x_1. Finally, Eq. (2.11) implies that also the hamiltonian function is continuous at t_1 because this time is free and $w(\cdot, \cdot)$ does not explicitly depend on t. Therefore we have

$0 \le t < t_1$

$\lambda_2(t) = \lambda_2(0) - \lambda_1(0)t$

$0 \le t \le t_1$

$\lambda_1(t) = \lambda_1(0)$

$x_1(t) = -\dfrac{\lambda_2(0)}{2}t^2 + \dfrac{\lambda_1(0)}{6}t^3$

$x_2(t) = -\lambda_2(0)t + \dfrac{\lambda_1(0)}{2}t^2$

$t_1 < t \le t_f$

$\lambda_2(t) = \lambda_2^+(t_1) - \lambda_1(0)(t - t_1)$

$t_1 \le t \le t_f$

$\lambda_1(t) = \lambda_1(0)$

$x_1(t) = x_1(t_1) + x_2(t_1)(t - t_1) + $
$\qquad -\dfrac{\lambda_2^+(t_1)}{2}(t - t_1)^2 + \dfrac{\lambda_1(0)}{6}(t - t_1)^3$

$x_2(t) = x_2(t_1) - \lambda_2^+(t_1)(t - t_1) + $
$\qquad + \dfrac{\lambda_1(0)}{2}(t - t_1)^2$

The values of the five parameters $\lambda_1(0)$, $\lambda_2(0)$, t_1, $\lambda_2^+(t_1)$ and t_f can be computed through the system of five equations resulting from feasibility (the three equations $x_2(t_1) = 1$, $x_1(t_f) = 0$, $x_2(t_f) = 0$), the transversality condition at the final time and the continuity at t_1 of the hamiltonian function, namely

$$H^-(t_1) = 1 + \frac{(u^-(t_1))^2}{2} + \lambda_1(t_1)x_2(t_1) + \lambda_2^-(t_1)u^-(t_1) = $$

$$= 1 - \frac{(\lambda_2^-(t_1))^2}{2} + \lambda_1(0) = 1 + \frac{(u^+(t_1))^2}{2} + \lambda_1(t_1)x_2(t_1) + \lambda_2^+(t_1)u^+(t_1) = $$

$$= 1 - \frac{(\lambda_2^+(t_1))^2}{2} + \lambda_1(0) = H^+(t_1)$$

The last equation implies

$$\frac{(\lambda_2^-(t_1))^2}{2} = \frac{(\lambda_2^+(t_1))^2}{2}$$

We find two solutions S_1 and S_2

$S_1:$ $\quad \lambda_1(0) = 2.00,$ $\quad \lambda_2(0) = 1.41,$ $\quad \lambda_2^+(t_1) = 2.45,$ $\quad t_1 = 1.93,$ $\quad t_f = 2.45$

$S_2:$ $\quad \lambda_1(0) = 3.00,$ $\quad \lambda_2(0) = 1.41,$ $\quad \lambda_2^+(t_1) = 2.83,$ $\quad t_1 = 1.41,$ $\quad t_f = 2.83$

The value (4.90) of the performance index corresponding to the first solution is smaller than the one relevant to the second solution (5.66), so that Table 9.1 reports the results of the first solution.

The plots of $x_1(\cdot)$ and $x_2(\cdot)$ are shown in Figs. 9.9 and 9.10 with dotted lines. The final time is denoted with t_{f_6} and the intermediate time t_1 with t_{1_6}.

As a conclusion we can notice that the values of the performance index are related in an easily predictable manner. In fact

(*i*) Problem 9.7-2 is less constrained than Problem 9.7-1. Consistently, $J_2 < J_1$.

(*ii*) Problem 9.7-3 and Problem 9.7-5 are less constrained than Problem 9.7-1. Consistently, $J_3 < J_1$ and $J_5 < J_1$.

(*iii*) Problem 9.7-4 and Problem 9.7-6 are less constrained than Problem 9.7-2. Consistently, $J_4 < J_2$ and $J_6 < J_2$.

(*iv*) Problem 9.7-4 is less constrained than Problem 9.7-3. Consistently, $J_4 < J_3$.
(*v*) Problem 9.7-6 is less constrained than Problem 9.7-5. Consistently, $J_6 < J_5$. ∎

We now discuss a fairly interesting problem where its solution, actually optimal, does not satisfy the NC. This seemingly contradictory statement originates from the fact that the solution does not comply with the assumptions concerning times reported in Sect. 2.3.1, point (b), and Sect. 2.3.3.

Problem 9.8

Let

$$x(0) = \begin{bmatrix} 0 \\ 0 \end{bmatrix}, \quad x(t_f) \in \overline{S}_f = \{x|\alpha_f(x) = 0\}, \quad \alpha_f(x) = x_2 - 1$$

$$J = \int_0^{t_f} l(x(t), u(t), t)\, dt, \quad l(x, u, t) = 1 + \frac{u^2}{2}, \quad u(t) \in R, \ \forall t$$

$$w(x(t_1), t_1) = 0, \quad w(x, t) = x_1 - 1$$

where the intermediate time t_1 and the final one t_f are free.

The set of the admissible final states $\overline{S}_f$ is a regular variety because

$$\Sigma(x) = \frac{d\alpha_f(x)}{dx} = \begin{bmatrix} 0 & 1 \end{bmatrix}$$

and $\text{rank}(\Sigma(x)) = 1, \forall x \in \overline{S}_f$.

The hamiltonian function and the H-minimizing control are

$$H = 1 + \frac{u^2}{2} + \lambda_1 x_2 + \lambda_2 u, \quad u_h = -\lambda_2$$

so that Eq. (2.5a) and its general solution over an interval beginning at time t_0 are

$$\dot{\lambda}_1(t) = 0 \qquad\qquad\qquad \lambda_1(t) = \lambda_1(t_0)$$
$$\dot{\lambda}_2(t) = -\lambda_1(t) \qquad\qquad \lambda_2(t) = \lambda_2(t_0) - \lambda_1(t_0)(t - t_0)$$

In view of Eq. (2.10) $\lambda_1(\cdot)$ can be discontinuous at time t_1 because $w(\cdot, \cdot)$ explicitly depends on x_1. On the contrary, $\lambda_2(\cdot)$ is continuous at time t_1 because $w(\cdot, \cdot)$ does not explicitly depend on x_2. Finally, Eq. (2.11) implies that also the hamiltonian function is continuous at t_1 because this time is free and $w(\cdot, \cdot)$ does not explicitly depend on t. Therefore we can state that $\lambda_1(\cdot)$ is piecewise constant and the transversality condition at the final time can be written at the initial time yielding

$$H(0) = 1 + \frac{u^2(0)}{2} + \lambda_1(0)x_2(0) + \lambda_2(0)u(0) = 1 - \frac{\lambda_2^2(0)}{2} = 0$$

so that $\lambda_2(0) = \pm\sqrt{2}$.

By resorting to the notation (9.1), the orthogonality condition at the final time is

$$\begin{bmatrix} \lambda_1(t_f) \\ \lambda_2(t_f) \end{bmatrix} = \begin{bmatrix} \lambda_1^+(t_1) \\ \lambda_2(t_1) - \lambda_1^+(t_1)(t_f - t_1) \end{bmatrix} = \vartheta_f \left. \frac{d\alpha_f(x)}{dx} \right|'_{x=x(t_f)} = \vartheta_f \begin{bmatrix} 0 \\ 1 \end{bmatrix}$$

and we conclude that $\lambda_1^+(t_1) = 0$. This fact, together with the continuity of the hamiltonian function at t_1, namely

$$H^-(t_1) = 1 + \frac{u^2(t_1)}{2} + \lambda_1^-(t_1)x_2(t_1) + \lambda_2(t_1)u(t_1) = 1 - \frac{\lambda_2^2(t_1)}{2} + \lambda_1 0 x_2(t_1) =$$

$$= H^+(t_1) = 1 + \frac{u^2(t_1)}{2} + \lambda_1^+(t_1)x_2(t_1) + \lambda_2(t_1)u(t_1) = 1 - \frac{\lambda_2^2(t_1)}{2}$$

entails that $\lambda_1(0)x_2(t_1) = 0$. Therefore we must examine the four following *cases*.

Case 1. $\lambda_2(0) = \sqrt{2}$, $\lambda_1(0) = 0$.

Case 2. $\lambda_2(0) = \sqrt{2}$, $x_2(t_1) = 0$.

Case 3. $\lambda_2(0) = -\sqrt{2}$, $\lambda_1(0) = 0$.

Case 4. $\lambda_2(0) = -\sqrt{2}$, $x_2(t_1) = 0$.

which are now sequentially considered.

Case 1. $\lambda_2(0) = \sqrt{2}$, $\lambda_1(0) = 0$.

We have $x_1(t) = -\frac{\sqrt{2}}{2}t^2$, $0 \le t \le t_1$. A positive value of t_1 such that $x_1(t_1) = 1$ does not exist because $x_1(t) \le 0$.

Case 2. $\lambda_2(0) = \sqrt{2}$, $x_2(t_1) = 0$.

We obtain

$$x_2(t_1) = \frac{\lambda_1(0)}{2}t_1^2 - \sqrt{2}t_1, \quad x_1(t_1) = \frac{\lambda_1(0)}{6}t_1^3 - \frac{\sqrt{2}}{2}t_1^2$$

so that the requests $x_2(t_1) = 0$ and $x_1(t_1) = 1$ give $t_1^2 = -\frac{6}{\sqrt{2}}$. Apparently this result can not be accepted.

Case 3. $\lambda_2(0) = -\sqrt{2}$, $\lambda_1(0) = 0$.

We have

$$x_1(t_1) = \frac{\sqrt{2}}{2}t_1^2, \quad x_2(t_1) = \sqrt{2}t_1$$

If we require that $x_1(t_1) = 1$, we obtain from the first of these two equations $t_1^2 = \sqrt{2}$. We also have $x_2(t) = x_2(t_1) + \sqrt{2}(t - t_1) = \sqrt{2}t$, $t_1 \le t \le t_f$, so that, by enforcing

Fig. 9.11 Problem 9.8. The functions $J^o(\cdot)$ (*solid line*) and $t_f^o(\cdot)$ (*dotted line*)

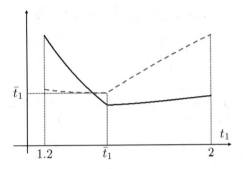

feasibility, it follows that $x_2(t_f) = \sqrt{2}t_f = 1$ which entails $t_f = 1/\sqrt{2} < t_1$. Therefore this case must be discarded as well.

Case 4. $\lambda_2(0) = -\sqrt{2}$, $x_2(t_1) = 0$.

We have, for $0 \le t \le t_1$,

$$x_1(t) = \frac{\sqrt{2}}{2}t^2 + \frac{\lambda_1(0)}{6}t^3, \quad x_2(t) = \sqrt{2}t + \frac{\lambda_1(0)}{2}t^2$$

By requiring that $x_2(t_1) = 0$ and $x_1(t_1) = 1$, we obtain

$$\lambda_1(0)t_1 = -\frac{2}{\sqrt{2}}, \quad t_1 = \sqrt{\frac{6}{\sqrt{2}}}$$

and, subsequently, $\lambda_2(t_1) = \sqrt{2}$. Then, for $t \ge t_1$, we have $x_2(t) = -\sqrt{2}(t - t_1)$ and the request $x_2(t_f) = 1$ can not be satisfied because $t_f > t_1$.

Nevertheless the problem admits a solution.

In order to prove this statement we consider, for the moment, an auxiliary problem where t_1 is given. By applying the Hamilton–Jacoby theory (see [15]) it is not difficult to ascertain that an optimal solution can be found for each $t_1 > 0$. The corresponding values $J^o(\cdot)$, $t_f^o(\cdot)$ of the performance index and the final time are plotted in Fig. 9.11 as functions of t_1. The function $J^o(\cdot)$ attains its minimum value at $t_1^o = \bar{t}_1 := 1.5$ and, correspondingly, the final time is $t_f^o = \bar{t}_1$. The NC failed to be satisfied because the obtained value of t_1^o is not smaller than t_f^o (see Sect. 2.3.1, point (b), and Sect. 2.3.3). ∎

Chapter 10
Punctual and Global Constraints

Abstract In this chapter we consider punctual and global constraints which concern the values taken on by the state and control variables at each time instants inside the control interval. First we consider punctual and global equality constraints and assume that the controlled system possesses more than one control variable. This assumption is needed because, otherwise, the problems will be either trivial or unsolvable. Then we tackle punctual and global inequality constraints. The main features of the relevant problems are: (1) the initial state is always given; (2) the final state is either given or partially specified or constrained to belong to a set which is not a regular variety; (3) the final time is given in one problem and is free in all the remaining ones; (4) other complex constraints (punctual and isolated equality constraints, integral inequality constraints) are also occasionally present.

In this chapter we consider punctual and global constraints. We recall that their denomination refers to the fact that they concern the values taken on by the state and control variables at each time instants inside the control interval. These constraints have been formally described in Sect. 2.3.1 (point (c)), whereas the relevant way of handling them has been outlined in Sect. 2.3.4.

Consistently with the discussion there, we first consider problems with punctual and global equality constraints (Sect. 10.1) and then those with punctual and global inequality constraints (Sect. 10.2).

10.1 Punctual and Global Equality Constraints

The systems considered in this section are a slight generalization of the classical double integrator: in fact they possess more than one control variable. According to this assumption (which is needed because, otherwise, we will be concerned with either trivial or unsolvable problems), the control u is a vector with two components in Problem 10.1 (where the absolute value of the second component is constrained) and in Problem 10.2 (where both components are unconstrained). The control u is a vector with three components in Problem 10.3 (where only the first component is

© Springer International Publishing Switzerland 2017

A. Locatelli, *Optimal Control of a Double Integrator*, Studies in Systems,
Decision and Control 68, DOI 10.1007/978-3-319-42126-1_10

unconstrained). Finally, the punctual and global equality constraint of Problem 10.4 is handled by expressing one of the two components of u as a function of the other one.

We make specific reference to the material of Sect. 2.3.4.

Problem 10.1

Two control variables act on the system which is described by

$$\dot{x}_1(t) = x_2(t)$$
$$\dot{x}_2(t) = u_1(t) + \beta u_2(t)$$

where $\beta > 0$ is a given parameter. Furthermore

$$x(t_0) = \begin{bmatrix} 0 \\ 0 \end{bmatrix}, \quad t_0 = 1, \quad x(t_f) = \begin{bmatrix} 1 \\ 0 \end{bmatrix}, \quad t_f = 2$$

$$J = \int_{t_0}^{t_f} l(x(t), u(t), t)\, dt, \quad l(x, u, t) = \frac{u_1^2}{2t}$$

$$w(x(t), u(t), t) = 0, \quad \forall t, \quad w(x, u, t) = u_1 u_2$$

$$u(t) = \begin{bmatrix} u_1(t) \\ u_2(t) \end{bmatrix}, \quad u(t) \in \overline{U}, \quad \forall t, \quad \overline{U} = \{u | u_1 \in R, \; -1 \le u_2 \le 1\}$$

Notice that the function $w(\cdot, \cdot, \cdot)$ prevents the contemporary use of both the control variables.

We take into account the punctual and global equality constraint by adding a term to the customary hamiltonian function. Thus we have

$$H_e = \frac{u_1^2}{2t} + \lambda_1 x_2 + \lambda_2 (u_1 + \beta u_2) + \mu u_1 u_2$$

The minimization of H_e requires $u_1 = -\lambda_2 t$ if $u_2 = 0$, whereas $u_2 = -\text{sign}(\lambda_2)$ if $u_1 = 0$.

Equation (2.5a) and its general solution over an interval beginning at time t_0 are

$$\dot{\lambda}_1(t) = 0 \qquad\qquad \lambda_1(t) = \lambda_1(t_0)$$
$$\dot{\lambda}_2(t) = -\lambda_1(t) \qquad\qquad \lambda_2(t) = \lambda_2(t_0) - \lambda_1(t_0)(t - t_0)$$

The function $\lambda_2(\cdot)$ is linear with respect to t: thus its sign can change at most once and, consistently, we must face the four following *cases*.

Case 1. $u_1(\cdot) = 0$

Case 2. $u_2(\cdot) = 0$

Case 3. $u_1(t) = 0$, $u_2(t) \neq 0$, $1 \le t < \tau$; $u_1(t) \neq 0$, $u_2(t) = 0$, $\tau < t \le t_f$ where the time τ must be determined and belongs to the interval (t_0, t_f)

Fig. 10.1 Problem 10.1.
State trajectories when
$u_1(\cdot) = 0$ and $u_2(\cdot) = 1$
(*solid lines*) or $u_2(\cdot) = -1$
(*dashed lines*)

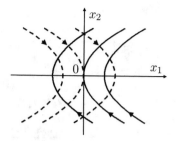

Case 4. $u_1(t) \neq 0$, $u_2(t) = 0$, $1 \leq t < \tau$; $u_1(t) = 0$, $u_2(t) \neq 0$, $\tau < t \leq t_f$ where the time τ must be determined and belongs to the interval (t_0, t_f)

Case 1. $u_1(\cdot) = 0$

Necessarily, $u_2(t) = 1$, $1 \leq \tau$ and $u_2(t) = -1$, $\tau < t \leq t_f$, where $0 < \tau < t_f$. In fact the state trajectories are

$$x_1 = \frac{x_2^2}{2\beta} + k \quad \text{if} \quad u_2(\cdot) = 1$$

$$x_1 = -\frac{x_2^2}{2\beta} + k \quad \text{if} \quad u_2(\cdot) = -1$$

and they are shown in Fig. 10.1. Thus we conclude that different $u_2(\cdot)$, namely $u_2(\cdot) = 1$, $u_2(\cdot) = -1$, $u_2(t) = -1$, $1 \leq t < \tau$ and $u_2(t) = 1$, $\tau < t \leq t_f$, are not feasible. Therefore we have

$1 \leq t < \tau$	$\tau < t \leq t_f$
$u_2(t) = 1$	$u_2(t) = -1$
$1 \leq t \leq \tau$	$\tau \leq t \leq t_f$
$x_1(t) = \beta \dfrac{(t - t_0)^2}{2}$	$x_1(t) = x_1(\tau) + x_2(\tau)(t - \tau) - \dfrac{\beta(t - \tau)^2}{2}$
$x_2(t) = \beta(t - t_0)$	$x_2(t) = x_2(\tau) - \beta(t - \tau)$

It is easy to ascertain that the constraint on $x(t_f)$, namely $x_1(t_f) = 1$ and $x_2(t_f) = 0$, requires

$$-\beta\tau^2 + 2\beta\tau - \frac{\beta}{2} = 1, \quad 2\beta\tau - \beta = 0$$

These equations can be satisfied only if $\beta = 4$ and $\tau = 1.5$. The corresponding value J_1 of the performance index is zero.

Case 2. $u_2(\cdot) = 0$

We have

$$u_1(t) = -\lambda_2(t)t = \lambda_1(t_0)(t - t_0)t - \lambda_2(t_0)t$$

$$x_1(t) = \frac{\lambda_1(t_0)t_0^3 + 3\lambda_2(t_0)t_0^2}{6}(t - t_0) - \frac{\lambda_1(t_0)t_0 + \lambda_2(t_0)}{6}(t^3 - t_0^3) +$$

$$+ \frac{\lambda_1(t_0)}{12}(t^4 - t_0^4)$$

$$x_2(t) = -\frac{\lambda_1(t_0)t_0 + \lambda_2(t_0)}{2}(t^2 - t_0^2) + \frac{\lambda_1(t_0)}{3}(t^3 - t_0^3)$$

By requiring the satisfaction of the constraint on $x(t_f)$, we obtain

$$\lambda_1(t_0) = -\frac{108}{13}, \quad \lambda_2(t_0) = -\frac{60}{13}$$

and the corresponding value of the performance index is $J_2 = 4.15$.

Case 3. $u_1(t) = 0, u_2(t) \neq 0, 1 \leq t < \tau, u_1(t) \neq 0, u_2(t) = 0, \tau < t \leq t_f, \tau$ free

We can state that $u_2(t) = 1, t < \tau$ because of the nature of the problem. Furthermore in order that $u_2(t) = 0, t > \tau$ it must be $\lambda_2(\tau) = \lambda_2(t_0) - \lambda_1(t_0)(\tau - t_0) = 0$ and, consequently, $\lambda_2(t) = -\lambda_1(t_0)(t - \tau)$ and $u_1(t) = \lambda_1(t_0)(t - \tau)t, t > \tau$. Therefore

$0 \leq t < \tau$ $\tau < t \leq t_f$

$u_1(t) = 0, \ u_2(t) = 1$ $u_1(t) = \lambda_1(t_0)(t - \tau)t, \ u_2(t) = 0$

$0 \leq t \leq \tau$ $\tau \leq t \leq t_f$

$x_1(t) = \dfrac{\beta(t - t_0)^2}{2}$ $x_1(t) = x_1(\tau) + x_2(\tau)(t - \tau) +$

$$+\lambda_1(t_0)\left(\frac{t^4 - \tau^4}{12} - \tau\frac{t^3 - \tau^3}{6} + \tau^3\frac{t - \tau}{6}\right)$$

$x_2(t) = \beta(t - t_0)$ $x_2(t) = x_2(\tau) + \lambda_1(t_0)\left(\dfrac{t^3 - \tau^3}{3} - \tau\dfrac{t^2 - \tau^2}{2}\right)$

By enforcing one of the two feasibility conditions, namely $x_2(t_f) = 0$, we obtain

$$\lambda_1(t_0) = -\frac{6\beta(\tau - t_0)}{\tau^3 - 12\tau + 16}$$

Subsequently, if we introduce this result into the second feasibility condition ($x_1(t_f) = 1$) and recall that $t_0 = 1$ and $t_f = 2$, we conclude that the time τ is a root of the polynomial

$$p(z) := \beta z^2 + (2 - 9\beta)z + 8(\beta + 1)$$

It is easy to ascertain that this polynomial possesses a root belonging to the interval (t_0, t_f) only if $\beta > 2$. The function $\tau(\cdot)$ which gives $\tau(\beta)$ within this interval is shown

Fig. 10.2 Problem 10.1. The
function $\tau(\cdot)$ when the
control function is the one of
Case 3 (*solid line*) and the
one of *Case 4* (*dashed line*)

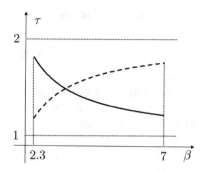

Fig. 10.3 Problem 10.1. The
functions J_2 (*dash-dotted
line*), $J_3(\cdot)$ (*solid line*), J_4
(*dashed line*)

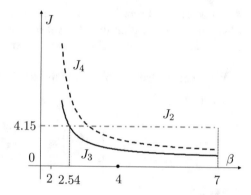

in Fig. 10.2, whether the function $J_3(\cdot)$ which supplies the value of the performance
index when $\beta > 2$ is plotted in Fig. 10.3.

Case 4. $u_1(t) \neq 0$, $u_2(t) = 0$, $1 \leq t < \tau$, $u_1(t) = 0$, $u_2(t) \neq 0$, $\tau < t \leq t_f$, τ free

We can state that $u_1(t) = -\lambda_2(t)t = (\lambda_2(t_0) - \lambda_1(t_0)(t - t_0))t > 0, t < \tau$, because
of the nature of the problem. Furthermore in order that $u_2(t) \neq 0$, $t > \tau$ it must
be $\lambda_2(\tau) = \lambda_2(t_0) - \lambda_1(t_0)(\tau - t_0) = 0$ and, consequently, $\lambda_2(t) = -\lambda_1(t_0)(t - \tau)$
and $u_1(t) = \lambda_1(t_0)(t - \tau)t$, $t < \tau$. Therefore

$$0 \leq t < \tau \qquad\qquad\qquad \tau < t \leq t_f$$
$$u_1(t) = \lambda_1(t_0)(t - \tau)t, \; u_2(t) = 0 \qquad u_1(t) = 0, \; u_2(t) = -1$$
$$0 \leq t \leq \tau \qquad\qquad\qquad \tau \leq t \leq t_f$$

$$x_1(t) = \lambda_1(t_0)\left(\frac{t^4 - t_0^4}{12}+\right.$$
$$\left.\frac{(t - t_0)(3\tau - 2) - \tau(t^3 - t_0^3)}{6}\right)$$

$$x_1(t) = x_1(\tau) + x_2(\tau)(t - \tau)+$$
$$-\frac{\beta}{2}(t - \tau)^2$$

$$x_2(t) = \lambda_1(t_0)\left(\frac{t^3 - t_0^3}{3} - \tau\frac{t^2 - t_0^2}{2}\right) \qquad x_2(t) = x_2(\tau) - \beta(t - \tau)$$

By enforcing one of the two feasibility conditions, namely $x_2(t_f) = 0$, we obtain

$$\lambda_1(t_0) = -\frac{6\beta(2-\tau)}{\tau^3 - 3\tau + 2}$$

Subsequently, if we introduce this result into the second feasibility condition ($x_1(t_f) = 1$) and recall that $t_0 = 1$ and $t_f = 2$, we conclude that the time τ is a root of the polynomial

$$r(z) := 2\beta z^2 + (2 - 3\beta)z + 4 - 2\beta$$

which possesses a root inside the interval $(1, 2)$ only if $\beta > 2$. The function $\tau(\cdot)$ which gives $\tau(\beta)$ within this interval is shown in Fig. 10.2, whether the function $J_4(\cdot)$ which supplies the value of the performance index when $\beta > 2$ is plotted in Fig. 10.3.

We conclude that the solution of the problem depends on β as follows.

If $\beta \in (0, 2.54)$

$$u(t) = \begin{bmatrix} u_1(t) \\ u_2(t) \end{bmatrix} = \begin{bmatrix} -\dfrac{108}{13}(t-1)t + \dfrac{60}{13}t \\ 0 \end{bmatrix}$$

If $\beta = 2.54$

$$u(t) = \begin{bmatrix} u_1(t) \\ u_2(t) \end{bmatrix} = \begin{cases} \begin{bmatrix} -\dfrac{108}{13}(t-1)t + \dfrac{60}{13}t \\ 0 \end{bmatrix} \\ \text{or} \\ \begin{bmatrix} 0 \\ 1 \end{bmatrix}, & 0 \le t < \tau \\ \begin{bmatrix} 0 \\ -1 \end{bmatrix}, & \tau < t \le 2 \\ \tau = \text{positive root of } q(\cdot) \end{cases}$$

If $\beta \in (2.54, 4)$

$$u(t) = \begin{bmatrix} u_1(t) \\ u_2(t) \end{bmatrix} = \begin{cases} \begin{bmatrix} 0 \\ 1 \end{bmatrix}, & 0 \le t < \tau \\ \begin{bmatrix} 0 \\ -1 \end{bmatrix}, & \tau < t \le 2 \\ \tau = \text{positive root of } q(\cdot) \end{cases}$$

If $\beta = 4$

$$u(t) = \begin{bmatrix} u_1(t) \\ u_2(t) \end{bmatrix} = \begin{cases} \begin{bmatrix} 0 \\ 1 \end{bmatrix}, & 0 \leq t < 1.5 \\ \begin{bmatrix} 0 \\ -1 \end{bmatrix}, & 1.5 < t \leq 2 \end{cases}$$

If $\beta > 4$

$$u(t) = \begin{bmatrix} u_1(t) \\ u_2(t) \end{bmatrix} = \begin{cases} \begin{bmatrix} 0 \\ 1 \end{bmatrix}, & 0 \leq t < \tau \\ \begin{bmatrix} 0 \\ -1 \end{bmatrix}, & \tau < t \leq 2 \\ \tau = \text{positive root of } q(\cdot) \end{cases}$$

Finally two remarks are worth to be done. The first one concerns the value of the performance index: it decreases as β increases, both when the control function is the one of *Case 3* and when the control function is the one of *Case 4*. This fact is easily predictable because the effectiveness of the second control component increases with β. The second one refers to the time τ. When β increases it decreases or increases corresponding to *Case 3* or *Case 4*. Indeed according to the form of the performance index, larger values of β make more convenient lowering the amplitude of the control u_1 while contemporarily extending the lasting of its use. Once more this fact is made possible by the greater effectiveness of the control u_2. A further insight is gained if we look at Fig. 10.4 where we have plotted (a) the function $u_1(\cdot)$ corresponding to *Case 3* and (b) to *Case 4*. We have chosen two different values of β which lie on the opposite side of 2.54 (the value where *Case 2* and *Case 3* are equally good), namely $\beta_1 = 2.3$ and $\beta_2 = 5$: it is apparent that the effort required to the first component of the control decreases as β increases.   ∎

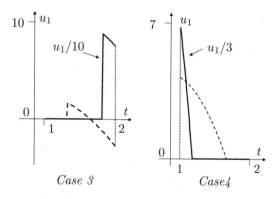

Fig. 10.4 Problem 10.1. The function u_1 when $\beta = \beta_1$ (*solid line*) and $\beta = \beta_2$ (*dashed line*)

Problem 10.2

Two control variables act on the system which is described by

$$\dot{x}_1(t) := x_2(t)$$
$$\dot{x}_2(t) := \beta_1 u_1(t) + \beta_2 u_2(t)$$

Furthermore

$$x(t_0) = \begin{bmatrix} 0 \\ 0 \end{bmatrix}, \quad t_0 = 1, \quad x(t_f) = \begin{bmatrix} 1 \\ 0 \end{bmatrix},$$

$$J = \int_{t_0}^{t_f} l(x(t), u(t), t)\, dt, \quad l(x, u, t) = \left(\frac{u_1^2}{2t} + \frac{u_2^2}{2}t \right), \quad u_1(t) \in R, \ u_2(t) \in R, \forall t$$

$$w(x(t), u(t), t) = 0, \ \forall t, \quad w(x, u, t) = u_1 u_2$$

In the above equations the parameters t_f, β_1, β_2, $\beta_2 > \beta_1$, are given and positive. We also assume that $t_f > \tau$, where $\tau := \beta_2/\beta_1$. Notice that the function $w(\cdot, \cdot, \cdot)$ prevents the contemporary use of both the control variables.

We take into account the punctual and global equality constraint by adding a term to the customary hamiltonian function. Thus we have

$$H_e = \frac{u_1^2}{2t} + \frac{u_2^2}{2}t + \lambda_1 x_2 + \lambda_2(\beta_1 u_1 + \beta_2 u_2) + \mu u_1 u_2$$

and the H_e-minimizing control is

$$u_{h_1} = -(\lambda_2 \beta_1 + \mu u_2)t$$

$$u_{h_2} = -\frac{\lambda_2 \beta_2 + \mu u_1}{t}$$

Therefore we can state that $u_{h_1} = -\lambda_2 \beta_1 t$, if $u_2 = 0$, whereas $u_{h_2} = -\frac{\lambda_2 \beta_2}{t}$, if $u_1 = 0$. By comparing the values of the hamiltonian function H_e in the two cases, we conclude that the minimum of H_e occurs when $u_1 = 0$ if $t < \tau$, whereas the minimum occurs when $u_2 = 0$ if $t > \tau$. These results are correct provided that $\lambda_2 \neq 0$.

Equation (2.5a) and its general solution over an interval beginning at time t_0 are

$$\dot{\lambda}_1(t) = 0 \qquad\qquad \lambda_1(t) = k_1 \qquad\qquad k_1 := \lambda_1(t_0)$$
$$\dot{\lambda}_2(t) = -\lambda_1(t) \qquad \lambda_2(t) = k_2 - k_1(t - t_0) \quad k_2 := \lambda_2(t_0)$$

Hence $\lambda_2(\cdot)$ can be zero at most at an isolated time instant. In fact if $\lambda_2(\cdot) = 0$ we also have $u_1(\cdot) = u_2(\cdot) = 0$ and the constraint on $x(t_f)$ can not be satisfied. Notice that the resulting use of the first or second component of the control completely agrees with the form of the performance index. Indeed $l(\cdot, \cdot, \cdot)$ suggests exploiting $u_2(\cdot)$ at the beginning of the control interval and $u_1(\cdot)$ later on. Thus we have, when

$t \in [t_0, \tau)$ (recall that $\tau = \beta_2/\beta_1 > 1 = t_0$),

$$u_1(t) = 0, \quad u_2(t) = -\frac{\beta_2[k_2 - k_1(t - t_0)]}{t}$$

$$x_1(t) = \beta_2^2 \left\{ \frac{k_1(t - t_0)^2}{2} - (k_2 + k_1 t_0) \left[t \left(\ln(t) - \ln(t_0) \right) - (t - t_0) \right] \right\}$$

$$x_2(t) = \beta_2^2 [k_1(t - t_0) - (k_2 + k_1 t_0)(\ln(t) - \ln(t_0))]$$

whereas, when $t \in (\tau, t_f]$,

$$u_1(t) = -\beta_1 t[k_2 - k_1(t - t_0)], \quad u_2(t) = 0$$
$$x_1(t) = x_1(\tau) + x_2(\tau)(t - \tau) +$$
$$+ \beta_1^2 \left[\frac{k_1(t^4 - 4\tau^3 t + 3\tau^4)}{12} - \frac{(k_2 + k_1 t_0)(t^3 - 3\tau^2 t + 2\tau^3)}{6} \right]$$
$$x_2(t) = x_2(\tau) + \beta_1^2 \left[\frac{k_1(t^3 - \tau^3)}{3} - \frac{(k_2 + k_1 t_0)(t^2 - \tau^2)}{2} \right]$$

By enforcing feasibility ($x_1(t_f) = 1$, $x_2(t_f) = 0$), we obtain a system of two linear equations for the two unknowns k_1, k_2. The state motion and the control are plotted in Fig. 10.5 when $\beta_1 = 1$, $\beta_2 = 2$, $t_f = 8$ and, accordingly, $\tau = 2$. Notice the discontinuity of $u(\cdot)$ at $t = \tau$. ∎

Problem 10.3
The control variable is a vector with three components. The second and third of them can not be contemporarily used. More precisely, we have

$$\dot{x}_1(t) := x_2(t)$$
$$\dot{x}_2(t) := u_1(t) + u_2(t) + u_3(t)$$

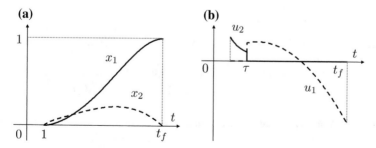

(a) **(b)**

Fig. 10.5 Problem 10.2: **a** State motion: x_1 (*solid line*) and x_2 (*dashed line*); **b** $u_1(\cdot)$ (*dashed line*) and $u_2(\cdot)$ (*solid line*)

and

$$x(0) = \begin{bmatrix} 0 \\ 0 \end{bmatrix}, \quad x(t_f) \in \overline{S}_f = \{x | \alpha_f(x) = 0\}, \quad \alpha_f(x) = x_1 - 10$$

$$J = \int_0^{t_f} l(x(t), u(t), t) \, dt, \quad l(x, u, t) = \left(1 + \frac{u_1^2}{2} + 0.1(u_2 - u_3)\right)$$

$$w(x(t), u(t), t) = 0, \quad \forall t, \quad w(x, u, t) = u_2 u_3$$

$$u(t) = \begin{bmatrix} u_1(t) \\ u_2(t) \\ u_3(t) \end{bmatrix} \in \overline{U}, \quad \forall t, \quad \overline{U} = \{u | u_1 \in R, \ 0 \le u_2 \le 1, \ -1 \le u_3 \le 0\}$$

where the final time t_f is free.

The set $\overline{S}_f$ of the admissible final states is a regular variety because

$$\Sigma_f(x) = \frac{d\alpha_f(x)}{dx} = \begin{bmatrix} 1 & 0 \end{bmatrix}$$

and rank$(\Sigma_f(x)) = 1, \forall x \in \overline{S}_f$.

We take into account the punctual and global equality constraint by adding a term to the customary hamiltonian function. Thus we have

$$H_e = 1 + \frac{u_1^2}{2} + 0.1(u_2 - u_3) + \lambda_1 x_2 + \lambda_2(u_1 + u_2 + u_3) + \mu(u_2 u_3)$$

and the H_e-minimizing control is

$$u_{h_1} = -\lambda_2$$

$$u_{h_2} = \begin{cases} 0 \text{ if } 0.1 + \lambda_2 + \mu u_3 > 0 \\ 1 \text{ if } 0.1 + \lambda_2 + \mu u_3 < 0 \end{cases}$$

$$u_{h_3} = \begin{cases} 0 \text{ if } -0.1 + \lambda_2 + \mu u_2 < 0 \\ -1 \text{ if } -0.1 + \lambda_2 + \mu u_2 > 0 \end{cases}$$

Equation (2.5a) and its general solution over an interval beginning at time 0 are

$$\dot{\lambda}_1(t) = 0 \qquad\qquad \lambda_1(t) = \lambda_1(0)$$
$$\dot{\lambda}_2(t) = -\lambda_1(t) \qquad \lambda_2(t) = \lambda_2(0) - \lambda_1(0)t$$

and the orthogonality condition at the final time

$$\begin{bmatrix} \lambda_1(t_f) \\ \lambda_2(t_f) \end{bmatrix} = \begin{bmatrix} \lambda_1(0) \\ \lambda_2(0) - \lambda_1(0)t_f \end{bmatrix} = \vartheta_f \frac{d\alpha_f(x)}{dx}\bigg|'_{x=x(t_f)} = \begin{bmatrix} 1 \\ 0 \end{bmatrix}$$

implies $\lambda_2(t_f) = 0$, so that $\lambda_2(t) = \lambda_1(0)(t_f - t)$ and $\lambda_2(\cdot)$ has always the same sign which is not positive. In fact the nature of the problem, particularly the constraint on $x_1(t_f)$, implies that a function $u_1(\cdot)$ with negative values is surely not optimal. Thus $\lambda_1(0) \leq 0$. A similar remark leads to stating that $u_3(\cdot) = 0$. Therefore we have

$$u_2(t) = \begin{cases} 0 \text{ if } 0.1 + \lambda_2(t) > 0 \\ 1 \text{ if } 0.1 + \lambda_2(t) < 0 \end{cases}$$

and $u_2(\cdot)$ can switch at most once. We notice that $\lambda_1(0) \neq 0$ because, otherwise, $\lambda_2(\cdot) = 0$ and $u_2(\cdot) = 0$ $(0.1 + \lambda_2(t) > 0)$.

We also notice that the switching (if it exists) must necessarily be from 1 to 0. In fact the switching from 0 to 1 can not take place because $0.1 + \lambda_2(t_f) = 0.1 > 0$. Similarly, it is not possible that $u_2(\cdot) = 1$.

Thus only the two following cases must be considered.

Case 1. $u_2(\cdot)$ switches from 1 to 0 at time τ

Case 2. $u_2(\cdot) = 0$

In both cases the transversality condition at the final time can be written at the initial time because the problem is time-invariant and it results

$$H_e(0) = 1 + \frac{u_1^2(0)}{2} + 0.1(u_2(0) - u_3(0)) + \lambda_1(0)x_2(0) +$$
$$+\lambda_2(0)(u_1(0) + u_2(0) + u_3(0)) + \mu u_2(0)u_3(0) =$$
$$= 1 + (0.1 + \lambda_2(0))u_2(0) - \frac{\lambda_2(0)^2}{2} = 0$$

This equation implies

Case 1	*Case 2*
$u_2(0) = 1,$	$u_2(0) = 0$
$\lambda_2(0) = 1 \pm \sqrt{3.2},$	$\lambda_2(0) = \pm\sqrt{2}$

Consequently, $\lambda_2(0) = 1 - \sqrt{3.2}$ in the first case and $\lambda_2(0) = -\sqrt{2}$ in the second case, because we have already established that $t_f\lambda_1(0) = \lambda_2(0)$ and $\lambda_1(0) < 0$.

The two cases are now examined.

Case 1. $u_2(\cdot)$ switches from 1 to 0 at time τ

We enforce feasibility $(x_1(t_f) = 10)$, the switching condition for $u_2(\cdot)$ $(0.1 + \lambda_2(\tau) = 0)$ and, obviously, recall that $\lambda_1(0) = \lambda_2(0)/t_f = -\sqrt{3.2}/t_f$. It results $t_f = t_{f_1} := 3.64$, $\tau = 3.18$, $\lambda_1(0) = -0.22$ $J = J_1 := 4.33$.

Case 2. $u_2(\cdot) = 0$

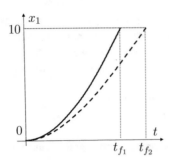
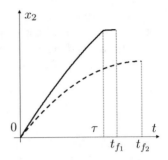

Fig. 10.6 Problem 10.3: State motions when $u_2(\cdot)$ switches (*solid line*) and does not switch (*dashed line*)

We enforce feasibility and recall that $\lambda_1(0) = \lambda_2(0)/t_f = -\sqrt{2}/t_f$. It results $t_f = t_{f_2} := 4.61$, $\lambda_1(0) = -0.31$ and $J = J_2 := 6.14$.

We choose the solution corresponding to *Case 1* because $J_2 > J_1$. The state motion is plotted in Fig. 10.6 when $u_2(\cdot)$ switches (solid line) and when it does not switch (dashed line). Notice that the derivative of $x_2(\cdot)$ is discontinuous at τ when switching occurs.   ∎

Problem 10.4

Two control variables act on the system which is described by

$$\dot{x}_1(t) := x_2(t)$$
$$\dot{x}_2(t) := u_1(t) + u_2(t)$$

Furthermore we have

$$x(0) = \begin{bmatrix} 0 \\ 0 \end{bmatrix}, \quad x(t_f) = \begin{bmatrix} 1 \\ 0 \end{bmatrix}, \quad J = \int_0^{t_f} l(x(t), u(t), t)\,dt, \quad l(x, u, t) = \frac{u_1^2}{2}$$
$$u_1(t \in R, \ u_2(t) \in R, \ \forall t, \quad w(x(t), u(t), t) = 0, \ \forall t, \quad w(x, u, t) = u_2 + \beta t^n$$

In these equations the parameters $\beta \geq 0$ and $n \geq 0$, n integer, are given, whereas the final time t_f is free.

Thanks to the form of $w(\cdot, \cdot, \cdot)$ we can deal with the constraint in an ad-hoc and very convenient way. Indeed the constraint simply requires that $u_2(t) = -\beta t^n$. Thus the problem at hand can be legitimately stated with reference to the new system

$$\dot{x}_1(t) := x_2(t)$$
$$\dot{x}_2(t) := u_1(t) - \beta t^n$$

and the punctual and global equality constraint is not present any more. Consequently, the hamiltonian function and the H-minimizing control are

$$H = \frac{u_1^2}{2} + \lambda_1 x_2 + \lambda_2 (u_1 - \beta t^n), \quad u_{1h} = -\lambda_2$$

Equation (2.5a) and its general solution over an interval beginning at time 0 are

$$\dot{\lambda}_1(t) = 0 \qquad\qquad \lambda_1(t) = \lambda_1(0)$$
$$\dot{\lambda}_2(t) = -\lambda_1(t) \qquad \lambda_2(t) = \lambda_2(0) - \lambda_1(0)t$$

The transversality condition at the final time (to be necessarily written at this time because the problem is not time-invariant) is

$$H(t_f) = \frac{u_1^2(t_f)}{2} + \lambda_1(t_f)x_2(t_f) + \lambda_2(t_f)(u_1(t_f) - \beta t_f^n) = -\frac{\lambda_2^2(t_f)}{2} - \lambda_2(t_f)\beta t_f^n = 0$$

where we have taken into account both the form of $u_h(\cdot)$ and the constraint on $x_2(t_f)$. The state motion can easily be computed

$$x_1(t) = -\frac{\lambda_2(0)}{2}t^2 + \frac{\lambda_1(0)}{6}t^3 - \frac{\beta}{(n+1)(n+2)}t^{n+2}$$

$$x_2(t) = -\lambda_2(0)t + \frac{\lambda_1(0)}{2}t^2 - \frac{\beta}{n+1}t^{n+1}$$

It is straightforward to ascertain that the transversality condition and feasibility $(x_1(t_f) = 1, x_2(t_f) = 0)$ can not be satisfied when $\beta = 0$. On the contrary, this does not happen if $\beta > 0$.

The parameters t_f, $\lambda_1(0)$, $\lambda_2(0)$ resulting from various choices of β and n are reported in Table 10.1

Table 10.1 Problem 10.4. The parameters t_f, λ_{10}, λ_{20} corresponding to various choices of β and n

		$\beta = 0.1$	$\beta = 1$	$\beta = 10$
$n = 0$	t_f	7.75	2.45	0.77
	$\lambda_1(0)$	−0.03	−0.82	−25.82
	$\lambda_2(0)$	−0.20	−2.00	−20.00
$n = 1$	t_f	3.91	1.82	0.84
	$\lambda_1(0)$	−0.10	−1.00	−10.00
	$\lambda_2(0)$	−0.39	−1.82	−8.43
$n = 5$	t_f	1.97	1.42	1.02
	$\lambda_1(0)$	−0.50	−1.34	−3.60
	$\lambda_2(0)$	−0.99	−1.90	−3.67
$n = 15$	t_f	1.39	1.21	1.06
	$\lambda_1(0)$	−1.23	−1.85	−2.77
	$\lambda_2(0)$	−1.71	−2.24	−2.94

Fig. 10.7 Problem 10.4:
State motion when
$\beta = 10$, $n = 15$ (*solid line*)
and $\beta = 0.1$, $n = 0$ (*dotted line*)

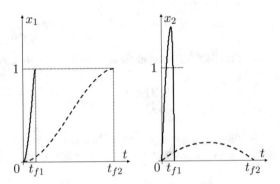

We notice that the final time t_f decreases both when, for a given n, β increases and, for a given β, n increases. This outcome is consistent with the fact that, corresponding to the particular initial and final states, the second component u_2 of the control can be legitimately considered as a disturbance which becomes more and more significant when β or n increase. The state motion corresponding to the pairs ($\beta = 0.1$, $n = 0$) and ($\beta = 10$, $n = 15$) are plotted in Fig. 10.7. We notice that the choice of the second pair is much more demanding (see $x_2(\cdot)$). ∎

10.2 Punctual and Global Inequality Constraints

Here the controlled system is again the classical integrator. Its initial state is always given in the forthcoming problems. On the contrary, the final state is given in the first four of them (Problems 10.5–10.8), is constrained to belong to a set which is not a regular variety in Problem 10.9 and is partially specified in Problem 10.10 (the first component is given, whereas the second one is free). The final time is given in Problem 10.8 and is free in all the remaining ones.

Only a punctual and global inequality constraint is present in Problems 10.5 and 10.6, whereas punctual and isolated equality constraints are also present in Problems 10.7 and 10.8. These constraints must be satisfied at one (Problem 10.7) or two (Problem 10.8) times.

An integral inequality constraint is also present in the second one of the two subproblems which constitute Problem 10.10.

The function $w(\cdot, \cdot, \cdot)$ (which specifies the punctual and global inequality constraints) explicitly depends on time only in Problems 10.8 and 10.9.

In the discussion below we make reference to the material of Sect. 2.3.4. Furthermore it is convenient to refer to Chap. 9 when dealing with Problems 10.7 and 10.8, where punctual and isolated equality constraints are present as well. Finally, we make use of the content of Sect. 2.2.3 when facing Problem 10.9, where also a non standard constraint on the final state must be satisfied.

Problem 10.5

Let

$$x(0) = \begin{bmatrix} 0 \\ 0 \end{bmatrix}, \quad x(t_f) = \begin{bmatrix} 2 \\ 0 \end{bmatrix}, \quad J = \int_0^{t_f} l(x(t), u(t), t)\, dt, \quad l(x, u, t) = 1$$

$$u(t) \in \overline{U}, \ \forall t, \ \overline{U} = \{u| -1 \le u \le 1\}$$

$$w(x(t), u(t), t) \le 0, \ \forall t, \ w(x, u, t) = x_2 - 1$$

where the final time is (obviously) free.

The hamiltonian function and the H-minimizing control are

$$H = 1 + \lambda_1 x_2 + \lambda_2 u, \quad u_h = -\text{sign}(\lambda_2)$$

so that Eq. (2.5a) and its general solution over an interval beginning at time t_0 are

$$\dot{\lambda}_1(t) = 0 \qquad\qquad \lambda_1(t) = \lambda_1(t_0)$$
$$\dot{\lambda}_2(t) = -\lambda_1(t) \qquad\quad \lambda_2(t) = \lambda_2(t_0) - \lambda_1(t_0)(t - t_0)$$

First, we consider Problem 10.5-1 where the punctual and global inequality constraint is not present, that is $w(\cdot, \cdot, \cdot) = 0$, or, equivalently, the request $x_2(\cdot) \le 1$ is ignored. Then we discuss Problem 10.5-2 where the constraint is present.

Problem 10.5-1 The constraint is not present, that is $w(\cdot, \cdot, \cdot) = 0$
The function $\lambda_2(\cdot)$ is linear with respect to t: therefore the control is a piecewise constant function which switches at most once from 1 to -1 or viceversa. The transversality condition at the final time, written at $t = 0$, because the problem is time-invariant, yields

$$H(0) = 1 + \lambda_1(0)x_2(0) + \lambda_2(0)u(0) = 1 - \lambda_2(0)\text{sign}(\lambda_2(0)) = 0$$

so that $\lambda_2(0) = \pm 1$. The choice $\lambda_2(0) = 1$ must be discarded. Indeed it implies that the control is negative at the beginning and the state moves away from the required final value, with a consequent increase of the length of the control interval, i.e. of the performance index. On the other hand, $u(\cdot) = 1$ is not feasible: as a consequence, the control must switch from 1 to -1 at a suitable time t_c. Thus

$0 \le t < t_c$	$t_c < t \le t_f$
$u(t) = 1$	$u(t) = -1$
$0 \le t \le t_c$	$t_c \le t \le t_f$
$x_1(t) = \dfrac{t^2}{2}$	$x_1(t) = \dfrac{t_c^2}{2} + t_c(t - t_c) - \dfrac{(t - t_c)^2}{2}$
$x_2(t) = t$	$x_2(t) = t_c - (t - t_c)$

Fig. 10.8 Problem 10.5.
State trajectories: **a** without
the constraint $x_2(\cdot) \leq 1$ and
b with the constraint
$x_2(\cdot) \leq 1$

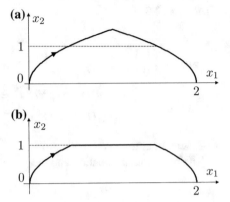

By enforcing feasibility $(x_1(t_f) = 2, x_2(t_f) = 0)$, we easily obtain

$$t_c = \sqrt{2}, \quad t_f = t_{f_l} := 2\sqrt{2}$$

The corresponding state trajectory and the function $u(\cdot)$ are shown in Figs. 10.8a and
10.9, respectively. Notice that the constraint $x_2(\cdot) \leq 1$ is violated.

Problem 10.5-2 $x_2(\cdot) \leq 1$
The shape of the trajectory relevant to Problem 10.5-1 indicates that the constraint is
binding over an interval $[t_1, t_2]$, namely $x_2(t) = 1, t_1 \leq t \leq t_2$. The function $w(\cdot, \cdot, \cdot)$
does not explicitly depend on u: therefore we proceed to its total differentiation with
respect to time and obtain

Fig. 10.9 Problem 10.5. The
function $u(\cdot)$ when the
constraint $x_2(\cdot) \leq 1$ is
ignored (*solid line*) and when
it is taken into account
(*dashed line*)

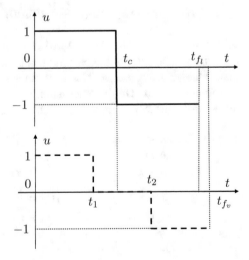

$$\frac{dw(x(t), u(t), t)}{dt} = \dot{x}_2(t) = \widehat{w}(x(t), u(t), t) = u(t)$$

As a result, we add the new constraint $u(t) \leq 0$, $t \in (t_1, t_2)$ to the punctual and isolated equality constraint $w(x(t_1), u(t_1), t_1) = x_2(t_1) - 1 = 0$, where t_1 is free.

In view of Eq. (2.10) the function $\lambda_2(\cdot)$ can be discontinuous at time t_1 because $w(\cdot, \cdot, \cdot)$ explicitly depends on x_2. On the contrary, Eqs. (2.10) and (2.11) imply that both the hamiltonian function and $\lambda_1(\cdot)$ are continuous at this time because $w(\cdot, \cdot, \cdot)$ explicitly depends neither on time nor on x_1 and t_1 is free. Thus, by resorting to the notation (9.1), we have

$$\lambda_1^-(t_1) = \lambda_1^+(t_1) = \lambda_1(0)$$
$$H^-(t_1) = 1 + \lambda_1^-(t_1)x_2(t_1) - \lambda_2^-(t_1)\text{sign}(\lambda_2^-(t_1)) =$$
$$= 1 + \lambda_1(0) - \lambda_2^-(t_1)\text{sign}(\lambda_2^-(t_1)) = H^+(t_1) =$$
$$= 1 + \lambda_1^+(t_1)x_2(t_1) + \lambda_2^+(t_1)u^+(t_1) = 1 + \lambda_1(0)$$

These equations imply $\lambda_2^-(t_1) = \lambda_2(0) - \lambda_1(0)t_1 = 0$. Furthermore $\lambda_2(t_2) = 0$ because $\lambda_2(\cdot)$ is continuous at t_2 and the value 0 is the only one consistent with

$$\lim_{t \to t_2^-} u(t) = 0$$

Therefore $\lambda_2(t) = -\lambda_1(0)(t - t_2)$, $t \geq t_2$.

As a conclusion we have

$0 \leq t < t_1$	$t_1 < t < t_2$	$t_2 < t \leq t_f$
$u(t) = 1$	$u(t) = 0$	$u(t) = -1$
$0 \leq t \leq t_1$	$t_1 \leq t \leq t_2$	$t_2 \leq t \leq t_f$
$x_1(t) = \dfrac{t^2}{2}$	$x_1(t) = x_1(t_1) +$	$x_1(t) = x_1(t_2) +$
	$+ x_2(t_1)(t - t_1)$	$+ x_2(t_2)(t - t_2) - \dfrac{(t - t_2)^2}{2}$
$x_2(t) = t$	$x_2(t) = x_2(t_1)$	$x_2(t) = x_2(t_2) - (t - t_2)$

By enforcing feasibility ($x_2(t_1) = 1$, $x_1(t_f) = 2$, $x_2(t_f) = 0$), we obtain

$$t_1 = 1, \quad t_2 = 2, \quad t_f = t_{f_v} := 3$$

The transversality condition at the final time can be exploited to compute $\lambda_1(0)$ and then $\lambda_2(0)$.

The state trajectory and the function $u(\cdot)$, corresponding to the resulting solution, are shown in Figs. 10.8b and 10.9, respectively.

We notice that the presence of the punctual and global inequality constraint implies the increase of the length of the control interval, that is a larger value of the performance index. ∎

Problem 10.6

Let

$$x(0) = \begin{bmatrix} 0 \\ 0 \end{bmatrix}, \quad x(t_f) = \begin{bmatrix} 2 \\ 0 \end{bmatrix}$$

$$J = \int_0^{t_f} l(x(t), u(t), t)\, dt, \quad l(x, u, t) = 8 + \frac{u^2}{2}, \quad u(t) \in R, \ \forall t$$

$$w(x(t), u(t), t) \le 0, \ \forall t, \quad w(x, u, t) = x_2 - 1$$

where the final time t_f is free.

The hamiltonian function and the H-minimizing control are

$$H = 8 + \frac{u^2}{2} + \lambda_1 x_2 + \lambda_2 u, \quad u_h = -\lambda_2$$

As previously done, it is interesting to evaluate the effect of the punctual and global inequality constraint. To this end, we first assume that the constraint is not present or, in other words, that $w(\cdot, \cdot, \cdot) = 0$ (Problem 10.6-1). Then we introduce the request $x_2(\cdot) \le 1$ (Problem 10.6-2).

In both cases, Eq. (2.5a) and its general solution over an interval beginning at time t_0 are

$$\dot\lambda_1(t) = 0 \qquad\qquad \lambda_1(t) = \lambda_1(t_0)$$
$$\dot\lambda_2(t) = -\lambda_1(t) \qquad\qquad \lambda_2(t) = \lambda_2(t_0) - \lambda_1(t_0)(t - t_0)$$

Problem 10.6-1 $w(\cdot, \cdot, \cdot) = 0$

The transversality condition at the final time, written at the initial time because the problem is time-invariant, yields

$$H(0) = 8 + \frac{u^2(0)}{2} + \lambda_1(0)x_2(0) + \lambda_2(0)u(0) = 8 - \frac{\lambda_2^2(0)}{2} = 0$$

and $\lambda_2(0) = \pm 4$.

Furthermore we have

$$x_1(t) = -\lambda_2(0)\frac{t^2}{2} + \lambda_1(0)\frac{t^3}{6}, \quad x_2(t) = -\lambda_2(0)t + \lambda_1(0)\frac{t^2}{2}$$

and feasibility ($x_1(t_f) = 2$, $x_2(t_f) = 0$) is attained if

$$\lambda_2(0) = -\frac{12}{t_f^2}, \quad \lambda_1(0) = \frac{2\lambda_2(0)}{t_f}$$

Fig. 10.10 Problem 10.6.
State trajectory when the
constraint $x_2(\cdot) \leq 1$ is
ignored (*solid line*) and when
is taken into account (*dashed
line*)

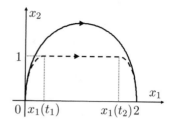

Fig. 10.11 Problem 10.6.
The function $u(\cdot)$ when the
constraint $x_2(\cdot) \leq 1$ is
ignored (*solid line*) and when
the constraint $x_2(\cdot) \leq 1$ is
taken into account (*dashed
line*)

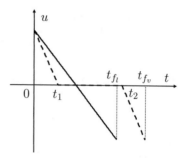

Therefore $\lambda_2(0) < 0$ and

$$\lambda_2(0) = -4, \quad t_f = t_{f_l} := \sqrt{3}, \quad \lambda_1(0) = -\frac{8}{\sqrt{3}}, \quad u(t) = 4 - \frac{8}{\sqrt{3}}t$$

The state trajectory and the control function are shown (solid lines) in Figs. 10.10
and 10.11, respectively.

We notice that the constraint $w(x(t), u(t), t) \leq 0, \; \forall t$ has been violated.

Problem 10.6-2 $x_2(\cdot) \leq 1$
The shape of the state trajectory relevant to Problem 10.6-1 indicates that the constraint is binding over an interval $[t_1, t_2]$, namely $x_2(t) = 1, t_1 \leq t \leq t_2$. The function
$w(\cdot, \cdot, \cdot)$ does not explicitly depend on u: therefore we proceed to its total differentiation with respect to time and obtain

$$\frac{d(w(x, t))}{dt} = \dot{x}_2(t) = \widehat{w}(x(t), u(t), t) = u(t)$$

As a result, we add the new constraint $u(t) \leq 0, \; t \in (t_1, t_2)$ to the punctual and
isolated equality constraint $w(x(t_1), u(t_1), t_1) = x_2(t_1) - 1 = 0$, where t_1 is free.

In view of Eq. (2.10) the function $\lambda_2(\cdot)$ can be discontinuous at time t_1 because
$w(\cdot, \cdot, \cdot)$ explicitly depends on x_2. On the contrary, Eqs. (2.10) and (2.11) imply that
both the hamiltonian function and $\lambda_1(\cdot)$ are continuous at this time because $w(\cdot, \cdot, \cdot)$
explicitly depends neither on time nor on x_1 and t_1 is free. Thus, by resorting to the
notation (9.1), we have

$$\lambda_1^-(t_1) = \lambda_1^+(t_1) = \lambda_1(0)$$

$$H^-(t_1) = 8 + \frac{(u^-(t_1))^2}{2} + \lambda_1^-(t_1)x_2(t_1) - \lambda_2^-(t_1)u^-(t_1) =$$

$$= 8 + \lambda_1(0) - \frac{(\lambda_2^-(t_1))^2}{2} = H^+(t_1) =$$

$$= 8 + \frac{(u^+(t_1))^2}{2} + \lambda_1^+(t_1)x_2(t_1) + \lambda_2^+(t_1)u^+(t_1) = 8 + \lambda_1(0)$$

These equations imply

$$\lambda_2^-(t_1) = \lambda_2(0) - \lambda_1(0)t_1 = 0$$

Furthermore $\lambda_2(t_2) = 0$ because $\lambda_2(\cdot)$ is continuous at t_2 and

$$0 = \lim_{t \to t_2^+} u(t) = - \lim_{t \to t_2^+} \lambda_2(t)$$

The transversality condition at the final time, written at the initial time because the hamiltonian function is continuous and the problem is time-invariant, is the same as in Problem 10.6-1, so that $\lambda_2(0) = \pm 4$ again. We also have

$0 \le t < t_1$	$t_1 < t < t_2$	$t_2 \le t \le t_f$
$u(t) = -\lambda_2(t) =$	$u(t) = 0$	$u(t) = -\lambda_2(t) =$
$= -\lambda_2(0) + \lambda_1(0)t$		$= \lambda_1(0)(t - t_2)$
$0 \le t \le t_1$	$t_1 \le t \le t_2$	$t_2 \le t \le t_f$
$x_1(t) = -\dfrac{\lambda_2(0)}{2}t^2+$	$x_1(t) = x_1(t_1)+$	$x_1(t) = x_1(t_2)+$
$+\dfrac{\lambda_1(0)}{6}t^3$	$+x_2(t_1)(t - t_1)$	$+x_2(t_2)(t - t_2)+$
		$+\dfrac{\lambda_1(0)}{6}(t - t_2)^3$
$x_2(t) = -\lambda_2(0)t+$	$x_2(t) = x_2(t_1)$	$x_2(t) = x_2(t_2)+$
$+\dfrac{\lambda_1(0)}{2}t^2$		$+\dfrac{\lambda_1(0)}{2}(t - t_2)^2$

If we recall that $\lambda_2(0) = \pm 4$, $\lambda_2(0) - \lambda_1(0)t_1 = 0$ and enforce feasibility ($x_2(t_1) = 1$, $x_1(t_f) = 2$, $x_2(t_f) = 0$), we get from the above equations

$$\lambda_2(0) = -4, \quad \lambda_1(0) = -8, \quad x_1(t_1) = \frac{1}{3}, \quad x_1(t_2) = \frac{5}{3}$$

$$t_1 = \frac{1}{2}, \quad t_2 = \frac{11}{6}, \quad t_f = t_{f_v} := \frac{7}{3}$$

The state trajectory and the control function are shown (dashed lines) in Figs. 10.10 and 10.11, respectively.

Notice that the length of the control interval increases when the constraint is enforced.   ∎

A punctual and isolated equality constraint is also present in the forthcoming problem.

Problem 10.7

Let

$$x(0) = \begin{bmatrix} 0 \\ 0 \end{bmatrix}, \quad x(t_f) = \begin{bmatrix} 1 \\ 0 \end{bmatrix}$$

$$J = \int_0^{t_f} l(x(t), u(t), t)\, dt, \quad l(x, u, t) = 1 + \frac{u^2}{2}, \quad u(t) \in R, \; \forall t$$

$$w^{(1)}(x(\tau), \tau) = 0, \quad w^{(1)}(x, \tau) = \begin{bmatrix} x_1 + 1 \\ x_2 \end{bmatrix}$$

$$w^{(2)}(x(t), u(t), t) \leq 0, \; \forall t, \quad w^{(2)}(x, u, t) = \begin{bmatrix} x_2 - 0.5 \\ -x_2 - 0.5 \end{bmatrix}$$

The intermediate time τ and the final one t_f are free.

The hamiltonian function and the H-minimizing control are

$$H = 1 + \frac{1}{2}u^2 + \lambda_1 x_2 + \lambda_2 u, \quad u_h = -\lambda_2$$

whereas Eq. (2.5a) and its general solution over an interval beginning at time t_0 are

$$\dot{\lambda}_1(t) = 0 \qquad\qquad \lambda_1(t) = \lambda_1(t_0)$$
$$\dot{\lambda}_2(t) = -\lambda_1(t) \qquad\qquad \lambda_2(t) = \lambda_2(t_0) - \lambda_1(t_0)(t - t_0)$$

We can handle the (vector) punctual and isolated equality constraint by making reference to the material presented in Sect. 2.3.3 and Chap. 9. However, a simple thought (the problem is time-invariant) allows us to conclude that the control interval can be subdivided into two subintervals, the first one relevant to values of t smaller than τ, the second to values greater than τ.

In order to perform meaningful comparisons, we discuss two cases. The first one deals with a problem (Problem 10.7-1) where the punctual and global inequality constraint is not present ($w^{(2)}(\cdot, \cdot, \cdot) = 0$), whereas the second considers a problem (Problem 10.7-2) where such constraint is present.

Problem 10.7-1 $w^{(2)}(\cdot, \cdot, \cdot) = 0$

First, we notice that Eqs. (2.10) and (2.11) imply that the hamiltonian function is continuous at τ because $w^{(1)}(\cdot, \cdot)$ does not explicitly depend on t and τ is free. On the contrary, Eq. (2.10) allows both $\lambda_1(\cdot)$ and $\lambda_2(\cdot)$ to be discontinuous at τ because $w^{(1)}(\cdot, \cdot)$ explicitly depends on x_1 and x_2. Therefore the transversality condition at the final time can be written at the initial time because the hamiltonian function is continuous and the problem is time-invariant. We obtain

$$H(0) = 1 + \frac{1}{2}u^2(0) + \lambda_1(0)x_2(0) + \lambda_2(0)u(0) = 1 - \frac{1}{2}\lambda_2^2(0) = 0$$

and $\lambda_2(0) = \pm\sqrt{2}$. It is fairly obvious that the control must be negative at the beginning (see the constraint on the state at time τ): thus $\lambda_2(0) = \sqrt{2}$.

Corresponding to the first subinterval ($0 \le t \le \tau$), we have

$$x_1(t) = \frac{\lambda_1(0)}{6}t^3 - \frac{\lambda_2(0)}{2}t^2, \quad x_2(t) = \frac{\lambda_1(0)}{2}t^2 - \lambda_2(0)t$$

and, by enforcing feasibility ($x_1(\tau) = -1, x_2(\tau) = 0$), we find

$$\tau = \tau_l := 2.06, \quad \lambda_1(0) = 1.37$$

We proceed in a similar way when considering the second subinterval ($\tau \le t \le t_f$). It is expedient to refer to the time t^* which ranges over the interval $[0, t_f^*], t_f^* := t_f - \tau$. Thanks to an analysis which is completely analogous to the one carried on above, we conclude that $\lambda_2^*(0) = -\sqrt{2}$ and compute the state motion

$$x_1^*(t^*) = -1 + \frac{\lambda_1^*(0)}{6}(t^*)^3 - \frac{\lambda_2^*(0)}{2}(t^*)^2, \quad x_2^*(t^*) = \frac{\lambda_1^*(0)}{2}(t^*)^2 - \lambda_2^*(0)t^*$$

By enforcing feasibility ($x_1(t_f^*) = 1, x_2(t_f^*) = 0$), we obtain

$$t_f^* = 2.91, \quad \lambda_1^*(0) = -0.97, \quad t_f = t_{f_l} := \tau_l + t_f^* = 4.97$$

and

$$u(t) = -\lambda_2(0) + \lambda_1(0)t, \quad 0 \le t < \tau_l$$
$$u(t) = -\lambda_2^*(0) + \lambda_1^*(0)(t - \tau_l), \quad \tau_l < t \le t_{f_l}$$

We notice that $\lambda_1(\cdot)$ and $\lambda_2(\cdot)$, if considered over the whole interval $[0, t_{f_l}]$, are discontinuous at τ_l, in accordance to the material of Sect. 2.3.3.

The state trajectory corresponding to the above parameters $\lambda_1(0)$, $\lambda_2(0)$, $\lambda_1^*(0)$, $\lambda_2^*(0)$, τ_l, t_{f_l} are plotted in Fig. 10.12 (dashed line).

Fig. 10.12 Problem 10.7. State trajectories when the constraint $w^{(2)}(\cdot, \cdot, \cdot) \le 0$ is ignored (*dashed line*) and when it is taken into account (*solid line*)

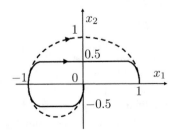

Problem 10.7-2 $w^{(2)}(x, u, t) = \begin{bmatrix} x_2 - 0.5 \\ -x_2 - 0.5 \end{bmatrix}$

The shape of the state trajectory relevant to Problem 10.7-1 indicates that the second component of the constraint is binding over an interval $T_1 := [t_1, t_2]$ (namely $x_2(t) = -0.5$, $t \in T_1$), whereas the first component is binding over an interval $T_2 := [t_3, t_4]$ (namely $x_2(t) = 0.5$, $t \in T_2$). The function $w^{(2)}(\cdot, \cdot, \cdot)$ does not explicitly depend on u so that we proceed to its total differentiation with respect to time and obtain

$$\frac{dw^{(2)}(x(t), u(t), t)}{dt} = \begin{bmatrix} \dot{x}_2(t) \\ -\dot{x}_2(t) \end{bmatrix} = \widehat{w}^{(2)}(x, u, t) = \begin{bmatrix} u(t) \\ -u(t) \end{bmatrix}$$

As it was done in solving Problem 10.7-1, we can subdivide the control interval into two subintervals $T_a := [0, \tau_v]$ and $T_b := [\tau_v, t_{f_v}]$, where t_{f_v} is the final time and τ_v is the time when the punctual and isolated equality constraint is satisfied, namely $w^{(1)}(x(\tau_v), \tau_v) = 0$. Then we can make the following statements.

(i) The constraints $-u(t) \leq 0$, $t_1 < t < t_2$ and $w_2^{(2)}(x(t_1), u(t_1), t_1) = -x_2(t_1) - 0.5 = 0$, where t_1 is free, must be taken into account with reference to the first subinterval T_a. Furthermore Eqs. (2.10) and (2.11) imply that the hamiltonian function and $\lambda_1(\cdot)$ are continuous at t_1. In fact $w_2^{(2)}(\cdot, \cdot, \cdot)$ explicitly depends neither on x_1 nor on t and t_1 is free. On the contrary, $\lambda_2(\cdot)$ can be discontinuous at t_1 because $w_2^{(2)}(\cdot, \cdot, \cdot)$ explicitly depends on x_2. Finally, the transversality condition at the final time can be written at $t = 0$ because the hamiltonian function is continuous and the problem is time-invariant.

(ii) The constraints $u(t) \leq 0$, $t_3 < t < t_4$ and $w_1^{(2)}(x(t_3), u(t_3), t_3) = x_2(t_3) - 0.5 = 0$, where t_3 is free, must be taken into account with reference to the second subinterval T_b. Eqs. (2.10) and (2.11) imply that the hamiltonian function and $\lambda_1(\cdot)$ are continuous at t_3. In fact $w_1^{(2)}(\cdot, \cdot, \cdot)$ explicitly depends neither on x_1 nor on t and t_3 is free. On the contrary, $\lambda_2(\cdot)$ can be discontinuous at t_3 because $w_1^{(2)}(\cdot, \cdot, \cdot)$ explicitly depends on x_2. Finally, the transversality condition at the final time can be written at the initial time of the subinterval T_b, namely at $t = \tau_v$ because the hamiltonian function is continuous and the problem is time-invariant.

More specifically, the transversality condition relevant to the subinterval T_a

$$H(0) = 1 + \frac{u^2(0)}{2} + \lambda_1(0)x_2(0) + \lambda_2(0)u(0) = 1 - \frac{\lambda_2^2(0)}{2} = 0$$

supplies $\lambda_2(0) = \pm\sqrt{2}$. By exploiting the notation (9.1), the continuity of the hamiltonian function at t_1

$$H^-(t_1) = 1 - \frac{(\lambda_2^-(t_1))^2}{2} + \lambda_1^-(t_1)x_2(t_1) = H^+(t_1) = 1 + \lambda_1^+(t_1)x_2(t_1)$$

implies (recall that $\lambda_1^-(t_1) = \lambda_1^+(t_1) = \lambda_1(0)$ because $\lambda_1(\cdot)$ is continuous) $\lambda_2^-(t_1) = \lambda_2(0) - \lambda_1(0)t_1 = 0$. These facts and the satisfaction of the constraint on $x_2(t_1)$

$$x_2(t_1) = -\lambda_2(0)t_1 + \frac{\lambda_1(0)}{2}t_1^2 = -\frac{1}{2}$$

lead to

$$\lambda_2(0) = 1.41, \quad \lambda_1(0) = 2, \quad t_1 = 0.71, \quad x_1(t_1) = -0.24$$

At time t_2, when the constraint is no more binding, $u(t_2) = 0 = -\lambda_2(t_2)$ so that, for $t \geq t_2$, we have

$$\lambda_2(t) = -\lambda_1(0)(t - t_2)$$
$$x_1(t) = x_1(t_2) - \frac{1}{2}(t - t_2) + \frac{\lambda_1(0)}{6}(t - t_2)^3$$
$$x_2(t) = -\frac{1}{2} + \frac{\lambda_1(0)}{2}(t - t_2)^2$$

where

$$x_1(t_2) = x_1(t_1) - \frac{1}{2}(t_2 - t_1)$$

By enforcing feasibility ($x_1(\tau) = -1$, $x_2(\tau) = 0$), we obtain

$$t_2 = 1.76, \quad \tau = \tau_v := 2.47$$

We proceed in the same way when dealing with the subinterval T_b. Thus we consider a problem defined on a system with state x^* which is function of a fictitious time t^* which ranges over the interval $[t_0^*, t_f^*]$. The control interval begins at $t_0^* = 0$ and ends at the free final time $t_f^* = t_{f_v} - \tau_v$. The initial and final states are given, because

$$x^*(t_0^*) = \begin{bmatrix} -1 \\ 0 \end{bmatrix}, \quad x^*(t_f^*) = \begin{bmatrix} 1 \\ 0 \end{bmatrix}$$

The punctual and global inequality constraint $x_2^*(t^*) \leq 0.5$, $\forall t^*$ originates from the first component of $w(\cdot, \cdot, \cdot)$. With reference to the fictitious time we have

$$\lambda_2^*(0) = -1.41, \quad \lambda_1^*(0) = -2, \quad t_3^* = 0.71, \quad t_4^* = 6.24, \quad t_f^* = 4.47, \quad x_1^*(t_3^*) = -0.76$$

The times t_3, t_4, t_f are simply obtained by adding τ_v to the above values. Furthermore it is obvious that $x_1(t_3) = x_1^*(t_3^*)$, whereas $\lambda_1^*(0)$ and $\lambda_2^*(0)$ have to be meant as the limits of $\lambda_1(\cdot)$ and $\lambda_2(\cdot)$ when t tends to τ_v from the right, so that $\lambda_1^+(\tau_v) = \lambda_1^*(0)$ and $\lambda_2^+(\tau_v) = \lambda_2^*(0)$. As a conclusion, we have

$$t_f = 4.94, \quad t_3 = 3.18, \quad t_4 = 4.23, \quad x_1(t_3) = 0.24$$

Fig. 10.13 Problem 10.7.
State motion and control
function when the constraint
$w^{(2)}(\cdot, \cdot, \cdot) \leq 0$ is ignored
(*dashed line*) and when it is
taken into account (*solid
line*)

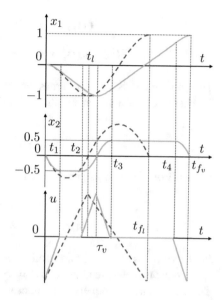

Notice that $\lambda_1(\cdot)$ and $\lambda_2(\cdot)$ are discontinuous at $t = \tau_v$. In fact

$$\lim_{t \to \tau_v^-} \lambda_1(t) = \lambda_1(0) = 2 \neq \lim_{t \to \tau_v^+} \lambda_1(t) = \lambda_1^*(0) = -2$$

$$\lim_{t \to \tau_v^-} \lambda_2(t) = \lambda_1(0)(\tau_v - t_2) = 1.41 \neq \lim_{t \to \tau_v^+} \lambda_2(t) = \lambda_2^*(0) = -1.41$$

We remark that the same results can be obtained if the problem is tackled without
splitting it into two subproblems.

The state trajectory corresponding to the solution above is plotted in Fig. 10.12
(solid line). The state motion and the control function for both Problems 10.7-1 and
10.7-2 are shown in Fig. 10.13 (dashed line for the first problem, continuous line for
the second one). We notice the remarkable differences of the two state motions, the
controls and the final times.   ∎

In the forthcoming problem we must also satisfy two punctual and isolated equality
constraints.

Problem 10.8
Let

$$x(0) = \begin{bmatrix} -1 \\ 0 \end{bmatrix}, \quad x(t_f) = \begin{bmatrix} 0 \\ 0 \end{bmatrix}$$

$$J = \int_0^{t_f} l(x(t), u(t), t)\, dt, \quad l(x, u, t) = \frac{u^2}{2}, \quad u(t) \in R, \; \forall t$$

$$w^{(1)}(x(\tau_1), \tau_1) = 0, \quad w^{(1)}(x, \tau) = \begin{bmatrix} x_1 \\ x_2 - 1 \end{bmatrix}$$

$$w^{(2)}(x(\tau_2), \tau_2) = 0, \quad w^{(2)}(x, \tau) = \begin{bmatrix} x_1 - 1 \\ x_2 \end{bmatrix}$$

$$w^{(3)}(x(t), u(t), t) \le 0, \quad \forall t, \quad w^{(3)}(x, u, t) = t - x_2 - t_f$$

The final time t_f is given, whereas the intermediate times τ_1 and τ_2, $\tau_1 < \tau_2$ are free. The hamiltonian function and the H-minimizing control are

$$H = \frac{1}{2}u^2 + \lambda_1 x_2 + \lambda_2 u, \quad u_h = -\lambda_2$$

so that Eq. (2.5a) and its general solution over an interval beginning at time t_0 are

$$\begin{aligned} \dot\lambda_1(t) &= 0 & \lambda_1(t) &= \lambda_1(t_0) \\ \dot\lambda_2(t) &= -\lambda_1(t) & \lambda_2(t) &= \lambda_2(t_0) - \lambda_1(t_0)(t - t_0) \end{aligned}$$

First, we observe that, if the constraint defined by $w^{(3)}(\cdot, \cdot, \cdot)$ becomes binding at $\tau_3 < t_f$, x_2 must be negative at τ_3 ($x_2(\tau_3) = \tau_3 - t_f$), besides being zero at the final time. These facts and the sequence of values which must be assumed by the state variables suggest that the interval of time when $w^{(3)}(x(t), u(t), t) = 0$ ends at t_f. Therefore we also conclude that $\tau_3 > \tau_2$. The solution is specified by the nine parameters $\lambda_{10} := \lambda_1(0), \lambda_{20} := \lambda_2(0), \tau_1, \tau_2, \tau_3, \lambda_{11} := \lambda_1^+(\tau_1), \lambda_{21} := \lambda_2^+(\tau_1), \lambda_{12} := \lambda_1^+(\tau_2), \lambda_{22} := \lambda_2^+(\tau_2)$, where we have exploited the notation (9.1). It is obvious that the values $\lambda_{ij}, i = 1, 2, j = 0, 1, 2$ are expedient in defining the control function within the intervals $[0, \tau_1), (\tau_1, \tau_2), (\tau_2, \tau_3)$. Furthermore we have $u(t) = 1, \tau_3 < t \le t_f$ because $w^{(3)}(\cdot, \cdot, \cdot)$ does not explicitly depend on u and its total derivative with respect to time is $1 - u$. In view of Eqs. (2.10) and (2.11), we impose the continuity of the hamiltonian function at $t = \tau_1, t = \tau_2$ ($w^{(1)}(\cdot, \cdot)$ and $w^{(2)}(\cdot, \cdot)$ do not explicitly depend on time and both τ_1 and τ_2 are free), namely

$$H^-(\tau_1) = H^+(\tau_1), \quad H^-(\tau_2) = H^+(\tau_2)$$

where

$$H^-(\tau_1) = \frac{(u^-(\tau_1))^2}{2} + \lambda_1^-(\tau_1)x_2(\tau_1) + \lambda_2^-(t_1)u^-(t_1) =$$
$$= -\frac{(\lambda_{20} - \lambda_{10}\tau_1)^2}{2} + \lambda_{10}$$
$$H^+(\tau_1) = \frac{(u^+(\tau_1))^2}{2} + \lambda_{11}x_2(\tau_1) + \lambda_{21}u^+(t_1) = -\frac{\lambda_{21}^2}{2} + \lambda_{11}$$
$$H^-(\tau_2) = \frac{(u^-(\tau_2))^2}{2} + \lambda_1^-(\tau_2)x_2(\tau_2) + \lambda_2^-(\tau_2)u^-(\tau_2) = -\frac{(\lambda_2^-(\tau_2))^2}{2} =$$
$$= -\frac{(\lambda_{21} - \lambda_{11}(\tau_2 - \tau_1))^2}{2}$$
$$H^+(\tau_2) = \frac{(u^+(\tau_2))^2}{2} + \lambda_{12}x_2(\tau_2) + \lambda_{22}u^+(t_2) = -\frac{\lambda_{22}^2}{2}$$

Moreover, if we take into account that

$$
\begin{aligned}
0 \le t < \tau_1, \quad & u(t) = -\lambda_2(t) = -\lambda_{20} + \lambda_{10} t \\
\tau_1 < t < \tau_2, \quad & u(t) = -\lambda_2(t) = -\lambda_{21} + \lambda_{11}(t - \tau_1) \\
\tau_2 < t < \tau_3, \quad & u(t) = -\lambda_2(t) = -\lambda_{22} + \lambda_{12}(t - \tau_2) \\
\tau_3 < t \le t_f, \quad & u(t) = 1, \quad x_2(t) = t - t_f
\end{aligned}
$$

the satisfaction of the constraints on the state variables requires that the following six equations hold

$$
x_2(\tau_1) = \frac{\lambda_{10}}{2} \tau_1^2 - \lambda_{20} \tau_1 = 1
$$

$$
x_1(\tau_1) = -1 + \frac{\lambda_{10}}{6} \tau_1^3 - \frac{\lambda_{20}}{2} \tau_1^2 = 0
$$

$$
x_2(\tau_2) = 1 + \frac{\lambda_{11}}{2} (\tau_2 - \tau_1)^2 - \lambda_{21}(\tau_2 - \tau_1) = 0
$$

$$
x_1(\tau_2) = \frac{\lambda_{11}}{6} (\tau_2 - \tau_1)^3 - \frac{\lambda_{21}}{2} (\tau_2 - \tau_1)^2 = 1
$$

$$
x_2(\tau_3) = \frac{\lambda_{12}}{2} (\tau_3 - \tau_2)^2 - \lambda_{22}(\tau_3 - \tau_2) = \tau_3 - t_f
$$

$$
x_1(t_f) = x_1(\tau_3) + \frac{1}{2}(t_f^2 - \tau_3^2) - t_f(t_f - \tau_3) = 0
$$

where

$$
x_1(\tau_3) = x_1(\tau_2) + \frac{\lambda_{12}}{6} (\tau_3 - \tau_2)^3 - \frac{\lambda_{22}}{2} (\tau_3 - \tau_2)^2
$$

Finally, Eqs. (2.10) and (2.11) imply that at time τ_3 both $\lambda_2(\cdot)$ and the hamiltonian function can be discontinuous because $w^{(3)}(\cdot, \cdot, \cdot)$ explicitly depends on time and x_2, the time τ_3 being free. On the contrary, $\lambda_1(\cdot)$ is continuous at this time because $w^{(3)}(\cdot, \cdot, \cdot)$ does not explicitly depend on x_1. Thus we have

$$
\begin{aligned}
\begin{bmatrix} \lambda_1^-(\tau_3) \\ \lambda_2^-(\tau_3) \end{bmatrix} &= \begin{bmatrix} \lambda_{12} \\ \lambda_2^-(\tau_3) \end{bmatrix} = \begin{bmatrix} \lambda_1^+(\tau_3) \\ \lambda_2^+(\tau_3) \end{bmatrix} + \left. \frac{\partial w^{(3)}(x, u(\tau_3), \tau_3)}{\partial x} \right|'_{x=x(\tau_3)} \mu = \\
&= \begin{bmatrix} \lambda_{12} \\ \lambda_2^+(\tau_3) \end{bmatrix} + \begin{bmatrix} 0 \\ -1 \end{bmatrix} \mu
\end{aligned}
$$

with reference to $\lambda(\cdot)$ and

$$
\begin{aligned}
H^-(\tau_3) &= \frac{(u^-(\tau_3))^2}{2} + \lambda_1^-(\tau_3) x_2(\tau_3) + \lambda_2^-(\tau_3) u^-(\tau_3) = \\
&= -\frac{(\lambda_2^-(\tau_3))^2}{2} + \lambda_{12}(\tau_3 - t_f) = H^+(\tau_3) - \left. \frac{\partial w^{(3)}(x(\tau_3), u(\tau_3), t)}{\partial t} \right|_{t=\tau_3} \mu =
\end{aligned}
$$

Table 10.2 Problem 10.8. The values of the parameters which define the solution corresponding to two values of t_f

t_f	λ_{10}	λ_{12}	λ_{20}	λ_{21}	λ_{22}	τ_1	τ_2	τ_3
2	−439.61	1588.70	−66.36	−59.36	66.36	0.29	0.57	0.61
5	−1.06	1.45	−1.46	−0.09	1.46	1.46	2.91	4.61

$$= \frac{(u^+(\tau_3))^2}{2} + \lambda_1^+(\tau_3)x_2(\tau_3) + \lambda_2^+(\tau_3)u^+(\tau_3) - \mu =$$
$$= \frac{1}{2} + \lambda_{12}(\tau_3 - t_f) + \lambda_2^+(\tau_3) - \mu$$

with reference to the hamiltonian function. We obtain from the last two sets of equations,
$$\lambda_2^-(\tau_3) = -1$$

where $\lambda_2^-(\tau_3) = \lambda_{22} - \lambda_{12}(\tau_3 - \tau_2)$. In summary, we have nine equations for the nine (unknown) parameters which define the solution: their detailed handling, though not completely trivial, is not worth to be reported here.

We have chosen two values for the final time, namely $t_f = 2$ and $t_f = 5$ which lead to the solutions reported in Table 10.2 and to the state trajectories plotted in Fig. 10.14. The values of λ_{11} are not shown in the quoted table since they are equal to λ_{10}. Notice that an increase of the value of the final time implies a smoother shape of the corresponding trajectory. The state motions corresponding to the two considered values of the final time are shown in Figs. 10.15 and 10.16. Consistently with the shapes of the state trajectories, a rather large increase of the absolute values of the second state variable is caused by the choice of the smaller final time.

The parameters λ_{ij} take on significantly different values according to the choice of the final time. Thus we expect that the control functions are significantly different as well. Figure 10.17 strengthens this fact, particularly at the beginning of the control interval.

If we now choose $t_f = 10$, it is easy to ascertain that the constraint is never binding. In fact when $t_f = 10$ the above equations do not admit a solution. As a matter of

Fig. 10.14 Problem 10.8. State trajectories when $t_f = 5$ (*dashed line*) and when $t_f = 2$ (*solid line*)

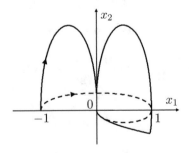

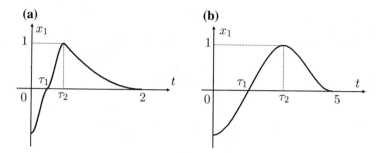

Fig. 10.15 Problem 10.8. The function $x_1(\cdot)$ when **a** $t_f = 2$ and **b** $t_f = 5$

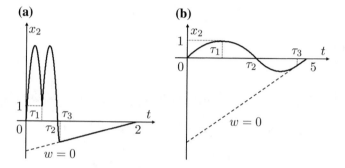

Fig. 10.16 Problem 10.8. The function $x_2(\cdot)$ when **a** $t_f = 2$ and **b** $t_f = 5$

Fig. 10.17 Problem 10.8. The function $u(\cdot)$ when $t_f = 5$ (*dashed line*) and when $t_f = 2$ (*solid line*)

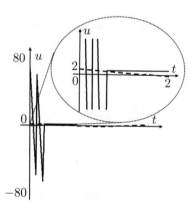

fact, if we consider the problem without such constraint, we find that the resulting motion of the second state variable complies with the request $w^{(3)}(x(t), u(t), t) \le 0$, $\forall t \in [0, t_f]$, as it clearly appears in Fig. 10.18. ∎

Fig. 10.18 Problem 10.8.
The function $x_2(\cdot)$ when
$t_f = 10$ and the constraint
$w^{(3)}(\cdot, \cdot, \cdot) \le 0$ is ignored
(*straight line r*). The
constraint is not violated

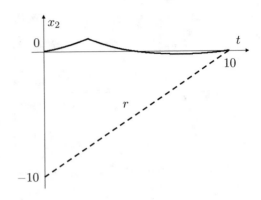

Problem 10.9

Let

$$x(0) = \begin{bmatrix} 0 \\ 5 \end{bmatrix}, \quad J = \int_0^{t_f} l(x(t), u(t), t)\, dt, \ l(x, u, t) = 1 + \frac{u^2}{2}, \ u(t) \in R, \ \forall t$$

$$x(t_f) \in \widehat{S}_f = \{x | x \in S_f, \ x_1 \le b_1\}, \ b_1 = 0$$

$$S_f = \overline{S}_f = \{x | \alpha_f(x) = 0\}, \ \alpha_f(x) = x_1 + x_2$$

$$w(x(t), u(t), t) \le 0, \ \forall t, \ w(x, u, t) = t - x_2 - 1$$

The final time t_f is free.

The set $\overline{S}_f$ is a regular variety because

$$\Sigma_f(x) = \frac{d\alpha_f(x)}{dx} = \begin{bmatrix} 1 & 1 \end{bmatrix}$$

and rank$(\Sigma_f(x)) = 1, \ \forall x \in \overline{S}_f$.

The hamiltonian function and the H-minimizing control are

$$H = 1 + \frac{u^2}{2} + \lambda_1 x_2 + \lambda_2 u, \quad u_h = -\lambda_2$$

Therefore Eq. (2.5a) and its general solution over an interval beginning at time t_0 are

$$\dot{\lambda}_1(t) = 0 \qquad\qquad \lambda_1(t) = \lambda_1(t_0)$$
$$\dot{\lambda}_2(t) = -\lambda_1(t) \qquad \lambda_2(t) = \lambda_2(t_0) - \lambda_1(t_0)(t - t_0)$$

First, we suppose that the punctual and global inequality constraint is absent (Problem 10.9-1), then we assume that it is present (Problem 10.9-2).

Problem 10.9-1 We ignore the constraint $w(\cdot, \cdot, \cdot) \leq 0$
Consistently with the material presented in Sect. 2.2.3, we first let $x(t_f) \in \overline{S}_f$, that
is we do not take into account the request $x_1(t_f) \leq 0$.

We must impose feasibility and both the orthogonality and transversality conditions at the final time. They lead to three equations for the unknown parameters $\lambda_1(0)$, $\lambda_2(0)$, t_f which specify the solution. To this end we observe that the state motion is

$$x_1(t) = 5t - \frac{\lambda_2(0)}{2}t^2 + \frac{\lambda_1(0)}{6}t^3$$

$$x_2(t) = 5 - \lambda_2(0)t + \frac{\lambda_1(0)}{2}t^2$$

the orthogonality condition is

$$\begin{bmatrix} \lambda_1(t_f) \\ \lambda_2(t_f) \end{bmatrix} = \begin{bmatrix} \lambda_1(0) \\ \lambda_2(0) - \lambda_1(0)t_f \end{bmatrix} = \vartheta_f \left. \frac{d\alpha_f(x)}{dx} \right|'_{x=x(t_f)} = \vartheta_f \begin{bmatrix} 1 \\ 1 \end{bmatrix}$$

and, finally, the transversality condition, written at the initial time because the problem is time-invariant, is

$$H(0) = 1 + \frac{u^2(0)}{2} + \lambda_1(0)x_2(0) + \lambda_2(0)u(0) = 1 - \frac{\lambda_2^2(0)}{2} + 5\lambda_1(0) = 0$$

The last two relations together with the one relevant to feasibility, namely

$$x_1(t_f) + x_2(t_f) = 5t_f - \frac{\lambda_2(0)}{2}t_f^2 + \frac{\lambda_1(0)}{6}t_f^3 + 5 - \lambda_2(0)t_f + \frac{\lambda_1(0)}{2}t_f^2 = 0$$

constitute a system of three equations which leads to

$$\lambda_2(0) = \frac{5}{1 + t_f} \pm \sqrt{\frac{25}{(1 + t_f)^2} + 2}$$

The problem data imply $u(0) < 0$, so that $\lambda_2(0)$ must be positive. Consequently we have

$$\lambda_1(0) = 0.39, \quad \lambda_2(0) = 2.44, \quad t_f = t_{f_1} := 5.18$$

and the value of the performance index is $J = J_1 := 11.27$. The state trajectory and the control function are plotted (dash-double-dotted lines) in Figs. 10.19 and 10.20, respectively. We conclude that the constraint $x_1(t_f) \leq 0$ can not be ignored, so that we require that $x_1(t_f) = x_2(t_f) = 0$. Thus we impose the transversality condition, identical to the one reported above, and feasibility

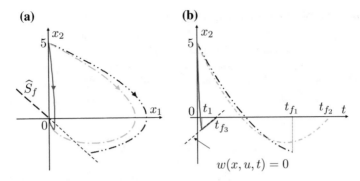

Fig. 10.19 Problem 10.9. **a** State trajectories; **b** the function $x_2(\cdot)$. The *dash-double-dotted lines* and *dash-single-dotted lines* refer to Problem 10.9-1 when $x(t_f) \in \overline{S}_f$ and when $x(t_f) \in \widehat{S}_f$, respectively. The *solid lines* refer to Problem 10.9-2

Fig. 10.20 Problem 10.9. The function $u(\cdot)$. The *dash-double-dotted lines* refer to Problem 10.9-1 when $x(t_f) \in \overline{S}_f$ and when $x(t_f) \in \widehat{S}_f$, respectively. The *solid line* refers to Problem 10.9-2 and its values have been divided by 10

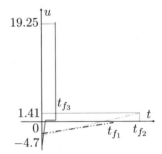

$$x_2(t_f) = 5 - \lambda_2(0)t_f + \frac{\lambda_1(0)}{2}t_f^2 = 0$$

$$x_1(t_f) = 5t_f - \frac{\lambda_2(0)}{2}t_f^2 + \frac{\lambda_1(0)}{6}t_f^3 = 0$$

where we have taken into account the previously found expressions for the state motion. From these three equations we obtain (recall that we have already established that $u(0) < 0$)

$$\lambda_1(0) = 0.6, \quad \lambda_2(0) = 2.83, \quad t_f = t_{f_2} := 7.07$$

The value of the performance index is now $J_2 := 14.14$. We notice that $J_2 > J_1$ as a consequence of the presence of the further constraint on $x_1(t_f)$. The state trajectory and the control function are plotted (dash-single-dotted lines) in Figs. 10.19 and 10.20, respectively.

Problem 10.9-2 $w(\cdot, \cdot, \cdot) \leq 0$

In view of the preceding discussion we assume $x_1(t_f) = x_2(t_f) = 0$. We proceed to the total differentiation with respect to time of $w(\cdot, \cdot, \cdot)$ because it does not explicitly depend on u and obtain

$$\frac{dw(x(t), u(t), t)}{dt} = \widehat{w}(x(t), u(t), t) = 1 - u(t)$$

The punctual and global inequality constraint can be binding over the interval $T :=$ $[t_1, t_2]$, namely $x_2(t) = t - 1$, $t \in T$, if $w(x(t_1), u(t_1), t_1) = 0$, t_1 is free and $1 - u(t) \leq 0$, $t \in T$. In view of Eqs. (2.10) and (2.11), $\lambda_1(\cdot)$ is continuous at t_1 whereas both the hamiltonian function and $\lambda_2(\cdot)$ can be discontinuous at this time. In fact $w(\cdot, \cdot, \cdot)$ does not explicitly depend on x_1, but it explicitly depends on x_2 and t and the time t_1 is free. The solution is defined by the six unknown parameters t_1, t_2, t_f, $\lambda_1(0)$, $\lambda_2(0)$, $\lambda_2^+(t_1)$, where we have exploited the notation (9.1). These parameters can be computed by solving the equations resulting from feasibility, the transversality condition at the final time, the constraints between the hamiltonian function and $\lambda_2(\cdot)$ imposed by Eqs. (2.10) and (2.11) at t_1 and those between $u(\cdot)$ and $\lambda_2(\cdot)$ at t_2. We observe that the transversality condition can be written for $t > t_1$ since the problem is time-invariant and the hamiltonian function is continuous when $t > t_1$. More in detail we have

$$x_2(t_1) = t_1 - 1$$
$$x_1(t_f) = 0$$
$$x_2(t_f) = 0$$
$$H^+(t_1) = 0$$
$$\lambda_2^-(t_1) = \lambda_2^+(t_1) - \mu$$
$$H^-(t_1) = H^+(t_1) - \mu$$
$$-\lambda_2(t_2) = u(t_2)$$

We have written seven rather than six equations because we have introduced the parameter μ in the fifth and sixth of them. Together with these equations we consider those which express how the parameters to be determined influence $\lambda_2(\cdot)$, $x(\cdot)$ and the hamiltonian function. More precisely, we have:

(*i*) for the first time interval

$$0 \leq t < t_1, \quad \lambda_2(t) = \lambda_2(0) - \lambda_1(0)t, \quad u(t) = -\lambda_2(t)$$

$$0 \leq t \leq t_1, \quad x(t) = \begin{bmatrix} 5t - \dfrac{\lambda_2(0)}{2}t^2 + \dfrac{\lambda_1(0)}{6}t^3 \\ 5 - \lambda_2(0)t + \dfrac{\lambda_1(0)}{2}t^2 \end{bmatrix}$$

(*ii*) for the second time interval (when the constraint is binding)

$$t_1 < t \le t_2, \quad \lambda_2(t) = \lambda_2^+(t_1) - \lambda_1(0)(t - t_1), \quad u(t) = 1$$

$$t_1 \le t \le t_2, \quad x(t) = \begin{bmatrix} x_1(t_1) - (t - t_1) + \dfrac{1}{2}(t^2 - t_1^2) \\ t - 1 \end{bmatrix}$$

(*iii*) for the third (last) time interval

$$t_2 \le t \le t_f, \quad \lambda_2(t) = -1 - \lambda_1(0)(t - t_2), \quad u(t) = -\lambda_2(t)$$

$$x(t) = \begin{bmatrix} x_1(t_2) + x_2(t_2)(t - t_2) + \dfrac{1}{2}(t - t_2)^2 + \dfrac{\lambda_1(0)}{6}(t - t_2)^3 \\ x_2(t_2) + t - t_2 + \dfrac{\lambda_1(0)}{2}(t - t_2)^2 \end{bmatrix}$$

(*iv*) for the hamiltonian function at t_1

$$H^-(t_1) = 1 - \frac{1}{2}(\lambda_2^-(t_1))^2 + \lambda_1(0)x_2(t_1)$$

$$H^+(t_1) = \frac{3}{2} + \lambda_1(0)x_2(t_1) + \lambda_2^+(t_1)$$

We obtain

$$\lambda_1(0) = 192, \quad \lambda_2(0) = 47, \quad \lambda_2^+(t_1) = 142.5$$
$$t_1 = 0.25, \quad t_2 = 0.9974, \quad t_f = t_{f_3} := 0.9996$$

The state trajectory and the control function are plotted (solid lines) in Figs. 10.19 and 10.20. The values of $u(\cdot)$ have been divided by 10.

We notice that the punctual and global inequality constraint requires that the second state variable is greater than a quantity which increases with the elapsed time: consequently, a very fast approach to the (imposed) final state results. Thus a demanding control action is needed and the performance index is now $J = J_3 := 113.34$ which is a significantly larger value than those previously found. ∎

Problem 10.10

In the two forthcoming problems the goal is to transfer the state from the same initial state to the same set of admissible final states and to comply with the same punctual and global inequality constraint. This has to be achieved without requiring large control actions and long control intervals. The two problem statements reflect these objectives by introducing different performance indexes and adding an integral inequality constraint in one of them. More precisely, we have, for both problems,

$$x(0) = \begin{bmatrix} 0 \\ 0 \end{bmatrix}, \quad x(t_f) \in \overline{S}_f = \{x \mid \alpha_f(x) = 0\}, \quad \alpha_f(x) = x_1 - x_{1f}$$

$$w(x(t), u(t), t) \le 0, \quad \forall t, \quad w(x, u, t) = x_2 - 2$$

where $x_{1f} > 0$ is given. The set $\overline{S}_f$ is a regular variety because

$$\Sigma_f(x) = \frac{d\alpha_f(x)}{dx} = \begin{bmatrix} 1 & 0 \end{bmatrix}$$

and $\mathrm{rank}(\Sigma_f(x)) = 1, \forall x \in \overline{S}_f$.

Problem 10.10-1

Let

$$J = \int_0^{t_f} l(x(t), u(t), t)\, dt, \quad l(x, u, t) = t^2 + \frac{u^2}{2}, \quad u(t) \in R, \quad \forall t$$

where the final time t_f is obviously free. The hamiltonian function and the H-minimizing control are

$$H = t^2 + \frac{u^2}{2} + \lambda_1 x_2 + \lambda_2 u, \quad u_h = -\lambda_2$$

Therefore Eq. (2.5a) and its general solution over an interval beginning at time t_0 are

$$\begin{aligned} \dot{\lambda}_1(t) &= 0 & \lambda_1(t) &= \lambda_1(t_0) \\ \dot{\lambda}_2(t) &= -\lambda_1(t) & \lambda_2(t) &= \lambda_2(t_0) - \lambda_1(t_0)(t - t_0) \end{aligned}$$

The orthogonality condition at the final time

$$\begin{bmatrix} \lambda_1(t_f) \\ \lambda_2(t_f) \end{bmatrix} = \vartheta_f \left. \frac{d\alpha_f(x)}{dx} \right|'_{x=x(t_f)} = \vartheta_f \begin{bmatrix} 1 \\ 0 \end{bmatrix}$$

gives $\lambda_2(t_f) = 0$, so that the transversality condition at the final time is

$$H(t_f) = t_f^2 + \frac{u^2(t_f)}{2} + \lambda_1(t_f)x_2(t_f) + \lambda_2(t_f)u(t_f) =$$

$$= t_f^2 - \frac{\lambda_2^2(t_f)}{2} + \lambda_1(t_f)x_2(t_f) = t_f^2 + \lambda_1(t_f)x_2(t_f) = 0$$

First, we ignore the punctual and global inequality constraint on $x_2(\cdot)$ (Problem 10.10-1a), then we take it into account (Problem 10.10-1b). In both problems we let $\lambda_{i0} := \lambda_i(0), i = 1, 2$.

Problem 10.10-1a The constraint on $x_2(\cdot)$ is not present
The orthogonality condition and the expression of $\lambda_2(\cdot)$ give $\lambda_2(t) = \lambda_{10}(t_f - t)$, so that

$$x_1(t) = \lambda_{10}\left(\frac{1}{6}t^3 - \frac{t_f}{2}t^2\right)$$

$$x_2(t) = \lambda_{10}\left(\frac{1}{2}t^2 - t_f t\right)$$

By enforcing feasibility $(x_1(t_f) = x_{1f})$ and the transversality condition, we first compute t_f, λ_{10} and then λ_{20}, $x_2(t_f)$. More precisely, we get

$$\lambda_{10} = \lambda_{10_{1a}} := -\sqrt{2}, \quad t_f = t_{f_{1a}} := \sqrt[3]{\frac{3x_{1f}}{\sqrt{2}}}, \quad \lambda_{20} = \lambda_{20_{1a}} := \lambda_{10_{1a}}t_{f_{1a}}$$

$$x_2(t_f) = x_{2f_{1a}} := -\frac{\lambda_{20_{1a}}t_{f_{1a}} + \lambda_{10_{1a}}}{2}t_{f_{1a}}^2$$

When $x_{1f} = 10$, the values of these parameters are reported in Table 10.3 and the state motion is plotted in Fig. 10.21 (solid line). It is apparent that the constraint on $x_2(\cdot)$ has been violated.

Problem 10.10-1b $w(x(t), u(t), t) \leq 0$, $\forall t$, namely $x_2(t) \leq 2$, $\forall t$

The solution of Problem 10.10-1a indicates that the constraint is binding over a time interval T_τ beginning at τ whenever $x_{2f1a} > 2$. In fact Fig. 10.21 shows that, when the constraint is ignored, $x_2(\cdot)$ attains its maximum at $t = t_{f1a}$, that is, by exploiting the previously found expressions, when

$$x_{1f} > \frac{4}{3}\sqrt{2\sqrt{2}} = 2.24$$

Suppose that this inequality holds. The function $w(\cdot, \cdot, \cdot)$ does not explicitly depend on u, so that we proceed to its total differentiation with respect to time and obtain

$$\frac{dw(x(t), u(t), t)}{dt} = \widehat{w}(x(t), u(t), t) = u(t)$$

Fig. 10.21 Problem 10.10-1. State motions when the constraint $x_2(\cdot) \leq 2$ is ignored (*solid line*) and when it is taken into account (*dashed line*)

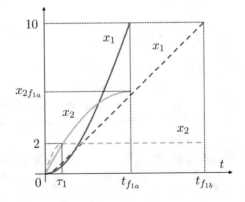

Therefore the problem at hand includes the following two constraints

$$\widehat{w}(x(t), u(t), t) \leq 0, \quad \widehat{w}(x, u, t) = u, \quad t \in T_\tau$$
$$w(x(\tau), u(\tau), \tau) = 0, \quad \tau \text{ free}$$

This implies that $u(\cdot) = 0$ when the inequality constraint is binding and, in view of Eq. (2.10), the function $\lambda_2(\cdot)$ can be discontinuous at τ because $w(\cdot, \cdot, \cdot)$ explicitly depends on x_2. On the contrary, thanks to Eqs. (2.10) and (2.11), $\lambda_1(\cdot)$ and the hamiltonian function are continuous at τ. In fact $w(\cdot, \cdot, \cdot)$ explicitly depends neither on x_1 nor on t and τ is free. As a consequence, $\lambda_1(\cdot) = \lambda_{10}$ and

$$H^-(\tau) = \tau^2 + \frac{(u^-(\tau))^2}{2} + \lambda_1^-(\tau)x_2(\tau) + \lambda_2^-(\tau)u^-(\tau) = \tau^2 - \frac{(\lambda_2^-(\tau))^2}{2} + 2\lambda_{10} =$$
$$= \tau^2 + \frac{(u^+(\tau))^2}{2} + \lambda_1^+(\tau)x_2(\tau) + \lambda_2^+(\tau)u^+(\tau) = \tau^2 + 2\lambda_{10} = H^+(\tau)$$

where we have used the notation (9.1). From this equation we deduce that

$$\lambda_2^-(\tau) = \lambda_{20} - \lambda_{10}\tau = 0$$

At the end of the interval T_τ (where the constraint is binding) the functions $\lambda_2(\cdot)$ and $u(\cdot)$ are continuous. There the control is zero and equal to $-\lambda_2$, so that also $\lambda_2(\cdot)$ must be zero at this time besides being zero at the final time (recall the orthogonality condition). Thus we can assume that the constraint is binding $\forall t \geq \tau$ and, consequently,

$$0 \leq t < \tau \qquad\qquad \tau < t \leq t_f$$
$$u(t) = \lambda_{10}t - \lambda_{20} \qquad u(t) = 0$$
$$0 \leq t \leq \tau \qquad\qquad \tau \leq t \leq t_f$$
$$x_1(t) = \frac{\lambda_{10}}{6}t^3 - \frac{\lambda_{20}}{2}t^2 \qquad x_1(t) = x_1(\tau) + 2(t - \tau)$$
$$x_2(t) = \frac{\lambda_{10}}{2}t^2 - \lambda_{20}t \qquad x_2(t) = 2$$

and

$$x_2(\tau) = 2 = \frac{\lambda_{10}}{2}\tau^2 - \lambda_{20}\tau$$
$$\lambda_2^-(\tau) = 0 = \lambda_{20} - \lambda_{10}\tau$$
$$H(t_f) = t_f^2 + 2\lambda_{10} = 0$$
$$x_1(t_f) = x_{1f} = \frac{\lambda_{10}}{6}\tau^3 - \frac{\lambda_{20}}{2}\tau^2 + 2(t_f - \tau)$$

Table 10.3 Problems 10.10-1 and 10.10-2. The parameters which define the solution

	Problem 10.10-1		Problem 10.10-2		Problem 10.10-2	
	–		$\beta = 4$		$\beta = 12$	
	$w = 0$	$w = x_2 - 2$	$w = 0$	$w = x_2 - 2$	$w = 0$	$w = x_2 - 2$
λ_{10}	-1.41	-13.43	-2.50	-13.64	-0.83	-12.87
λ_{20}	-3.91	-7.33	-8.37	-9.09	-1.93	-2.86
λ_{30}	–	–	3.12	1.52	0.35	0.16
t_f	2.77	5.18	3.35	5.22	2.32	5.07
τ	–	0.55	–	0.67	–	0.22
$x_2(t_f)$	5.42	2.00	4.48	2.00	6.46	2.00
J	14.14	51.27	12.50	47.47	4.17	43.55

From the four equations above (the last one of them stems from feasibility) we can compute the four parameters λ_{10}, λ_{20}, τ, t_f which specify the solution. More in detail, the first three equations supply t_f, τ e λ_{20} as functions of λ_{10}, namely

$$t_f = t_{f_{1b}} := \sqrt{-2\lambda_{10}}, \quad \tau = \tau_1 := \sqrt{-\frac{4}{\lambda_{10}}}, \quad \lambda_{20} = \lambda_{20_{1b}} := \lambda_{10}\tau_1$$

If we substitute these expressions into the fourth equation, we determine the value of λ_{10}, to be denoted with $\lambda_{10_{1b}}$. When $x_{1f} = 10$ we obtain the results in Table 10.3, whereas the state motion is plotted in Fig. 10.21 (dashed line). We notice the remarkable increase of the final time due to the presence of the constraint on $x_2(\cdot)$. A similar effect can be ascertained on the value of performance index which changes from $J = J_{1a} := 14.14$ to $J = J_{1b} := 51.27$.

Problem 10.10-2
Here the goal of limiting the control action is expressed by giving an upper bound to the integral of $\dfrac{u^2(t)}{2}$ rather than inserting such a term into the performance index, as done in Problem 10.10 -1. Consistently, let

$$J = \int_0^{t_f} l(x(t), u(t), t)\, dt, \quad l(x, u, t) = t^2$$

$$\int_0^{t_f} w_d(x(t), u(t), t)\, dt \le \overline{w}_d, \quad w_d(x, u, t) = \frac{u^2}{2}, \quad \overline{w}_d = \beta, \quad u(t) \in R, \ \forall t$$

where t_f is free and $\beta > 0$ is given.

We proceed as indicated in Sect. 2.3.2 and take into account the integral constraint by adding a new state variable x_3 which must comply with the equations

$$\dot{x}_3(t) = w_d(x(t), u(t), t) = \frac{u^2(t)}{2}, \quad x_3(0) = 0, \quad x_3(t_f) \leq \beta$$

In so doing, the system is *enlarged* because its state is now the three-dimensional vector $x_a = [\, x_1 \; x_2 \; x_3 \,]'$. As a consequence, the set of the admissible final states is

$$\widehat{S}_{af} = \left\{ x_a | x_a \in S_{af}, \; x_3 \leq b_3 \right\}, \; b_3 = \beta$$
$$S_{af} = \overline{S}_{af} = \left\{ x_a | \alpha_{af}(x_a) = 0 \right\}, \; \alpha_{af}(x_a) = x_1 - x_{1f}$$

where the set $\overline{S}_{af}$ is a regular variety because

$$\Sigma_{af}(x_a) = \frac{d\alpha_{af}(x_a)}{dx_a} = \begin{bmatrix} 1 & 0 & 0 \end{bmatrix}$$

and $\operatorname{rank}(\Sigma_{af}(x_a)) = 1, \forall x_a \in \overline{S}_{af}$.

The hamiltonian function and the H-minimizing control are

$$H = t^2 + \lambda_1 x_2 + \lambda_2 u + \lambda_3 \frac{u^2}{2}, \quad u_h = -\frac{\lambda_2}{\lambda_3}$$

provided that $\lambda_3 > 0$. Therefore Eq. (2.5a) and their general solution over an interval beginning at time t_0 are

$$\begin{aligned} \dot{\lambda}_1(t) &= 0 & \lambda_1(t) &= \lambda_1(t_0) \\ \dot{\lambda}_2(t) &= -\lambda_1(t) & \lambda_2(t) &= \lambda_2(t_0) - \lambda_1(t_0)(t - t_0) \\ \dot{\lambda}_3(t) &= 0 & \lambda_3(t) &= \lambda_3(t_0) \end{aligned}$$

As frequently done in the preceding discussion, we first assume that the punctual and global inequality constraint on $x_2(\cdot)$ is ignored (Problem 10.10-2a). On the contrary, this constraint is present in Problem 10.10-2b. In both cases $\lambda_3(\cdot)$ is a continuous function: this fact is apparent in the first problem whereas it is proved in the second problem. Thus, if the integral constraint is not binding, the orthogonality condition at the final time is

$$\begin{bmatrix} \lambda_1(t_f) \\ \lambda_2(t_f) \\ \lambda_3(t_f) \end{bmatrix} = \begin{bmatrix} \lambda_1(t_f) \\ \lambda_2(t_f) \\ \lambda_3(t_0) \end{bmatrix} = \vartheta_{af} \frac{d\alpha_{af}(x_a)}{dx_a} \bigg|'_{x_a = x_a(t_f)} = \begin{bmatrix} 1 \\ 0 \\ 0 \end{bmatrix}$$

and implies $\lambda_3(t_0) = 0$. Therefore the integral constraint is binding, $x_3(t_f) = \beta$ and the set of the admissible final states becomes

$$\overline{S}_{afM} := \left\{ x_a | \alpha_{afM_i}(x_a) = 0, \; i = 1, 2 \right\}, \; \alpha_{afM_1}(x_a) = x_1 - x_{1f}, \; \alpha_{afM_2}(x_a) = x_3 - \beta$$

where the set $\overline{S}_{afM}$ is a regular variety because

$$\Sigma_{afM}(x_a) = \frac{d\left[\begin{array}{c}\alpha_{afM_1}(x_a)\\ \alpha_{afM_2}(x_a)\end{array}\right]}{dx_a} = \left[\begin{array}{ccc}1 & 0 & 0\\ 0 & 0 & 1\end{array}\right]$$

and $\mathrm{rank}(\Sigma_{afM}(x_a)) = 2, \forall x_a \in \overline{S}_{afM}$.

The orthogonality condition at the final time is now

$$\left[\begin{array}{c}\lambda_1(t_f)\\ \lambda_2(t_f)\\ \lambda_3(t_f)\end{array}\right] = \left[\begin{array}{c}\lambda_1(t_f)\\ \lambda_2(t_f)\\ \lambda_3(t_0)\end{array}\right] = \vartheta_{afM_1}\frac{d\alpha_{afM_1}(x_a)}{dx_a}\bigg|'_{x_a = x_a(t_f)} +$$

$$+\vartheta_{afM_2}\frac{d\alpha_{afM_2}(x_a)}{dx_a}\bigg|'_{x_a = x_a(t_f)} = \left[\begin{array}{c}\vartheta_{afM_1}\\ 0\\ \vartheta_{afM_2}\end{array}\right]$$

and supplies $\lambda_2(t_f) = 0$. Therefore the transversality condition at the final time is

$$H(t_f) = t_f^2 + \lambda_1(t_f)x_2(t_f) + \lambda_2(t_f)u(t_f) + \lambda_3(t_f)\frac{u^2(t_f)}{2} =$$

$$= t_f^2 + \lambda_1(t_f)x_2(t_f) - \frac{\lambda_2^2(t_f)}{2\lambda_3(t_0)} = t_f^2 + \lambda_1(t_f)x_2(t_f) = 0$$

where we have taken into account the expression of u_h. This condition is the same for the two cases presented below.

Problem 10.10-2a We ignore the constraint on $x_2(\cdot)$, namely $w(\cdot, \cdot, \cdot) = 0$
We obtain

$$u(t) = \frac{1}{\lambda_{30}}(\lambda_{10}t - \lambda_{20})$$

$$x_1(t) = \frac{1}{\lambda_{30}}\left(\frac{\lambda_{10}}{6}t^3 - \frac{\lambda_{20}}{2}t^2\right)$$

$$x_2(t) = \frac{1}{\lambda_{30}}\left(\frac{\lambda_{10}}{2}t^2 - \lambda_{20}t\right)$$

$$x_3(t) = \frac{1}{6\lambda_{10}\lambda_{30}^2}\left((\lambda_{10}t - \lambda_{20})^3 + \lambda_{20}^3\right)$$

where $\lambda_{i0} := \lambda_i(0)$, $i = 1, 2, 3$, The three relations $x_1(t_f) = x_{1f}$, $x_3(t_f) = \beta$, $\lambda_2(t_f) = 0$ together with the transversality condition constitute a system of four equations for the four unknown parameters λ_{i0}, $i = 1, 2, 3$ and t_f. Its solution supplies

Fig. 10.22 Problem
10.10-2. State motion when
$\beta = 4$ and the constraint
$x_2(\cdot) \le 2$ is ignored (*solid
line*) or it is taken into
account (*dashed line*)

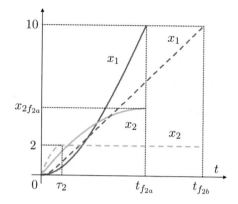

$$\lambda_{10} = \lambda_{10_{2a}} := -\frac{x_{1f}}{\beta}, \quad t_f = t_{f_{2a}} := \sqrt[3]{\frac{3x_{1f}^2}{2\beta}}, \quad \lambda_{20} = \lambda_{20_{2a}} := \lambda_{10_{2a}} t_{f_{2a}},$$

$$\lambda_{30} = \lambda_{30_{2a}} := \frac{\lambda_{10_{2a}}^2}{2}$$

Notice that $\lambda_{10} \ne 0$ and therefore $\lambda_{30} > 0$. We can now compute the second state
variable at the final time

$$x_2(t_{f_{2a}}) = x_{2f_{2a}} = -\frac{\lambda_{10_{2a}}}{2\lambda_{30_{2a}}} t_{f_{2a}}^2$$

The four parameters and $x_{2f_{2a}}$ are given in Table 10.3, corresponding to $x_{1f} = 10$ and
to two values of β. The relevant state motions are plotted in Figs. 10.22 and 10.23
(solid lines). The two values of β (4 and 12) have been chosen so that one of them is
smaller and the other one is greater than

Fig. 10.23 Problem
10.10-2. State motion when
$\beta = 12$ and the constraint
$x_2(\cdot) \le 2$ is ignored (*solid
line*) or it is taken into
account (*dashed line*)

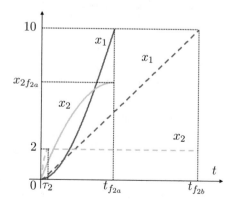

$$\overline{\beta} := \int_0^{t_{f_{1a}}} \frac{u_{1a}^2(t)}{2} \, dt = 7.07$$

where $u_{1a}(\cdot)$ is the control which results from the solution of Problem 10.10-1a.

Problem 10.10-**2b** The constraint on $x_2(\cdot)$ is taken into account, namely $w(x(t), u(t), t) \le 0$

The solution of Problem 10.10-2a suggests that the constraint is binding over an interval T_τ, beginning at τ, whenever $x_{2f_{2a}} > 2$. In fact Fig. 10.22 shows that, when the constraint is ignored, the maximum of $x_2(\cdot)$ is attained at $t = t_{f_{2a}}$ or, by exploiting what has been previously found, whenever

$$x_{1f} > \frac{32}{9\beta}$$

Assume that this inequality holds. The function $w(\cdot, \cdot, \cdot)$ does not explicitly depend on u, so that we proceed to its total differentiation with respect to time and obtain

$$\frac{dw(x(t), u(t), t)}{dt} = \widehat{w}(x(t), u(t), t) = u(t)$$

Therefore the problem at hand includes the following two constraints

$$\widehat{w}(x(t), u(t), t) \le 0, \quad \widehat{w}(x, u, t) := u, \quad t \in T_\tau$$
$$w(x(\tau), u(\tau), \tau) = 0, \quad \tau \text{ free}$$

Thus $u(\cdot) = 0$ when the inequality constraint is binding and, in view of Eq. (2.10), $\lambda_2(\cdot)$ can be discontinuous at τ because $w(\cdot, \cdot, \cdot)$ explicitly depends on x_2. On the contrary, Eqs. (2.10) and (2.11) imply that $\lambda_1(\cdot)$, $\lambda_3(\cdot)$ and the hamiltonian function are continuous at τ because $w(\cdot, \cdot, \cdot)$ explicitly depends neither on x_1 nor on x_3 nor on t and τ is free. Therefore $\lambda_1(\cdot) = \lambda_{10}$ and

$$H^-(\tau) = \tau^2 + \lambda_{10} x_2(\tau) + \lambda_2^-(\tau) u^-(\tau) + \lambda_{30} \frac{(u^-(\tau))^2}{2} =$$
$$= \tau^2 + 2\lambda_{10} - \frac{(\lambda_2^-(\tau))^2}{2\lambda_{30}} = \tau^2 + \lambda_{10} x_2(\tau) + \lambda_2^+(\tau) u^+(\tau) + \lambda_{30} \frac{(u^+(\tau))^2}{2} =$$
$$= \tau^2 + 2\lambda_{10} - \frac{(\lambda_2^+(\tau))^2}{2\lambda_{30}} = H^+(\tau)$$

where we have exploited the notation (9.1). From this equation we deduce that

$$\lambda_2^-(\tau) = \lambda_{20} - \lambda_{10}\tau = 0$$

At the end of the interval T_τ (where the constraint is binding) the functions $\lambda_2(\cdot)$ and $u(\cdot)$ are continuous. There the control is zero and equal to $-\lambda_2$, so that also $\lambda_2(\cdot)$ is zero at that time. On the other hand, $\lambda_2(\cdot)$ must be zero also at the final time (recall the orthogonality condition) so that we can assume that the constraint is binding $\forall t \geq \tau$. Therefore we have

$0 \leq t < \tau$

$$u(t) = \frac{\lambda_{10}t - \lambda_{20}}{\lambda_{30}}$$

$\tau < t \leq t_f$

$$u(t) = 0$$

$0 \leq t \leq \tau$

$$x_1(t) = \frac{\lambda_{10}t^3 - 3\lambda_{20}t^2}{6\lambda_{30}}$$

$$x_2(t) = \frac{\lambda_{10}t^2 - 2\lambda_{20}t}{2\lambda_{30}}$$

$$x_3(t) = \frac{(\lambda_{10}t - \lambda_{20})^3 + \lambda_{20}^3}{6\lambda_{10}\lambda_{30}^2}$$

$\tau \leq t \leq t_f$

$$x_1(t) = \frac{\lambda_{10}\tau^3 - 3\lambda_{20}\tau^2}{6\lambda_{30}} + 2(t - \tau)$$

$$x_2(t) = 2$$

$$x_3(t) = \frac{(\lambda_{10}\tau - \lambda_{20})^3 + \lambda_{20}^3}{6\lambda_{10}\lambda_{30}^2}$$

In view of the above results, the five parameters $\lambda_{10}, \lambda_{20}, \lambda_{30}, \tau, t_f$ which define the solution can be computed by solving the system of five equations

$$x_2(\tau) = 2$$
$$\lambda_2^-(\tau) = 0$$
$$x_1(t_f) = x_{1f}$$
$$x_3(t_f) = \beta$$
$$H(t_f) = t_f^2 + 2\lambda_{10} = 0$$

More explicitly, we have

$$\tau = \tau_2 := \frac{8}{3\beta}, \quad t_f = t_{f_{2b}} := \frac{3x_{1f} + 2\tau_2}{6}, \quad \lambda_{10} = \lambda_{10_{2b}} := -\frac{t_{f_{2b}}^2}{2}$$

$$\lambda_{20} = \lambda_{20_{2b}} := -\frac{\tau_2 t_{f_{2b}}^2}{2}, \quad \lambda_{30} = \lambda_{30_{2b}} := \frac{\tau_2^2 t_{f_{2b}}^2}{8}$$

The resulting values are shown in Table 10.3. The state motions corresponding to $\beta = 4$ and $\beta = 12$ are plotted in Figs. 10.22 and 10.23 (dashed lines).

The following remarks hold both when the punctual and global inequality constraint is ignored and when it is taken into account. First, the final time and the performance index decrease as β increases. Second, when $\beta = 4$ ($\beta = 12$) the final time is greater (smaller) than that of Problem 10.10-1. This is consistent with the fact that $4 < \bar{\beta}$ ($12 > \bar{\beta}$).   ∎

Chapter 11
Singular Arcs

Abstract Here we focus the attention on the issues raised by the presence of the so-called singular arcs. Accordingly, we always have a linear hamiltonian function and constraints on the absolute value of the control variable. Only simple constraints are present in all considered problems but in the second one where also a punctual and isolated constraint on the state variables is included.

In this chapter we focus the attention on the issues raised by the presence of the so-called singular arcs. We have presented them in Sect. 2.4 together with a way of handling them. Accordingly, we always have a linear hamiltonian function (a requirement for the existence of singular arcs) and constraints on the absolute value of the control variable. Furthermore the forthcoming problems exhibit the following features: (1) the final time is free and the performance indexes are equal in Problems 11.1 and 11.2; (2) the initial state is given and the final one is the origin in Problem 11.1, where no other constraints have to be fulfilled; (3) both the initial and final state are given in Problem 11.2 where also a punctual and isolated constraint on the state variables is included; (4) the final time and both the initial and final state are given in Problem 11.3; (5) the length of the control interval is free also in Problem 11.4, where the performance index includes a term which is a function of the final state; (6) the solution of Problem 11.2 is compared with that of a problem which has significantly different constraints on the control and state variables: despite of this, the solutions of them are substantially similar.

The discussion of the forthcoming problems is mainly based on Sect. 2.4, but it is expedient to recall the content of Sect. 2.2 together with the material in Sects. 2.3.3, 2.3.4 and the relevant problems presented in Chaps. 9 and 10.

Problem 11.1
Let

$$x(0) = x_0, \quad x(t_f) = \begin{bmatrix} 0 \\ 0 \end{bmatrix}$$

$$J = \int_0^{t_f} l(x(t), u(t), t) \, dt, \quad l(x, u, t) = \frac{1 + x_2^2}{2}$$

$$u(t) \in \overline{U}, \ \forall t, \ \overline{U} = \{u \mid -1 \le u \le 1\}$$

© Springer International Publishing Switzerland 2017
A. Locatelli, *Optimal Control of a Double Integrator*, Studies in Systems, Decision and Control 68, DOI 10.1007/978-3-319-42126-1_11

where the initial state x_0 is given, whereas the final time t_f is free.

The hamiltonian function is

$$H = \frac{1}{2} + \frac{x_2^2}{2} + \lambda_1 x_2 + \lambda_2 u$$

so that Eq. (2.5a) is

$$\dot{\lambda}_1(t) = 0$$
$$\dot{\lambda}_2(t) = -\lambda_1(t) - x_2(t)$$

and its solution over an interval beginning at time 0 gives $\lambda_1(t) = \lambda_1(0)$. When $\lambda_2 \neq 0$ the H-minimizing control is $u_h = -\text{sign}(\lambda_2)$. Nevertheless, we can not exclude that $\lambda_2(\cdot) = 0$ over an interval $[t_1, t_2]$ of non-zero length. In fact within such an interval it can well happen that

$$a(\lambda(t)) := \lambda_2(t) = 0$$
$$\frac{da(\lambda(t))}{dt} = \frac{d\lambda_2(t)}{dt} = -\lambda_1(0) - x_2(t) = 0$$
$$\frac{d^2 a(\lambda(t))}{dt^2} = \frac{d^2 \lambda_2(t)}{dt^2} = -\frac{d\lambda_1(t)}{dt} - \frac{dx_2(t)}{dt} = -u(t) = 0$$

where we have taken into account that $\lambda_1(\cdot) = \lambda_1(0)$. The conditions of Theorem 2.9 are satisfied: therefore if the solution includes a singular arc, the control can take on the three values $-1, 0, 1$, whereas, if the arc is not present, the value 0 must not be taken into consideration.

We notice that the problem is time-invariant, so that the transversality condition at the final time can be written at any time within the interval $[t_1, t_2]$. If the control is singular over such an interval (obviously assuming that it exists), we have

$$H(t) = \frac{1}{2} + \frac{x_2^2(t)}{2} + \lambda_1(t)x_2(t) + \lambda_2(t)u(t) = \frac{1}{2} - \frac{x_2^2(t)}{2} = 0$$

where it has been taken into account that $x_2(t) = -\lambda_1(0)$, $t \in [t_1, t_2]$. Therefore two singular arcs can exist and they are characterized by $x_2(t) = \pm 1$, $t \in [t_1, t_2]$.

Now we make reference to Fig. 11.1 where various state trajectories are plotted. They correspond to the above mentioned values of $u(\cdot)$ and are the curves defined by the equations (the check is straightforward)

$$\begin{cases} x_1 = -\dfrac{x_2^2}{2} + k_1, & \text{if } u(\cdot) = -1 \\ x_2 = k_2, & \text{if } u(\cdot) = 0 \\ x_1 = \dfrac{x_2^2}{2} + k_3, & \text{if } u(\cdot) = 1 \end{cases}$$

where k_i, $i = 1, 2, 3$, are constants.

Fig. 11.1 Problem 11.1.
State trajectories
corresponding to $u(\cdot) = 1$
(*dash-double-dotted line*),
$u(\cdot) = 0$ (*solid line*),
$u(\cdot) = -1$
(*dash-single-dotted line*)

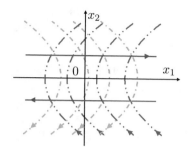

We conclude that the origin (the final state) can be reached only if during the last part of the control interval we have $u(\cdot) = 1$ or $u(\cdot) = -1$. The symmetry characteristic of the problem, in particular the one concerning possible singular arcs, allows us to only discuss the case where, at the end, the control is $u(\cdot) = 1$, so that the final part of the trajectory is a piece of the curve which passes trough the points $(0, a_4, a_1, a_6)$ and is defined by the equation $x_1 = \dfrac{x_2^2}{2}$ (see Fig. 11.2).

In order to identify the trajectories, among those shown in Fig. 11.2, which satisfy the NC, it is expedient letting the time flow in reverse direction starting from the final time t_f. Consistently, we define $\tau := t_f - t$ and add a tilde to the symbols of the involved functions, so that their dependence on the new variable τ is emphasized.

We notice that the sign of the derivatives with respect to τ of odd order is opposite to the one of the derivatives with respect to t of the same order. We also notice that the transversality condition at $\tau = 0$ is

$$\tilde{H}(0) = \frac{1}{2} + \frac{\tilde{x}_2^2(0)}{2} + \tilde{\lambda}_1(0)\tilde{x}_2(0) + \tilde{\lambda}_2(0)\tilde{u}(0) = \frac{1}{2} + \tilde{\lambda}_2(0) = 0$$

and requires $\tilde{\lambda}_2(0) = -\dfrac{1}{2}$, independently of the considered trajectory.

Fig. 11.2 Problem 11.1. The
NC are satisfied by the
trajectories $(a_{31}\text{-}a_2\text{-}a_1\text{-}0,$
$a_{32}\text{-}a_2\text{-}a_1\text{-}0), (a_5\text{-}a_4\text{-}0,$
$a_6\text{-}a_1\text{-}0)$, whereas they are
not satisfied by the trajectory
$(a_7\text{-}a_6\text{-}0)$

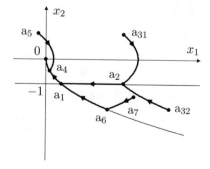

We now discuss the various *families* of trajectories which end at the origin (see Fig. 11.2). We say that two trajectories belong to the same family if they have the same shape.

Family 1. Trajectories like $(a_{31}\text{-}a_2\text{-}a_1\text{-}0)$

Let τ_1 and τ_2 be the times when the state is at point a_1 and a_2, respectively. The trajectory at hand satisfies the NC if

$$
\begin{array}{lll}
0 \leq \tau < \tau_1 & \tau_1 \leq \tau \leq \tau_2 & \tau > \tau_2 \\
\tilde\lambda_2(\tau) < 0 & \tilde\lambda_2(\tau) = 0 & \tilde\lambda_2(\tau) > 0 \\
\tilde x_2(\tau) = -\tau & \tilde x_2(\tau) = -1 & \tilde x_2(\tau) = -1 + (\tau - \tau_2)
\end{array}
$$

We notice that $\tau_1 = 1$ because $\tilde x_2(\tau) = -\tau$. Furthermore the trajectory includes a singular arc $(a_1\text{-}a_2)$ if

$$
\tilde\lambda_1(\tau_1) = -\tilde x_2(\tau_1) = 1
$$

When $0 \leq \tau \leq \tau_1$, we have

$$
\frac{d\tilde\lambda_2(\tau)}{d\tau} = \tilde\lambda_1(\tau) + \tilde x_2(\tau) = \tilde\lambda_1(0) + \tilde x_2(\tau)1 - \tau
$$

so that

$$
\tilde\lambda_2(\tau) = \tilde\lambda_2(0) + \tau - \frac{\tau^2}{2} = -\frac{1}{2} + \tau - \frac{\tau^2}{2}
$$

and also

$$
\tilde\lambda_2(\tau) \leq 0, \quad 0 \leq \tau \leq \tau_1
$$

Finally, when $\tau \geq \tau_2$, we have

$$
\frac{d\tilde\lambda_2(\tau)}{d\tau} = \tilde\lambda_1(\tau) + \tilde x_2(\tau) = 1 + \tilde x_2(\tau_2) + \tau - \tau_2 = \tau - \tau_2 \geq 0
$$

which implies $\tilde\lambda_2(\tau) \geq 0$, because $\tilde\lambda_2(\tau_2) = 0$.

Family 2. Trajectories like $(a_{32}\text{-}a_2\text{-}a_1\text{-}0)$

A procedure similar to the one exploited with reference to *Family* 1 leads to the conclusion that the trajectories of this *family* satisfy the NC.

Family 3. Trajectories like $(a_5\text{-}a_4\text{-}0)$

Let $\tau_4 < 1$ be the time when the state is at point a_4 (recall that $\tilde x_2(\tau) = -\tau, \tau \leq \tau_4$). As we have already found

$$
\frac{d\tilde\lambda_2(\tau)}{d\tau} = \tilde\lambda_1(0) + \tilde x_2(\tau)
$$

so that, in view of the previously computed value of $\tilde{\lambda}_2(0)$, we have

$$\tilde{\lambda}_2(\tau) = -\frac{1}{2} + \tilde{\lambda}_1(0)\tau - \frac{\tau^2}{2}$$

We get

$$\tilde{\lambda}_1(0) = \frac{\tau_4^2 + 1}{2\tau_4}$$

because $\tilde{\lambda}_2(\tau_4) = 0$ and, by recalling that $\tau_4 < 1$, we easily ascertain that

$$\left.\frac{d\tilde{\lambda}(\tau)}{d\tau}\right|_{\tau=\tau_4+d\tau} > 0$$

Therefore

$$\tau = \tau_4 + d\tau \Rightarrow \tilde{\lambda}_2(\tau) > 0 \Rightarrow \tilde{u}(\tau) = -1 \Rightarrow \tilde{x}_2(\tau) = -\tau_4 + (\tau - \tau_4)$$

and

$$\tilde{\lambda}_2(\tau) = \frac{(1-\tau_4)(\tau-\tau_4) + \tau_4(\tau-\tau_4)^2}{2\tau_4} > 0$$

We conclude that $\tilde{\lambda}_2$ can not change sign another time and the trajectories of this *family* satisfy the NC.

Family 4. Trajectories like (a₇-a₆-0)

Let $\tau_6 > 1$ be the time when the state is at the point a_6 and recall that along the trajectory a_6-0 we have $\tilde{x}_2(\tau) = -\tau$. Furthermore

$$\tilde{\lambda}_2(\tau) = -\frac{1}{2} + \tilde{\lambda}_1(0)\tau - \frac{\tau^2}{2}, \quad \tilde{\lambda}_2(\tau_6) = 0$$

where the second equation is the switching condition for the control to commute from -1 to 1. We obtain

$$\tau_6 = \tilde{\lambda}_1(0) \pm \sqrt{\tilde{\lambda}_1(0)^2 - 1}$$

and conclude that $\tilde{\lambda}_1(0) \geq 1$ because, otherwise, negative or complex values of τ_6 result. The smaller of the two admissible values of τ_6 is less than 1, so that the switching takes place prior to point a_1. The considered *family* of trajectories does not satisfy the NC.

In summary, at each time the value of the control variable depends on the values of the two state variables at the same time only and the solution can be implemented in a closed-loop form, that is by resorting to a regulator which supplies $u(t)$ as a

Fig. 11.3 Problem 11.1. The switching curve ϱ_1-B-0-C-ϱ_2 which defines the control law $u(x)$

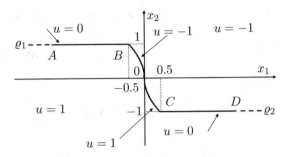

function of $x(t)$, as shown in Fig. 11.3. More in detail, we refer to the curve $A - B - C - D$ which appears in the quoted figure and is usually called *switching curve*. It is composed by the two half-straight-lines ϱ_1 and ϱ_2 passing through the points A, B and C, D, respectively, and by the two pieces B-0 and 0-C of the state trajectories through the origin which correspond to $u(\cdot) = -1$ and to $u(\cdot) = 1$, respectively. The two half-straight-lines are specified by the equations

$$\varrho_1 : \begin{cases} x_2 = 1 \\ x_1 \leq -\dfrac{1}{2} \end{cases} \qquad \varrho_2 : \begin{cases} x_2 = -1 \\ x_1 \geq \dfrac{1}{2} \end{cases}$$

If at time t the state is located above the switching curve or lies on its B-0 part, then $u(t) - 1$, if the state is located below the switching curve or lies on its 0-C part, then $u(t) = 1$, if the state lies on the half-straight-line ϱ_1 or on the half-straight-line ϱ_2, then $u(t) = 0$. These rules define the control law $u(x)$ which generates the solutions of the problem.   ∎

Problem 11.2
We now consider two problems which are stated in a substantially different way. In spite of this, their solutions deserve to be compared because they lead to a fairly similar behavior of the controlled system.

Problem 11.2-1
Let

$$x(0) = \begin{bmatrix} 2 \\ 0 \end{bmatrix}, \quad x(t_f) = \begin{bmatrix} 0 \\ 0 \end{bmatrix}, \quad J = \int_0^{t_f} l(x(t), u(t), t)\, dt, \quad l(x, u, t) = \frac{1 + x_2^2}{2}$$

$$w(x(\tau), \tau) = 0, \quad w(x, \tau) = \begin{bmatrix} x_1 - 5 \\ x_2 \end{bmatrix}, \quad u(t) \in \overline{U}, \ \forall t, \ \overline{U} = \{u | -1 \leq u \leq 1\}$$

where the intermediate and final times τ and t_f, $\tau < t_f$, are free.

A little thought allows us to conclude that the problem can legitimately be split into two very similar subproblems. The first of them (Problem 11.2-1a) refers to the time interval $0 \leq t \leq \tau$ whereas, the second one (Problem 11.2-1b) refers to the time interval $\tau \leq t \leq t_f$. In fact the initial and final state are given for both subproblems, the final time is free and the performance index does not explicitly depends on time. This is why we discuss in detail Problem 11.2-1a only and supply a less deep analysis of Problem 11.2-1b.

Problem 11.2-1a

We have

$$x(0) = \begin{bmatrix} 2 \\ 0 \end{bmatrix}, \ x(\tau) = \begin{bmatrix} 5 \\ 0 \end{bmatrix}, \ J = \int_0^\tau l(x(t), u(t), t) \, dt, \ l(x, u, t) = \frac{1 + x_2^2}{2}$$

$$u(t) \in \overline{U}, \ \forall t < \tau, \ \overline{U} = \{u \mid -1 \leq u \leq 1\}$$

where the final time τ is free. The hamiltonian function is

$$H = \frac{1}{2} + \frac{x_2^2}{2} + \lambda_1 x_2 + \lambda_2 u$$

so that Eq. (2.5a) is

$$\dot{\lambda}_1(t) = 0$$
$$\dot{\lambda}_2(t) = -\lambda_1(t) - x_2(t)$$

and its solution over an interval beginning at time 0 gives $\lambda_1(t) = \lambda_1(0)$. Furthermore the transversality condition at the final time, written at the initial time because the problem is time-invariant,

$$H(0) = \frac{1}{2} + \frac{x_2^2(0)}{2} + \lambda_1(0)x_2(0) + \lambda_2(0)u(0) = \frac{1}{2} + \lambda_2(0)u(0)$$

implies $\lambda_2(0) = \pm\frac{1}{2}$.

If $\lambda_2 \neq 0$ the H-minimizing control is $u_h = -\text{sign}(\lambda_2)$. Nevertheless we can not exclude that $\lambda_2(\cdot) = 0$ over an interval T_s of non-zero length. In fact within such an interval it can well happen that

$$a(\lambda(t)) := \lambda_2(t) = 0$$
$$\frac{da(\lambda(t))}{dt} = \frac{d\lambda_2(t)}{dt} = -\lambda_1(0) - x_2(t) = 0$$
$$\frac{d^2a(\lambda(t))}{dt^2} = \frac{d^2\lambda_2(t)}{dt^2} = -\frac{d\lambda_1(t)}{dt} - \frac{dx_2(t)}{dt} = -u(t) = 0$$

where we have taken into account that $\lambda_1(\cdot) = \lambda_1(0)$. The conditions of Theorem 2.9 are satisfied: therefore if the solution includes a singular arc, the control can take on the three values $-1, 0, 1$, whereas, if the arc is not present, the value 0 must not be taken into consideration. As we have already noticed, the problem is time-invariant, so that the transversality condition at the final time can be written at any time within the interval T_s. If the control is singular over such an interval (obviously assuming that it exists), we have

$$H(t) = \frac{1}{2} + \frac{x_2^2(t)}{2} + \lambda_1(t)x_2(t) + \lambda_2(t)u(t) = \frac{1}{2} - \frac{x_2^2(t)}{2} = 0$$

where it has been taken into account that $x_2(t) = -\lambda_1(0)$, $t \in T_s$. Therefore two singular arcs can exist and they correspond to $x_2(t) = \pm 1$, $t \in T_s$.

If we look at the state trajectory relevant to the above values of the control variable (see Fig. 11.1), we conclude that the singular arc to be taken into consideration is $x_2(t) = 1$, $t \in T_s$ because x_1 must increase, so that $u(t) = 1$, at least at the beginning of the control interval.

Assume that the singular solution occurs within the interval $T_s := [\tau_1, \tau_2]$. Again Fig. 11.1 suggests that the control is

$$u(t) = \begin{cases} 1, & 0 \le t < \tau_1 \\ 0, & \tau_1 < t < \tau_2 \\ -1, & \tau_2 < t \le \tau \end{cases}$$

Consequently, the functions $\lambda_2(\cdot)$, $x_1(\cdot)$, $x_2(\cdot)$ are

$$t \in [0, \tau_1] \quad \begin{cases} \lambda_2(t) = -\dfrac{1 - 2t + t^2}{2} \\ x_1(t) = \dfrac{4 + t^2}{2} \\ x_2(t) = t \end{cases}$$

$$t \in [\tau_1, \tau_2] \quad \begin{cases} \lambda_2(t) = 0 \\ x_1(t) = x_1(\tau_1) + (t - \tau_1) \\ x_2(t) = 1 \end{cases}$$

$$t \in [\tau_2, \tau] \quad \begin{cases} \lambda_2(t) = \dfrac{(t - \tau_2)^2}{2} \\ x_1(t) = x_1(\tau_2) + (t - \tau_2) + \dfrac{(t - \tau_2)^2}{2} \\ x_2(t) = 1 - t + \tau_2 \end{cases}$$

where, as far as $\lambda_2(\cdot)$ is concerned, we have taken into account that $x_2(\tau_1) = \tau_1 = 1$, $\lambda_2(0) = -\dfrac{1}{2}$, $\lambda_1(0) = -x_2(\tau_1) = -\tau_1$. By enforcing feasibility, namely $x_2(\tau) = 0$, $x_1(\tau) = 5$, we find

$$\tau_2 = 3, \quad \tau = 4, \quad J_s = \frac{10}{3}$$

where J_s is the value of the performance index. The state trajectory (curve (s-a-b-r)) is plotted in Fig. 11.4, whereas the function $\lambda_2(t)$, $t \in [0, \tau]$, is shown in Fig. 11.5.

Fig. 11.4 Problem 11.2-1.
The state trajectory. The
pieces *a-b* and *c-d*
correspond to the two
singular arcs

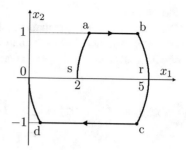

Fig. 11.5 Problem 11.2-1.
The function $\lambda_2(\cdot)$

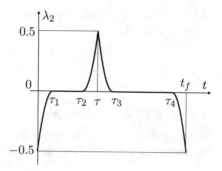

In order to complete the discussion of the problem, we also consider the possible existence of a solution which satisfies the NC and requires $a(\lambda(t)) = 0$ at isolated instants of time only.

Recall that $\lambda_2(0) = -\dfrac{1}{2}$ (transversality condition at $t = 0$), because, initially, the control is positive. Thus, at the beginning of the control interval, we have

$$x_2(t) = t, \quad \lambda_2(t) = -\frac{1}{2} - \lambda_1(0)t - \frac{t^2}{2}$$

and $\lambda_2(\cdot)$ can be zero at most twice because it is a parabola with the concavity pointing downwards. On the other hand, $\lambda_2(\cdot)$ must be positive at the end of the control interval (the control must be negative) so that at most only one time τ_c can exist where $\lambda_2(\tau_c) = 0$. Again from Fig. 11.1, we conclude that feasibility requires $x_1(\tau_c) = \dfrac{7}{2}$ and easily find

$$\tau = 2\tau_c = 2\sqrt{3}, \quad \lambda_1(0) = -\frac{2\sqrt{3}}{3}, \quad J_{ns} = 2\sqrt{3}$$

where J_{ns} is the value of the performance index.

Thus we have found a second solution which satisfies the NC. However the first one (which includes the singular arc) must be preferred because $J_s < J_{ns}$.

Problem 11.2-1b

We proceed in the same manner adopted in dealing with the first subproblem. Temporarily, we consider the problem which makes reference to the time t^* ranging over the interval $0 \leq t^* \leq t_f^*$. We have

$$x^*(0) = \begin{bmatrix} 5 \\ 0 \end{bmatrix}, \quad x(t_f^*) = \begin{bmatrix} 0 \\ 0 \end{bmatrix}, \quad J = \int_0^{t_f^*} l(x(t), u(t), t)\, dt^*, \quad l(x, u, t) = \frac{1 + (x_2^*)^2}{2}$$

$$u(t^*) \in \overline{U}, \ 0 < t^* \leq t_f^*, \ \overline{U} = \{u | -1 \leq u \leq 1\}$$

The transversality condition at the final time, written at the initial time because the problem is time-invariant,

$$H^*(0) = \frac{1}{2} + \frac{(x_2^*)^2(0)}{2} + \lambda_1^*(0)x_2^*(0) + \lambda_2^*(0)u^*(0) = \frac{1}{2} + \lambda_2^*(0)u^*(0) = 0$$

supplies $\lambda_2^*(0) = \dfrac{1}{2}$, being quite obvious that the control must be negative at the beginning. We ascertain that a singular arc corresponding to $x_2^*(t^*) = -1$ can exist within the interval $[\tau_3^*, \tau_4^*]$, so that the control function is

$$u^*(t^*) = \begin{cases} -1, & 0 \leq t^* < \tau_3^* \\ 0, & \tau_3^* < t^* < \tau_4^* \\ 1, & \tau_4^* < t^* < t_f^* \end{cases}$$

and, consequently, the functions $\lambda_2^*(\cdot)$ and $x_i^*(\cdot)$, $i = 1, 2$ have the form

$$t^* \in [0, \tau_3^*] \quad \begin{cases} \lambda_2^*(t^*) = \dfrac{1 - 2t^* + (t^*)^2}{2} \\[2mm] x_1^*(t^*) = \dfrac{10 - (t^*)^2}{2} \\[2mm] x_2^*(t^*) = -t^* \end{cases}$$

$$t^* \in [\tau_3^*, \tau_4^*] \quad \begin{cases} \lambda_2^*(t^*) = 0 \\ x_1^*(t^*) = x_1^*(\tau_3^*) - (t^* - \tau_3^*) \\ x_2^*(t^*) = -1 \end{cases}$$

$$t^* \in [\tau_4^*, t_f^*] \quad \begin{cases} \lambda_2^*(t^*) = -\dfrac{(t^* - \tau_4^*)^2}{2} \\[2mm] x_1^*(t^*) = x_1^*(\tau_4^*) - (t^* - \tau_4^*) + \dfrac{(t^* - \tau_4^*)^2}{2} \\[2mm] x_2^*(t^*) = -1 + (t^* - \tau_4^*) \end{cases}$$

where, with reference to $\lambda_2^*(\cdot)$, we have taken into account that $x_2^*(\tau_3^*) = -\tau_3^* = -1$, $\lambda_2^*(0) = \dfrac{1}{2}$, $\lambda_1^*(0) = -x_2^*(\tau_3^*) = \tau_3^*$. By enforcing feasibility, namely $x_2^*(t_f^*) = 0$, $x_1^*(t_f^*) = 0$, we get $\tau_4^* = 5$, $t_f^* = 6$.

Coming back to the time t we see that a singular arc can exist over the intervals $[\tau_3, \tau_4]$ and the final time is t_f, where

$$\tau_3 = \tau_3^* + \tau = 5, \quad \tau_4 = \tau_4^* + \tau = 9, \quad t_f = t_f^* + \tau = 10$$

If we restrain the attention to the time interval $[\tau, t_f]$, we have the state trajectory (curve r-c-d-0) shown in Fig. 11.4, whereas $\lambda_2(\cdot)$ is plotted in Fig. 11.5.

Also for this problem we can easily compute a different solution which satisfies the NC and is such that $a(\lambda(t)) = 0$ at isolated times only. However, as we have found for Problem 11.2-1b, this solution should not be adopted because the value of the performance index is larger.

Finally, we notice that we can deal with Problem 11.2-1 without partitioning it into two subproblems, provided that the material of Chap. 9 is taken into account. In so doing, we ascertain, in particular, the discontinuity of $\lambda_1(\cdot)$ at τ where, on the contrary, the hamiltonian function is continuous. We also observe that $\lambda_2(\cdot)$ is continuous at $t = \tau$ as well, in spite of the fact that its discontinuity is allowable.

The comparison of the solution above with the one of the forthcoming Problem 11.2-2 deserves attention because the corresponding behavior of the state trajectory is similar, although no singular arcs are included in it.

Problem 11.2-2

Some data are equal to those of Problem 11.2-1. More specifically, the initial and final states together with the constraint on the state at the intermediate time τ_v are the same. On the contrary, we now have a different performance index, a punctual and global inequality constraint and no limitations on the control variable. Consistently, we have

$$x(0) = \begin{bmatrix} 0 \\ 0 \end{bmatrix}, \quad x(t_f) = \begin{bmatrix} 0 \\ 0 \end{bmatrix}$$

$$w^{(1)}(x(\tau_v), \tau_v) = 0, \quad w^{(1)}(x, \tau) = \begin{bmatrix} x_1 - 5 \\ x_2 \end{bmatrix}$$

$$J = \int_0^{t_{fv}} l(x(t), u(t), t)\, dt, \quad l(x, u, t) = \frac{1 + u^2}{2}, \quad u(t) \in R, \ \forall t$$

$$w^{(2)}(x(t), u(t), t) \le 0, \ \forall t, \quad w^{(2)}(x, u, t) = \begin{bmatrix} -x_2 - 1 \\ x_2 - 1 \end{bmatrix}$$

where τ_v and t_{fv}, $\tau_v < t_{fv}$ are free. We first notice that the punctual and global inequality constraints are satisfied if and only if $|x_2(\cdot)| \le 1$ and that, also in this case, it is expedient splitting the problem into two subproblems, Problems 11.2-2a and 11.2-2b. The former concerns the time interval $0 \le t < \tau_v$, whereas the latter concerns the time interval $\tau_v < t \leq t_{fv}$. Both problems have the initial and final state given and the final time free.

Problem 11.2-2a

We have

$$x(0) = \begin{bmatrix} 2 \\ 0 \end{bmatrix}, \quad x(\tau_v) = \begin{bmatrix} 5 \\ 0 \end{bmatrix}$$

$$J = \int_0^{\tau_v} l(x(t), u(t), t) \, dt, \quad l(x, u, t) = \frac{1 + u^2}{2}, \quad u(t) \in R, \quad \forall t$$

$$w^{(2)}(x(t), u(t), t) \leq 0, \quad \forall t, \quad w^{(2)}(x, u, t) = \begin{bmatrix} -x_2 - 1 \\ x_2 - 1 \end{bmatrix}$$

The hamiltonian function, the H-minimizing control and Eq. (2.5a) are

$$H = \frac{1}{2} + \frac{u^2}{2} + \lambda_1 x_2 + \lambda_2 u, \quad u_h = -\lambda_2, \quad \begin{array}{l} \dot{\lambda}_1(t) = 0 \\ \dot{\lambda}_2(t) = -\lambda_1(t) \end{array}$$

It is apparent (see the values of the initial and final states) that we must consider the second component of $w^{(2)}(\cdot, \cdot, \cdot)$, so that in this subproblem we simply have $w_2^{(2)}(x(t), u(t), t) \leq 0, w_2^{(2)}(x, u, t) = x_2 - 1$. By recalling the material of Sect. 10.2 and, in particular, of Sect. 2.3.4, we first compute the total derivative of $w_2(\cdot, \cdot, \cdot)$ with respect to time because this function does not explicitly depend u. We obtain

$$\frac{dw_2^{(2)}(x(t), u(t), t)}{dt} = \widehat{w}_2^{(2)}(x(t), u(t), t) = u(t)$$

Therefore if we want to ascertain whether the constraint is binding over the interval $[\tau_{1v}, \tau_{2v}], 0 < \tau_{1v} \leq \tau_{2v} < \tau_v$ we can consider the problem with

$$\widehat{w}_2^{(2)}(x(t), u(t), t) \leq 0, \quad \widehat{w}_2^{(2)}(x, u, t) = u, \quad t \in [\tau_{1v}, \tau_{2v}]$$

$$w_2^{(2)}(x(\tau_{1v}), u(\tau_{1v}), \tau_{1v}) = 0, \quad w_2^{(2)}(x, u, t) = x_2 - 1$$

where τ_{1v} is free. In view of Eqs. (2.10), (2.11) the hamiltonian function and $\lambda_1(\cdot)$ are continuous at τ_{1v}, because $w_2(\cdot, \cdot, \cdot)$ explicitly depends neither on x_1 nor on t and τ_{1v} is free. On the contrary, $\lambda_2(\cdot)$ can be discontinuous at τ_{1v} because $w_2(\cdot, \cdot, \cdot)$ explicitly depends on x_2 (see Eq. (2.10)). Furthermore

$$u(t) = 0, \quad \tau_{1v} < t < \tau_{2v}$$

Thus the transversality condition at the final time can be written at the initial time (the problem is time-invariant) yielding

$$H(0) = \frac{1}{2} + \frac{u^2(0)}{2} + \lambda_1(0)x_2(0) + \lambda_2(0)u(0) = \frac{1}{2} - \frac{\lambda_2^2(0)}{2} = 0$$

and $\lambda_2(0) = \pm 1$. In view of the problem data, it is apparent that the control must be positive at the beginning, so that $\lambda_2(0) = -1$. By resorting to the notation (9.1), continuity at τ_{1v} of the hamiltonian function amounts to

$$H^-(\tau_{1v}) = \frac{1}{2} + \frac{(u^-(\tau_{1v}))^2}{2} + \lambda_1^-(\tau_{1v})x_2(\tau_{1v}) + \lambda_2^-(\tau_{1v})u^-(\tau_{1v}) =$$

$$= \frac{1}{2} - \frac{(\lambda_2^-(\tau_{1v}))^2}{2} + \lambda_1(0)x_2(\tau_{1v}) = \frac{1}{2} + \frac{(u^+(\tau_{1v}))^2}{2} + \lambda_1^+(\tau_{1v})x_2(\tau_{1v}) +$$

$$+ \lambda_2^+(\tau_{1v})u^+(\tau_{1v}) = \frac{1}{2} + \lambda_1(0)x_2(\tau_{1v}) = H^+(\tau_{1v})$$

where we have taken into account that $\lambda_1(\cdot)$ is continuous, namely

$$\lambda_1^-(\tau_{1v}) = \lambda_1(0) = \lambda_1^+(\tau_{1v})$$

We conclude that

$$\lambda_2^-(\tau_{1v}) = \lambda_2(0) - \lambda_1(0)\tau_{1v} = 0$$

This equation, together with the one relevant to the inequality constraint becoming binding at τ_{1v}, that is

$$x_2(\tau_{1v}) = -\lambda_2(0)\tau_{1v} + \frac{\lambda_1(0)\tau_{1v}^2}{2} = 1$$

constitutes a system of equations for the two unknowns $\tau_{1v}, \lambda_1(0)$ which gives

$$\lambda_1(0) = -\frac{1}{2}, \quad \tau_{1v} = 2$$

Then we have

$\tau_{1v} \leq t \leq \tau_{2v}$	$\tau_{2v} \leq t \leq \tau_v$
$u(t) = 0$	$u(t) = \lambda_1(0)(t - \tau_{2v})$
$x_1(t) = x_1(\tau_{1v}) + (t - \tau_{1v})$	$x_1(t) = x_1(\tau_{2v}) + (t - \tau_{2v}) + \frac{\lambda_1(0)}{6}(t - \tau_{2v})^3$
$x_2(t) = 1$	$x_2(t) = 1 + \frac{\lambda_1(0)}{2}(t - \tau_{2v})^2$

where $x_1(\tau_{1v})$ can easily be computed if the previous results and the equation $x_1(\tau_{2v}) = x_1(\tau_{1v}) + \tau_{2v} - \tau_{1v}$ are exploited.

By enforcing feasibility, we obtain

$$\tau_{2v} = \frac{7}{3}, \quad \tau_v = \frac{13}{3}$$

Problem 11.2-2b

Analogously to what we did while discussing Problem 11.2-1b, we can deal with the problem at hand as if it is defined with reference to the time t_v^* ranging over the interval $[0, t_{fv}^*]$ and then proceeding as it has been done just now. We have

$$x^*(0) = \begin{bmatrix} 5 \\ 0 \end{bmatrix}, \quad x^*(t_{fv}^*) = \begin{bmatrix} 0 \\ 0 \end{bmatrix}$$

$$J = \int_0^{t_{fv}^*} l(x^*(t_v^*), u^*(t_v^*), t_v^*)\, dt_v^*, \quad l(x^*, u^*, t_v^*) = \frac{1 + (u^*)^2}{2}, \quad u^*(t) \in R, \; \forall t$$

$$w^{(2)*}(x^*(t), u^*(t), t^*) \le 0, \; \forall t^*, \quad w^{(2)*}(x^*, u^*, t^*) = \begin{bmatrix} -x_2^* - 1 \\ x_2^* - 1 \end{bmatrix}$$

It is clear (see the data relevant to the initial and final state) that the first component of $w^{(2)*}(\cdot, \cdot, \cdot)$ must be considered, so that we only have

$$w_1^{(2)*}(x^*(t_v^*), u^*(t_v^*), t_v^*) \le 0, \quad w_1^{(2)*}(x^*, u^*, t_v^*) = -x_2^* - 1$$

In view of the material of Sect. 10.2 and, in particular, of Sect. 2.3.4 we first compute the total derivative of $w_1^{(2)*}(\cdot, \cdot, \cdot)$ with respect to time, because this function does not explicitly depend on u^*. We obtain

$$\frac{dw_1^{(2)*}(x^*(t_v^*), u^*(t_v^*), t_v^*)}{dt_v^*} = \widehat{w}_1^{(2)*}(x^*(t_v^*), u^*(t_v^*), t_v^*) = -u^*(t_v^*)$$

Therefore if we want to discuss whether the constraint is binding over the interval $[\tau_{3v}^*, \tau_{4v}^*], 0 < \tau_{3v}^* \le \tau_{4v}^* < t_{fv}^*$, we must consider a problem with

$$\widehat{w}_1^{(2)*}(x^*(t_v^*), u^*(t_v^*), t_v^*) \le 0, \quad \widehat{w}_1^{(2)*}(x^*, u^*, t_v^*) = -u^*, \; t_v^* \in [\tau_{3v}^*, \tau_{4v}^*]$$
$$w_1^{(2)*}(x^*(\tau_{3v}^*), u^*(\tau_{3v}^*), \tau_{3v}^*) = 0, \quad w_1^{(2)}(x^*, u^*, t_v^*) = -x_2^* - 1$$

where τ_{3v}^* is free. In view of Eqs. (2.10), (2.11), the hamiltonian function and $\lambda_1^*(\cdot)$ are continuous at τ_{3v}^*, because τ_{3v}^* is free and $w_1^*(\cdot, \cdot, \cdot)$ explicitly depends neither on x_1^* nor on t_v^*. On the contrary, $\lambda_2^*(\cdot)$ can be discontinuous at τ_{3v}^* because $w_1^*(\cdot, \cdot, \cdot)$ explicitly depends on x_2^* (see Eq. (2.10)). Furthermore

$$u^*(t_v^*) = 0, \quad \tau_{3v}^* < t_v^* < \tau_{4v}^*$$

The transversality condition at the final time, written at the initial time because the problem is time-invariant, is

$$H^*(0) = \frac{1}{2} + \frac{(u^*(0))^2}{2} + \lambda_1^*(0)x_2^*(0) + \lambda_2^*(0)u^*(0) = \frac{1}{2} - \frac{(\lambda_2^*(0))^2}{2} = 0$$

so that $\lambda_2^*(0) = \pm 1$. The problem data clearly imply that the control must be negative at the beginning. Thus $\lambda_2^*(0) = 1$. By exploiting the notation (9.1), the continuity of the hamiltonian function at time τ_{3v}^* yields

$$
\begin{aligned}
H^{*-}(\tau_{3v}^*) &= \frac{1}{2} + \frac{(u^{*-}(\tau_{3v}^*))^2}{2} + \lambda_1^{*-}(\tau_{3v}^*)x_2^*(\tau_{3v}^*) + \lambda_2^{*-}(\tau_{3v}^*)u^{*-}(\tau_{3v}^*) = \\
&= \frac{1}{2} - \frac{(\lambda_2^{*-}(\tau_{3v}^*))^2}{2} + \lambda_1^*(0)x_2^*(\tau_{3v}^*) = \\
&= \frac{1}{2} + \frac{(u^{*+}(\tau_{3v}^*))^2}{2} + \lambda_1^{*+}(\tau_{3v}^*)x_2^*(\tau_{3v}^*) + \lambda_2^{*+}(\tau_{3v}^*)u^{*+}(\tau_{3v}^*) = \\
&= \frac{1}{2} + \lambda_1^*(0)x_2^*(\tau_{3v}^*) = H^{*+}(\tau_{3v}^*)
\end{aligned}
$$

where we have taken into account that $\lambda_1^*(\cdot)$ is continuous, namely

$$
\lambda_1^{*-}(\tau_{3v}^*) = \lambda_1^*(0) = \lambda_1^{*+}(\tau_{3v}^*)
$$

We conclude that $\lambda_2^{*-}(\tau_{3v}^*) = \lambda_2^*(0) - \lambda_1^*(0)\tau_{3v}^* = 0$. This equation together with the one which follows from the inequality constraint becoming binding at τ_{3v}^*, i.e.

$$
x_2^*(\tau_{3v}^*) = -\lambda_2^*(0)\tau_{3v}^* + \frac{\lambda_1^*(0)(\tau_{3v}^*)^2}{2} = -1
$$

constitutes a system of equations for the two unknowns τ_{3v}^*, $\lambda_1^*(0)$. The solution is

$$
\lambda_1^*(0) = \frac{1}{2}, \quad \tau_{3v}^* = 2
$$

Furthermore we have

$$
\begin{aligned}
&\tau_{3v}^* \le t^* \le \tau_{4v}^* \\
&u(t) = 0 \\
&x_1^*(t^*) = x_1^*(\tau_{3v}^*) + t^* - \tau_{3v}^* \\
&x_2^*(t^*) = -1
\end{aligned}
\qquad
\begin{aligned}
&\tau_{4v}^* \le t^* \le t_{fv}^* \\
&u(t) = \lambda_1^*(0)(t^* - \tau_{4v}^*) \\
&x_1^*(t^*) = x_1^*(\tau_{4v}^*) + t^* - \tau_{4v}^* + \frac{\lambda_1^*(0)}{6}(t^* - \tau_{4v}^*)^3 \\
&x_2(t^*) = -1 + \frac{\lambda_1^*(0)}{2}(t^* - \tau_{4v}^*)^2
\end{aligned}
$$

where $x_1^*(\tau_{3v}^*)$ is easily computed from the results above and the equation $x_1^*(\tau_{4v}^*) = x_1^*(\tau_{3v}^*) + \tau_{4v}^* - \tau_{3v}^*$. By enforcing feasibility, we obtain

$$
\tau_{4v}^* = \frac{13}{3}, \quad t_{fv}^* = \frac{19}{3}
$$

Going back to time t, we conclude that

$$\tau_{3v} = \tau_{3v}^* + \tau_v = \frac{19}{3}, \quad \tau_{4v} = \tau_{4v}^* + \tau_v = \frac{26}{3}, \quad t_{fv} = t_{fv}^* + \tau_v = \frac{32}{3}$$

The state trajectories of Problem 11.2-1 (where singular arcs are present) and Problem 11.2-2 (where $|x_2(\cdot)| \le 1$) are shown in Fig. 11.6 with a dotted and solid line, respectively. The trajectories are fairly similar but we notice that the derivative dx_2/dx_1 is not discontinuous in the second case. We also notice that the intermediate times τ, τ_v (when the constraint becomes binding) as well as the final times t_f, t_{fv} are quite different. These differences are put into evidence in Fig. 11.7 where the control functions are plotted. The figure also points out the discontinuity (continuity) of $u(\cdot)$ in the first (second) case.   ∎

Fig. 11.6 Problem 11.2. The state trajectory of Problem 11.2-2 where $|x_2(\cdot)| \le 1$ (*solid line*) and of Problem 11.2-1 where singular arcs are present (*dotted line*)

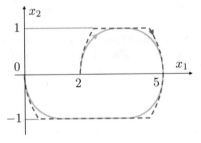

Fig. 11.7 Problem 11.2. The function $u(\cdot)$ of Problem 11.2-2 where $|x_2(\cdot)| \le 1$ (*solid line*) and of Problem 11.2-1 where singular arcs are present (*dotted line*)

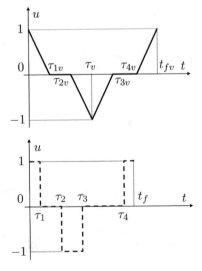

Problem 11.3
Let

$$x(0) = \begin{bmatrix} \beta \\ 0 \end{bmatrix}, \quad x(t_f) = \begin{bmatrix} 0 \\ 0 \end{bmatrix}, \quad J = \int_0^{t_f} l(x(t), u(t), t)\, dt, \quad l(x, u, t) = \frac{x_2^2}{2}$$

$$u(t) \in \overline{U}, \ \forall t, \ \overline{U} = \{u | -1 \le u \le 1\}$$

where the parameter β and the final time t_f are given. In order to prevent triviality, we assume $\beta \neq 0$, whereas we can also assume that it is positive because of obvious symmetry reasons.

We notice that the problem does not admit a solution corresponding to certain couples (t_f, β) because of the bounds on the control. In fact it is fairly obvious that the fastest way to bring the state from a given $x(0)$ to the origin simply consists in letting $u(\cdot) = -1$ at the beginning, and then switching to $u(\cdot) = 1$ (see also Chap. 7). Let τ be the switching time. Thus we have

$$0 \le t < \tau, \qquad\qquad\qquad \tau < t \le t_f$$
$$u(t) = -1, \qquad\qquad\qquad u(t) = 1$$
$$0 \le t \le \tau, \qquad\qquad\qquad \tau \le t \le t_f$$
$$x_1(t) = \beta - \frac{1}{2}t^2, \qquad\quad x_1(t) = \beta - \frac{1}{2}\tau^2 - \tau(t - \tau) + \frac{1}{2}(t - \tau)^2$$
$$x_2(t) = -t, \qquad\qquad\qquad x_2(t) = -\tau + (t - \tau)$$

If $x(t_f) = 0$, we obtain

$$\tau = \frac{t_f}{2}, \quad t_f = \bar{t}_f := 2\sqrt{\beta}$$

Therefore we can transfer the state to the origin only if $t_f \ge \bar{t}_f$. When this relation holds with the equality sign, only one feasible control exists and the problem is trivial, so that we assume $t_f > \bar{t}_f$.

The hamiltonian function is

$$H = \frac{x_2^2}{2} + \lambda_1 x_2 + \lambda_2 u$$

Consequently, Eq. (2.5a) is

$$\dot{\lambda}_1(t) = 0$$
$$\dot{\lambda}_2(t) = -\lambda_1(t) - x_2(t)$$

and its solution over an interval beginning at time 0 implies $\lambda_1(t) = \lambda_1(0)$.

The H-minimizing control, when $\lambda_2 \neq 0$, is $u_h = -\text{sign}(\lambda_2)$. Nevertheless, we can not exclude that $\lambda_2(\cdot) = 0$ over an interval T_s of non-zero length. In fact within such an interval it can well happen that

$$a(\lambda(t)) := \lambda_2(t) = 0$$

$$\frac{da(\lambda(t))}{dt} = \frac{d\lambda_2(t)}{dt} = -\lambda_1(0) - x_2(t) = 0$$

$$\frac{d^2 a(\lambda(t))}{dt^2} = \frac{d^2 \lambda_2(t)}{dt^2} = -\frac{d\lambda_1(t)}{dt} - \frac{dx_2(t)}{dt} = -u(t) = 0$$

where we have taken into account that $\lambda_1(\cdot) = \lambda_1(0)$. The conditions of Theorem 2.9 are satisfied: thus, if the solution includes a singular arc, the admissible control values are $-1, 0, 1$, whereas the value 0 should not be considered if such an arc is absent.

Let the singular solution occur within the interval $T_s := [\tau_1, \tau_2]$. In view of Fig. 11.1, we conclude that the control is

$$u(t) = \begin{cases} -1, & 0 \le t < \tau_1 \\ 0, & \tau_1 < t < \tau_2 \\ 1, & \tau_2 < t \le \tau \end{cases}$$

Now we can easily compute $\lambda_2(\cdot), x_1(\cdot), x_2(\cdot)$ as functions of the four parameters β, t_f, τ_1, τ_2: More precisely, we have

$$t \in [0, \tau_1] \qquad \begin{cases} \lambda_2(t) = \lambda_2(0) - \lambda_1(0)t + \dfrac{t^2}{2} \\ x_1(t) = \beta - \dfrac{t^2}{2} \\ x_2(t) = -t \end{cases}$$

$$t \in [\tau_1, \tau_2] \qquad \begin{cases} \lambda_2(t) = 0 \\ x_1(t) = x_1(\tau_1) + x_2(\tau_1)(t - \tau_1) \\ x_2(t) = x_2(\tau_1) \end{cases}$$

$$t \in [\tau_2, \tau] \qquad \begin{cases} \lambda_2(t) = -\lambda_1(0)(t - \tau_2) - x_2(\tau_2)(t - \tau_2) - \dfrac{(t - \tau_2)^2}{2} \\ x_1(t) = x_1(\tau_2) + x_2(\tau_2)(t - \tau_2) + \dfrac{(t - \tau_2)^2}{2} \\ x_2(t) = x_2(\tau_2) + (t - \tau_2) \end{cases}$$

If we recall that

$$x_1(\tau_1) = \beta - \frac{\tau_1^2}{2}, \quad x_2(\tau_1) = -\tau_1$$

$$x_1(\tau_2) = x_1(\tau_1) + x_2(\tau_1)(\tau_2 - \tau_1), \quad x_2(\tau_2) = -\tau_1$$

we obtain, by enforcing feasibility, namely

$$x_1(t_f) = \beta + \frac{1}{2}\tau_1^2 - \tau_1 t_f + \frac{1}{2}(t_f - \tau_2)^2 = 0$$

$$x_2(t_f) = t_f - \tau_1 - \tau_2 = 0$$

Fig. 11.8 Problem 11.3.
State trajectories when $\beta = 4$
and $t_f = 4$ (β-a-0); $t_f = 5$
(β-d-e-0); $t_f = 7$ (β-b-c-0)

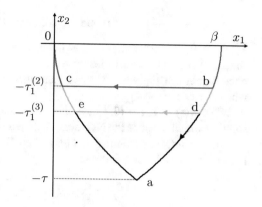

the values

$$\tau_1 = \frac{1}{2}\left(t_f - \sqrt{t_f^2 - 4\beta}\right), \quad \tau_2 = \frac{1}{2}\left(t_f + \sqrt{t_f^2 - 4\beta}\right)$$

Thus the solution is completely specified.

We can find $\lambda_1(0)$ and $\lambda_2(0)$ by noticing that $\lambda_1(0) = -x_2(t) = \tau_1$, $t \in [\tau_1, \tau_2]$ and $\lambda_2(\tau_1) = \lambda_2(0) - \lambda_1(0)\tau_1 + \dfrac{\tau_1^2}{2} = 0$.

We continue the discussion by assuming $\beta = 4$. The state trajectories corresponding to various values of the final time are plotted in Fig. 11.8.

When $t_f = 4$ (the lower bound for the problem to admit a solution) the trajectory does not include a singular arc, the control switches from -1 to 1 at $\tau = 2$ and $x_2(\tau) = -\tau$ (trajectory (β-a-0)).

When $t_f = 7$ a singular arc is included. The control switches from -1 to 0 at $\tau_1^{(2)} = 0.63$, from 0 to 1 at $\tau_2^{(2)} = 6.27$ and $x_2(\tau_1^{(2)}) = -\tau_1^{(2)}$ (trajectory (β-b-c-0)).

Finally, when $t_f = 5$ a singular arc is included. The control switches from -1 to 0 at $\tau_1^{(3)} = 1$, from 0 to 1 at $\tau_2^{(3)} = 4$ and $x_2(\tau_1^{(3)}) = -\tau_1^{(3)}$ (trajectory (β-d-e-0)).

We complete our analysis by checking whether, for some values of the final time $t_f > \bar{t}_f$, a solution exists which does not include a singular arc and satisfies the NC. We first assume $u(t) = -1$, $t \in [0, \tau_c)$, where $\tau_c > 0$ is the (first) time when the control switches to 1. Necessarily $\tau_c < t_f$ because $u(\cdot) = -1$ is not feasible. For $0 \le t < \tau_c$ we have

$$u(t) = -1, \quad x_2(t) = -t, \quad \lambda_2(t) = \lambda_2(0) - \lambda_1(0)t + \frac{t^2}{2}$$

with $\lambda_2(0) \ge 0$ because $u(0) = -1$. We also have $\dfrac{d\lambda_2(t)}{dt}\bigg|_{t=\tau_c} \le 0$ because $\lambda_2(t) > 0$, $t < \tau_c$, $\lambda_2(\tau_c) = 0$ and, for $t > \tau_c$,

$$u(t) = 1, \quad x_2(t) = x_2(\tau_c) + (t - \tau_c), \quad \lambda_2(t) = -(\lambda_1(0) + x_2(\tau_c))(t - \tau_c) - \frac{(t - \tau_c)^2}{2}$$

Therefore $\left.\dfrac{d^2\lambda_2(t)}{dt^2}\right|_{t=\tau_c} < 0$ and we conclude that non other time τ^* exists where $\lambda_2(\tau^*) = 0$. Consequently, $u(\cdot)$ switches only once from -1 to 1. In view of the shapes of the state trajectories corresponding to a control which switches at most once (again see Fig. 11.1) we can state that it is not possible to have $x(t_f) = 0$ whenever $t_f > \bar{t}_f$.

When $u(t) = 1$, $t \in [0, \tau_c)$, a similar discussion leads to the conclusion that a solution which satisfies the NC must include a singular arc if $t_f > \bar{t}_f$.

Now we consider a similar problem where the final state, rather than being given, is constrained to belong to the set

$$\overline{S}_f = \left\{x | \alpha_f(x) = 0\right\}, \quad \alpha_f(x) = x_1$$

which is a regular variety because

$$\Sigma_f(x) = \frac{d\alpha_f(x)}{dx} = \begin{bmatrix} 1 & 0 \end{bmatrix}$$

and $\mathrm{rank}(\Sigma_f(x)) = 1$, $\forall x \in \overline{S}_f$.

The forthcoming, relevant, discussion is essentially analogous to the previous one. In particular, we easily find that the final time must be not less than $\bar{t}_f = \sqrt{2\beta}$.

Indeed the fastest way to bring the system state to meet the x_2 axis is to set $u(\cdot) = -1$ which entails that the final time is $\bar{t}_f$ (see also Chap. 7).

When $t_f > \bar{t}_f$ the solution includes a singular arc.

More in detail, we have

$$u(t) = -1, \ 0 \le t < \tau, \quad u(t) = 0, \ \tau < t \le t_f$$

where τ (the time when the control switches and the state trajectory becomes coincident with a singular arc) is

$$\tau = t_f - \sqrt{t_f^2 - 2\beta} \qquad \blacksquare$$

Problem 11.4
Let

$$x(0) = \begin{bmatrix} 1 \\ 1 \end{bmatrix}, \quad x(t_f) \in \overline{S}_f = \left\{x | \alpha_f(x) = 0\right\}, \quad \alpha_f(x) = x_1$$

$$J = m(x(t_f), t_f) + \int_0^{t_f} l(x(t), u(t), t) \, dt,$$

$$m(x, t) = \beta x_2, \quad l(x, u, t) = 1 + \frac{x_2^2}{2}$$

$$u(t) \in \overline{U}, \ \forall t, \ \overline{U} = \{u | -1 \le u \le 1\}$$

where the final time t_f is free, whereas β is a given parameter belonging to the interval $(0, 1]$.

The set $\overline{S}_f$ of the admissible final states is a regular variety because

$$\Sigma_f(x) = \frac{d\alpha_f(x)}{dx} = \begin{bmatrix} 1 & 0 \end{bmatrix}$$

and $\mathrm{rank}(\Sigma_f(x)) = 1, \forall x \in \overline{S}_f$.

The hamiltonian function is

$$H = 1 + \frac{x_2^2}{2} + \lambda_1 x_2 + \lambda_2 u$$

so that Eq. (2.5a) is

$$\dot{\lambda}_1(t) = 0$$
$$\dot{\lambda}_2(t) = -\lambda_1(t) - x_2(t)$$

and its solution over an interval beginning at time 0 entails $\lambda_1(t) = \lambda_1(0)$.

If $\lambda_2 \neq 0$, the H-minimizing control is $u_h = -\mathrm{sign}(\lambda_2)$. Nevertheless, we can not exclude that $\lambda_2(\cdot) = 0$ over an interval of non-zero length. In fact within such an interval it can well happen that

$$a(\lambda(t)) := \lambda_2(t) = 0$$
$$\frac{da(\lambda(t))}{dt} = \frac{d\lambda_2(t)}{dt} = -\lambda_1(0) - x_2(t) = 0$$
$$\frac{d^2 a(\lambda(t))}{dt^2} = \frac{d^2 \lambda_2(t)}{dt^2} = -\frac{dx_2(t)}{dt} = -u(t) = 0$$

where we have taken into account that $\lambda_1(\cdot) = \lambda_1(0)$. The conditions of Theorem 2.9 are satisfied: therefore if the solution includes a singular arc, the control can take on the three values $-1, 0, 1$, whereas, if the arc is not present, the value 0 must not be taken into consideration.

Now we notice that the performance index penalizes large absolute values of x_2 during the control interval (which is desired to be of small lasting) and positive values of this variable at the final time. If we observe the shape of the state trajectories (see Fig. 11.1), we easily conclude that $u(0) = -1$.

The transversality condition at the final time, written at the initial time because the problem is time-invariant,

$$H(0) = 1 + \frac{x_2^2(0)}{2} + \lambda_1(0)x_2(0) + \lambda_2(0)u(0) = \frac{3}{2} + \lambda_1(0) - \lambda_2(0) = 0$$

implies $\lambda_2(0) = \frac{3}{2} + \lambda_1(0)$, whereas the orthogonality condition at the final time

$$\begin{bmatrix} \lambda_1(t_f) \\ \lambda_2(t_f) \end{bmatrix} - \frac{\partial m(x, t_f)}{\partial x}\Bigg|'_{x=x(t_f)} = \begin{bmatrix} \lambda_1(t_f) \\ \lambda_2(t_f) \end{bmatrix} - \begin{bmatrix} 0 \\ \beta \end{bmatrix} =$$

$$= \vartheta_f \frac{d\alpha_f(x)}{dx}\Bigg|'_{x=x(t_f)} = \vartheta_f \begin{bmatrix} 1 \\ 0 \end{bmatrix}$$

entails $\lambda_2(t_f) = \beta$.

By looking at Fig. 11.9 we can easily conclude that five cases must be considered with reference to the control

Case 1. $u(t) = -1, 0 \le t \le t_{f1}$

Case 2. $u(t) = -1, 0 \le t < \tau_2, u(t) = 0, \tau_2 < t \le t_{f2}$

Case 3. $u(t) = -1, 0 \le t < \tau_3, u(t) = 1, \tau_3 < t \le t_{f3}$

Case 4. $u(t) = -1, 0 \le t < \tau_{41}, u(t) = 0, \tau_{41} < t < \tau_{42}, u(t) = -1, \tau_{42} < t \le t_{f4}$

Case 5. $u(t) = -1, 0 \le t < \tau_{51}, u(t) = 0, \tau_{51} < t < \tau_{52}, u(t) = 1, \tau_{52} < t \le t_{f5}$

where the switching times $\tau_2 < t_{f2}, \tau_3 < t_{f3}, \tau_{41} < \tau_{42} < t_{f4}, \tau_{51} < \tau_{52} < t_{f5}$ and the final times $t_{fi}, i = 1, 2, \ldots, 5$ are all free.

We now discuss the five cases above.

Case 1. $u(t) = -1, 0 \le t \le t_{f1}, t_{f1}$ free

We have

$$0 \le t \le t_{f1}$$
$$u(t) = -1$$
$$x_1(t) = 1 + t - \frac{1}{2}t^2$$
$$x_2(t) = 1 - t$$
$$\lambda_2(t) = \lambda_2(0) - (\lambda_1(0) + 1)t + \frac{1}{2}t^2$$

Fig. 11.9 Problem 11.4.
State trajectories
corresponding to *Case 1.*
(a-b-c-d-e-f), *Case 2.*
(a-b-c-h), *Case 3.*
(a-b-c-d-e-l), *Case 4.*
(a-b-c-d-k-g), *Case 5.*
(a-b-j-i)

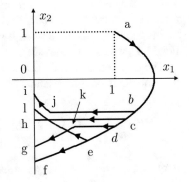

By enforcing feasibility ($x_1(t_{f1}) = 0$) and the fulfillment of the transversality and orthogonality conditions at the final time, we have

$$t_{f1} = 1 + \sqrt{3}, \quad \lambda_1(0) = \frac{5 - 2\beta}{2\sqrt{3}}, \quad \lambda_2(0) = \frac{3\sqrt{3} + 5 - 2\beta}{2\sqrt{3}}, \quad x_2(t_{f1}) = -\sqrt{3}$$

It is easy to ascertain that $\lambda_2(t) > 0, 0 \leq t \leq t_{f1}$, consistently with $u(\cdot) = -1$. The value of the performance index is

$$J_1(\beta) := \beta x_2(t_{f1}) + t_{f1} + \frac{(t_{f1} - 1)^3 + 1}{6}$$

Case 2. $u(t) = -1, 0 \leq t < \tau_2, u(t) = 0, \tau_2 < t \leq t_{f2}, \tau_2$ and t_{f2} free

We must disregard this case. In fact the orthogonality condition at the final time can not be satisfied because $\beta > 0$ and $\lambda_2(t) = 0, \tau < t \leq t_{f2}$ as a consequence of the last part of the state trajectory being a singular arc.

Case 3. $u(t) = -1, 0 \leq t < \tau_3, u(t) = 1, \tau_3 < t \leq t_{f3}, \tau_3$ and t_{f3} free

Also this case must be disregarded. In fact we have

$$
\begin{array}{ll}
0 \leq t < \tau_3 & \tau_3 < t \leq t_{f3} \\
u(t) = -1 & u(t) = 1 \\
0 \leq t \leq \tau_3 & \tau_3 \leq t \leq t_{f3} \\
x_1(t) = 1 + t - \dfrac{t^2}{2} & x_1(t) = x_1(\tau_3) + x_2(\tau_3)(t - \tau_3) + \dfrac{(t - \tau_3)^2}{2} \\
x_2(t) = 1 - t & x_2(t) = x_2(\tau_3) + t - \tau_3 \\
\lambda_2(t) = \lambda_2(0) - (\lambda_1(0) + 1)t + \dfrac{t^2}{2} & \lambda_2(t) = -(\lambda_1(0) + x_2(\tau_3))(t - \tau_3) - \dfrac{(t - \tau_3)^2}{2}
\end{array}
$$

where we have taken into account that the switching of the control requires $\lambda_2(\tau_3) = 0$. The four unknown parameters $\lambda_1(0), \lambda_2(0), \tau_3$ and t_{f3} are solution of the system of four equations $x_1(t_f) = 0$ (feasibility), $\lambda_2(\tau_3) = 0$ (switching), $\lambda_2(t_{f3}) = \beta$ (orthogonality at the final time) and $\lambda_2(0) = \lambda_1(0) + \dfrac{3}{2}$ (transversality). However it is not difficult to ascertain that no solution exists.

Case 4. $u(t) = -1, 0 \leq t < \tau_{41}, u(t) = 0, \tau_{41} < t < \tau_{42}, u(t) = -1, \tau_{42} < t \leq t_{f4}$, τ_{41}, τ_{42} and t_{f4} free

Here the control switches twice and a singular arc is present.

Thus we have to comply with feasibility, transversality and orthogonality conditions at the final time as well as with the relations $\lambda_1(0) = -x_2(\tau_{41}), \lambda_2(\tau_{41}) = 0$ deriving from the presence of the singular arc. Therefore

| $0 \le t < \tau_{41}$ | $u(t) = -1$ |

$0 \le t \le \tau_{41}$

$$\begin{cases} x_1(t) = 1 + t - \dfrac{t^2}{2} \\ x_2(t) = 1 - t \\ \lambda_2(t) = \lambda_2(0) - (\lambda_1(0) + 1)t + \dfrac{t^2}{2} \end{cases}$$

$\tau_{41} < t < \tau_{42}$ $\quad u(t) = 0$

$\tau_{41} \le t \le \tau_{42}$

$$\begin{cases} x_1(t) = x_1(\tau_{41}) + x_2(\tau_{41})(t - \tau_{41}) \\ x_2(t) = x_2(\tau_{41}) \\ \lambda_2(t) = 0 \end{cases}$$

$\tau_{42} < t \le t_{f4}$ $\quad u(t) = -1$

$\tau_{42} \le t \le t_{f4}$

$$\begin{cases} x_1(t) = x_1(\tau_{42}) + x_2(\tau_{42})(t - \tau_{42}) - \dfrac{(t - \tau_{42})^2}{2} \\ x_2(t) = x_2(\tau_{42}) - (t - \tau_{42}) \\ \lambda_2(t) = -(\lambda_1(0) + x_2(\tau_{42}))(t - \tau_{42}) + \dfrac{(t - \tau_{42})^2}{2} \end{cases}$$

From these equations and those relevant to transversality and orthogonality at the final time we find the parameters which specify the solution of the problem. More in detail, we get

$$\tau_{41} = 1 + \sqrt{2}, \quad \lambda_1(0) = \sqrt{2}, \quad \lambda_2(0) = \frac{3}{2} + \sqrt{2}$$

$$\tau_{42} = \frac{\dfrac{5}{2} + \sqrt{2} - 2\sqrt{\beta} - \beta}{\sqrt{2}}, \quad t_{f4} = \tau_{42} + \sqrt{2\beta}$$

$$x_2(\tau_{41}) = 1 - \tau_{41}, \quad x_2(\tau_{42}) = x_2(\tau_{41}), \quad x_2(t_{f4}) = x_2(\tau_{42}) - (t_{f4} - \tau_{42})$$

It is easy to ascertain that $\tau_{42} - \tau_{41} > 0$ only if $\beta < \bar{\beta} := 0.0505$: therefore a singular arc is present only if this inequality holds. Furthermore we find

$$J_4(\beta) := \beta x_2(t_{f4}) + t_{f4} + \frac{(\tau_{41} - 1)^3 + 1}{6} + \tau_{42} - \tau_{41} + \frac{(t_{f4} + \sqrt{2} - \tau_{42})^3 - \sqrt{8}}{6}$$

Case 5. $u(t) = -1, \, 0 \le t < \tau_{51}, \, u(t) = 0, \, \tau_{51} < t < \tau_{52}, \, u(t) = 1, \, \tau_{52} < t \le t_{f5}$, τ_{51}, τ_{52} and t_{f5} free

Here the control switches twice and a singular arc is present.

Thus we have to comply with feasibility and the transversality and orthogonality conditions at the final time as well as with the relations $\lambda_1(0) = -x_2(\tau_{51}), \lambda_2(\tau_{51}) = 0$ deriving from the presence of the singular arc. Therefore

$0 \le t < \tau_{51}$ $\qquad\qquad\qquad\qquad$ $u(t) = -1$

$0 \le t \le \tau_{51}$ $\qquad\qquad\qquad$ $\begin{cases} x_1(t) = 1 + t - \dfrac{t^2}{2} \\[2mm] x_2(t) = 1 - t \\[2mm] \lambda_2(t) = \lambda_2(0) - (\lambda_1(0) + 1)t + \dfrac{t^2}{2} \end{cases}$

$\tau_{51} < t < \tau_{52}$ $\qquad\qquad\qquad$ $u(t) = 0$

$\tau_{51} \le t \le \tau_{52}$ $\qquad\qquad$ $\begin{cases} x_1(t) = x_1(\tau_{51}) + x_2(\tau_{51})(t - \tau_{51}) \\ x_2(t) = x_2(\tau_{51}) \\ \lambda_2(t) = 0 \end{cases}$

$\tau_{52} < t \le t_{f5}$ $\qquad\qquad\qquad$ $u(t) = 1$

$\tau_{52} \le t \le t_{f5}$ $\qquad$ $\begin{cases} x_1(t) = x_1(\tau_{52}) + x_2(\tau_{52})(t - \tau_{52}) + \dfrac{(t - \tau_{42})^2}{2} \\[2mm] x_2(t) = x_2(\tau_{52}) + (t - \tau_{52}) \\[2mm] \lambda_2(t) = -(\lambda_1(0) + x_2(\tau_{52}))(t - \tau_{52}) - \dfrac{(t - \tau_{52})^2}{2} \end{cases}$

From these equations and the transversality condition at the final time we easily find

$$\tau_{51} = 1 + \sqrt{2}, \quad \lambda_1(0) = \sqrt{2}, \quad \lambda_2(0) = \frac{3}{2} + \sqrt{2}$$

and also

$$x_1(\tau_{51}) = \frac{1}{2}, \quad x_2(\tau_{51}) = -\sqrt{2}$$

Then

$$x_2(\tau_{52}) = x_2(\tau_{51}), \quad \lambda_2(t_{f5}) = -\frac{(t_{f5} - \tau_{52})^2}{2} < 0$$

and the orthogonality condition at the final time ($\lambda_2(t_{f5}) = \beta$) can not be satisfied because $\beta > 0$.

As a conclusion, when $\beta < \bar{\beta}$ we have two solutions of the problem which correspond to *Case 1.* and *Case 4.*, whether we have only the solution of *Case 1.* when $\beta \ge \bar{\beta}$. When $\beta = 0.02 < \bar{\beta}$ we have $J_1(\beta) = 3.7301$ and $J_2(\beta) = 3.7284$, so that the solution which incorporates a singular arc must be preferred. ∎

Chapter 12
Local Sufficient Conditions

Abstract We briefly mention an important topic of optimal control theory which is strictly connected to the necessary conditions framework and is a natural complement of it, namely the sufficient conditions issue. As in the (at least conceptually) related field of minimization of functions, we can distinguish between global and local conditions. Some of the most significant results for global conditions stem from the Hamilton–Jacoby theory: however they are not mentioned in the present treatment as they are rooted in a somehow unrelated context. On the contrary, we focus the attention on local conditions resulting from a variational approach closely related to the one underlying the achievements of the Maximum Principle. Indeed if we require that, once the first variation of the performance index is zero, its second variation is positive, then we end up with local sufficient conditions. We illustrate them with reference to four significant scenarios.

The aim of this chapter is to briefly mention an important topic of optimal control theory which is strictly connected to the *necessary conditions* framework and is a natural complement of it, namely the *sufficient conditions* issue. As in the (at least conceptually) related field of minimization of functions, we can distinguish between *global* and *local* conditions. Some of the most significant results for global conditions stem from the Hamilton–Jacoby theory (see, for instance, [7, 11, 15]). However they are not mentioned in the present treatment as they are rooted in a somehow unrelated context. On the contrary, we focus the attention on local conditions resulting from a variational approach closely related to the one underlying the achievements of the Maximum Principle. Indeed its necessary conditions are established by imposing that the first variation of the performance index, due to a specified set of perturbations of the claimed optimal solution at hand, is zero. If we require that, once the first variation of the performance index is zero, its second variation is positive, then we end up with sufficient conditions for a local minimum.

In the sequel we present local sufficient optimality conditions which are referred to as *weak* because they are deduced under the constraint that the allowed control perturbations $\delta u(\cdot)$ comply with the request

$$\sup_{t_0 \le t \le t_f} \sqrt{\delta u'(t)\delta u(t)} < \varepsilon$$

© Springer International Publishing Switzerland 2017
A. Locatelli, *Optimal Control of a Double Integrator*, Studies in Systems, Decision and Control 68, DOI 10.1007/978-3-319-42126-1_12

where $[t_0, t_f]$ is the control interval. The state perturbations caused by such control perturbations satisfy a relation of the same kind, provided that the function $f(\cdot, \cdot, \cdot)$ possesses suitable continuity properties as those mentioned below.

The results in the forthcoming sections are relevant to optimal control problems of the simplest kind, namely to problems characterized by the equations

$$\dot{x}(t) = f(x(t), u(t), t) \qquad (12.1a)$$

$$x(0) = x_0 \qquad (12.1b)$$

$$x(t_f) \in S_f \qquad (12.1c)$$

$$J = m(x(t_f, t_f) + \int_0^{t_f} l(x(t), u(t), t)dt, \ u(t) \in R^m \qquad (12.1d)$$

where x_0 is given, the initial time has been set to zero, the final time is either given or free and one of the two Assumptions 2.1 and 2.2 holds. Furthermore the considered problems are supposed to be non-pathological and in Eq. (12.1c) the set S_f is either the whole state space or a regular variety. Thus we have

$$S_f = \begin{cases} R^n \\ \text{or} \\ \overline{S}_f = \left\{ x \mid \alpha_{fi}(x) = 0, \ i = 1, 2, \ldots, q < n \right\} \end{cases} \qquad (12.2)$$

The functions $f(\cdot, \cdot, \cdot)$, $l(\cdot, \cdot, \cdot)$, $m(\cdot, \cdot)$, $\alpha_{fi}(\cdot)$, $i = 1, 2, \ldots, q$ are continuous together with all their first and second order derivatives.

We recall that the hamiltonian function for the (non-pathological) problem at hand is

$$H(x, u, t, \lambda) = l(x.u.t) + \lambda' f(x, u, t)$$

and make reference to Theorem 2.2 so that, in particular, $(x^o(\cdot), u^o(\cdot), t_f^o)$ is a triple which satisfies the NC of the Maximum Principle corresponding to $\lambda_0^o = 1$ and $\lambda^o(\cdot)$, solution of Eq. (2.5a). Moreover, by letting γ and δ be either x or u or t, we adopt the notation

$$f^o(t) = f(x^o(t), u^o(t), t), \quad f_\gamma^o(t) = \left. \frac{\partial f(x, u, t)}{\partial \gamma} \right|_{x=x^o(t), \, u=u^o(t)} \qquad (12.3a)$$

$$H^o(t) = H(x^o(t), u^o(t), t, \lambda^o(t)) \qquad (12.3b)$$

$$H_\gamma^o(t) = \left. \frac{\partial H(x, u, t, \lambda)}{\partial \gamma} \right|_{x=x^o(t), \, u=u^o(t), \, \lambda=\lambda^o(t)} \qquad (12.3c)$$

$$H_{\gamma\delta}^o(t) = \left. \frac{\partial}{\partial \gamma} \frac{\partial H(x, u, t, \lambda)}{\partial \delta} \right|_{x=x^o(t), \, u=u^o(t), \, \lambda=\lambda^o(t)} \qquad (12.3d)$$

$$\alpha(x) = \left[\alpha_{f1}(x) \ldots \alpha_{fq}(x) \right]', \quad \alpha_x^o = \left. \frac{d\alpha(x)}{dx} \right|_{x=x^o(t_f^o)} \qquad (12.3e)$$

$$\alpha^o_{xx} = \sum_{i=1}^{q} \vartheta_{fi} \frac{d^2\alpha_{fi}(x)}{dx^2}\bigg|_{x=x^o(t^o_f)} , \quad m^o_{xx} = \frac{d^2m(x,t^o_f)}{dx^2}\bigg|_{x=x^o(t^o_f)} \tag{12.3f}$$

Then we define

$$R(t) := H^o_{uu}(t), \quad A(t) := f^o_x(t) - f^o_u(t)R^{-1}(t)H^o_{ux}(t) \tag{12.4a}$$

$$B(t) := f^o_u(t), \quad Q(t) := H^o_{xx}(t) - H^o_{xu}(t)R^{-1}(t)H^o_{ux}(t) \tag{12.4b}$$

$$K(t) := B(t)R^{-1}(t)B'(t) \tag{12.4c}$$

$$S_1 := m^o_{xx} + \alpha^o_{xx}, \quad S_2 := \alpha^{o'}_x \tag{12.4d}$$

$$S_3 := S_1 f^o(t^o_f) + H^{o'}_x(t^o_f) \tag{12.4e}$$

$$S_4 := \left[f^{o'}(t^o_f)S_1 + H^o_x(t^o_f) \right] f^o(t^o_f) + H^o_t(t^o_f) \tag{12.4f}$$

In the forthcoming Sects. 12.1–12.4 we state *local sufficient optimality conditions in a weak sense* with reference to problems where the final time and state are characterized in one of the four following ways:

(i) $x(t_f) =$ partially given, $t_f =$ free
(ii) $x(t_f) =$ given, $t_f =$ given
(iii) $x(t_f) =$ free, $t_f =$ free
(iv) $x(t_f) =$ free, $t_f =$ given

With *partially given* we mean that the final state is neither free nor given, that is it must belong to the regular variety $\overline{S}_f$ specified in Eq. (12.2).

As we have previously done along the entire book, the relevant results (Theorems 12.1–12.4) are illustrated by assuming that the controlled system is a double integrator.

12.1 $x(t_f) =$ Partially Given, $t_f =$ Free

We consider optimal control problems defined by Eq. (12.1) where the final state $x(t_f)$ is partially given and the final time t_f is free. For such problems we have the following result.

Theorem 12.1

Let $(x^o(\cdot), u^o(\cdot), t^o_f)$ be a triple which satisfies the necessary optimality conditions stated in Theorem 2.2 under the assumption that $x(t_f)$ is partially given and t_f is free. Furthermore assume that $R(t) > 0$, $t \in [0, t^o_f]$ and there exist solutions of the differential equations

$$\dot{P}_1(t) = -P_1(t)A(t) - A'(t)P_1(t) + P_1(t)K(t)P_1(t) - Q(t)$$

$$\dot{P}_2(t) = -[A(t) - K(t)P_1(t)]' P_2(t)$$

$$\dot{P}_3(t) = -[A(t) - K(t)P_1(t)]' \, P_3(t)$$
$$\dot{P}_4(t) = P_3'(t)K(t)P_3(t)$$
$$\dot{P}_5(t) = P_2'(t)K(t)P_3(t)$$
$$\dot{P}_6(t) = P_2'(t)K(t)P_2(t)$$

with the boundary conditions

$$P_1(t_f^o) = S_1, \quad P_2(t_f^o) = S_2, \quad P_3(t_f^o) = S_3$$
$$P_4(t_f^o) = S_4, \quad P_5(t_f^o) = 0, \quad P_6(t_f^o) = 0$$

such that $P_4(t) > 0$, $t \in [0, t_f^o]$, $P_6(t) < 0$, $t \in [0, t_f^o)$. Then $(x^o(\cdot), u^o(\cdot), t_f^o)$ is a locally optimal triple, in a weak sense. ∎

We first notice that the sign definition of $R(\cdot)$ is not required by the NC. However, it is likely to be verified because it matches almost naturally with the minimization of the hamiltonian function with respect to u. We also note that the six differential equations for the functions $P_i(\cdot)$ can be integrated one at a time starting from the one relevant to $P_1(\cdot)$ which, by the way, is the only nonlinear one (actually it is a Riccati equation). The six unknown functions are three matrices (P_1, $n \times n$; P_2, $n \times q$; P_6, $q \times q$), two vectors (P_3, $n \times 1$; P_5, $q \times 1$) and one scalar (P_4).

Problem 12.1

Let

$$x(0) = \begin{bmatrix} 0 \\ 0 \end{bmatrix}, \quad x(t_f) \in \overline{S}_f = \{x | \alpha_f(x) = 0\}, \quad \alpha_f(x) = x_2 + 1, \quad u(t) \in R, \ \forall t$$

$$J = \int_0^{t_f} l(x(t), u(t), t) \, dt + m(x(t_f), t_f), \quad l(x, u, t) = 1 - x_2 + \frac{u^2}{2}, \quad m(x, t) = -x_1$$

where the final time is free. Notice that the set of the admissible final states S_f is a regular variety: in fact

$$\Sigma_f(x) = \frac{d\alpha_f(x)}{dx} = \begin{bmatrix} 0 & 1 \end{bmatrix}$$

and $rank(\Sigma_f(x)) = 1$, $\forall x \in S_f$.

The hamiltonian function and the H-minimizing control are

$$H = 1 - x_2 + \frac{u^2}{2} + \lambda_1 x_2 + \lambda_2 u, \quad u_h = -\lambda_2$$

so that Eq. (2.5a) and its general solution over an interval beginning at time 0 are

$$\dot{\lambda}_1(t) = 0 \qquad\qquad\qquad \lambda_1(t) = \lambda_{10}$$
$$\dot{\lambda}_2(t) = 1 - \lambda_1(t) \qquad\qquad \lambda_2(t) = \lambda_{20} + (1 - \lambda_{10})t$$

The transversality and orthogonality conditions at the final time, the former written at the initial time because the problem is time-invariant, are

$$H(0) = 1 - x_2(0) + \frac{u^2(0)}{2} + \lambda_1(0)x_2(0) + \lambda_2(0)u(0) = 1 - \frac{\lambda_{20}^2}{2}$$

and

$$\begin{bmatrix} \lambda_1(t_f) \\ \lambda_2(t_f) \end{bmatrix} - \frac{\partial m(x,t)}{\partial x}\bigg|'_{x=x(t_f)} = \begin{bmatrix} \lambda_{10} \\ \lambda_{20} + (1 - \lambda_{10})t_f \end{bmatrix} - \begin{bmatrix} -1 \\ 0 \end{bmatrix} =$$

$$= \vartheta_f \frac{d\alpha_f(x)}{dx}\bigg|'_{x=x(t_f)} = \vartheta_f \begin{bmatrix} 0 \\ 1 \end{bmatrix}$$

They supply

$$\lambda_{10} = -1, \quad \lambda_{20} = \pm\sqrt{2}$$

Thus the state motion is

$$x_1(t) = -\lambda_{20}\frac{t^2}{2} - \frac{t^3}{3}, \quad x_2(t) = -\lambda_{20}t - t^2$$

By enforcing feasibility ($x_2(t_f) = -1$) we obtain

$$t_f = \frac{-\lambda_{20} \pm \sqrt{6}}{2}, \quad \lambda_2(t_f) = \pm\sqrt{6}, \quad u(t_f) = \mp\sqrt{6}$$

Therefore we have found two solutions which satisfy the NC

$$\lambda_{20} = \sqrt{2} \qquad\qquad\qquad \lambda_{20} = -\sqrt{2}$$

$$t_f = t_{f1} := \frac{\sqrt{6} - \sqrt{2}}{2} \qquad\qquad t_f = t_{f2} := \frac{\sqrt{6} + \sqrt{2}}{2}$$

$$u(t) = -\sqrt{2} - 2t \qquad\qquad u(t) = \sqrt{2} - 2t$$

We now check whether they verify the local sufficiency conditions as well. With reference to Eq. (12.4) it is easy to verify that only S_4 depends on the adopted solution. In fact we have

$$R(t) = 1, \quad A(t) = \begin{bmatrix} 0 & 1 \\ 0 & 0 \end{bmatrix}, \quad B(t) = \begin{bmatrix} 0 \\ 1 \end{bmatrix}, \quad Q(t) = \begin{bmatrix} 0 & 0 \\ 0 & 0 \end{bmatrix}$$

$$K(t) = \begin{bmatrix} 0 & 0 \\ 0 & 1 \end{bmatrix}, \quad \alpha_{xx}(x) = \begin{bmatrix} 0 & 0 \\ 0 & 0 \end{bmatrix}, \quad S_1 = \begin{bmatrix} 0 & 0 \\ 0 & 0 \end{bmatrix}$$

$$S_2 = \begin{bmatrix} 0 \\ 1 \end{bmatrix}, \quad S_3 = \begin{bmatrix} 0 \\ -2 \end{bmatrix}, \quad S_4 = -2u(t_f)$$

Correspondingly, the solutions of the differential equations in the statement of Theorem 12.1 can easily be computed and we find

$$P_1(t) = \begin{bmatrix} 0 & 0 \\ 0 & 0 \end{bmatrix}, \quad P_2(t) = \begin{bmatrix} 0 \\ 1 \end{bmatrix}, \quad P_3(t) = \begin{bmatrix} 0 \\ -2 \end{bmatrix}$$
$$P_4(t) = P_4(0) + 4t, \quad P_5(t) = 2(t_f - t), \quad P_6(t) = t - t_f$$

where $P_4(0) = 2\sqrt{2}$ if $\lambda_{20} = \sqrt{2}$, whereas $P_4(0) = -2\sqrt{2}$ if $\lambda_{20} = -\sqrt{2}$. We conclude that the requirements of Theorem 12.1 on $P_4(\cdot)$ and $P_6(\cdot)$ are satisfied only when $\lambda_{20} = \sqrt{2}$.

We can gain further insight into the significance of the above conclusion by noticing that it is possible to obtain a globally optimal solution of the problem where the final time t_f is given and the performance index is

$$I = \int_0^{t_f} l^*(x(t), u(t), t)\, dt + m(x(t_f), t_f), \quad l^*(x, u, t) = -x_2 + \frac{u^2}{2}, \quad m(x, t) = -x_1$$

In fact a straightforward application of the Hamilton–Jacoby theory (see [15]) allows us to compute the (optimal) value $I^o(t_f)$ of the performance index as a function of t_f. It results

$$I^o(t_f) = \beta(t_f) + 2t_f + \frac{\beta^3(t_f) - (2t_f + \beta(t_f))^3}{12}, \quad \beta(t_f) = \frac{1 - t_f^2}{t_f}$$

For any given t_f the (minimum) value $J(t_f)$ of the performance index of the original problem is $J(t_f) = t_f + I^o(t_f)$. The plot of $J(\cdot)$ is shown in Fig. 12.1. We note that the first derivative of this function vanishes at the values of the final time which satisfy the NC. We also note that the sign of the second derivative agrees with the local sufficient conditions not being satisfied at t_{f2}. Finally, we observe that the original problem does not admit a solution even though we have been able to compute a locally optimal triple $(x^o(\cdot), u^o(\cdot), t_f^o)$. ∎

Fig. 12.1 Problem 12.1. The function $J(t_f)$

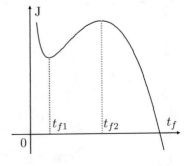

12.2 $x(t_f) =$ Given, $t_f =$ Given

We consider optimal control problems defined by Eq. (12.1) where both the final state $x(t_f)$ and time t_f are given. Obviously, the performance index is purely integral. For such problems we have the following result.

Theorem 12.2

Let $(x^o(\cdot), u^o(\cdot))$ be a pair which satisfies the necessary optimality conditions stated in Theorem 2.2 under the assumption that both $x(t_f)$ and t_f are given. Furthermore assume that $R(t) > 0$, $t \in [0, t_f]$ and there exist solutions of the differential equations

$$\dot{P}_1(t) = -P_1(t)A(t) - A'(t)P_1(t) + P_1(t)K(t)P_1(t) - Q(t)$$
$$\dot{P}_2(t) = -[A(t) - K(t)P_1(t)]' P_2(t)$$
$$\dot{P}_6(t) = P_2'(t)K(t)P_2(t)$$

with the boundary conditions

$$P_1(t_f) = S_1, \ P_2(t_f) = S_2, \ P_6(t_f) = 0$$

such that $P_6(t) < 0$, $t \in [0, t_f)$. Then $(x^o(\cdot), u^o(\cdot))$ is a locally optimal pair, in a weak sense. ∎

Problem 12.2

Let

$$x(0) = \begin{bmatrix} 0 \\ 0 \end{bmatrix}, \quad x(t_f) = \begin{bmatrix} 1 \\ 0 \end{bmatrix}, \quad u(t) \in R, \ \forall t$$

$$J = \int_0^{t_f} l(x(t), u(t), t) \, dt, \ l(x, u, t) = 2x_2 + \frac{u^2}{2}, \ t_f = 1$$

The hamiltonian function and the H-minimizing control are

$$H = 2x_2 + \frac{u^2}{2} + \lambda_1 x_2 + \lambda_2 u, \quad u_h = -\lambda_2$$

so that Eq. (2.5a) and its general solution over an interval beginning at time 0 are

$$\dot{\lambda}_1(t) = 0 \qquad\qquad \lambda_1(t) = \lambda_{10}$$
$$\dot{\lambda}_2(t) = -(2 + \lambda_1(t)) \qquad \lambda_2(t) = \lambda_{20} - (2 + \lambda_{10})t$$

Thus the state motion is

$$x_1(t) = (2 + \lambda_{10})\frac{t^3}{6} - \lambda_{20}\frac{t^2}{2}, \quad x_2(t) = (2 + \lambda_{10})\frac{t^2}{2} - \lambda_{20}t$$

and, by enforcing feasibility $(x_1(t_f) = 1, x_2(t_f) = 0)$ we ascertain that a control which satisfies the NC is

$$u(t) = (2 + \lambda_{10})t - \lambda_{20}, \quad \lambda_{10} = -10, \quad \lambda_{20} = -4$$

We now check whether the local sufficiency conditions are satisfied as well. With reference to Eq. (12.4) we have

$$R(t) = 1, \quad A(t) = \begin{bmatrix} 0 & 1 \\ 0 & 0 \end{bmatrix}, \quad B(t) = \begin{bmatrix} 0 \\ 1 \end{bmatrix}, \quad Q(t) = \begin{bmatrix} 0 & 0 \\ 0 & 0 \end{bmatrix}$$

$$K(t) = \begin{bmatrix} 0 & 0 \\ 0 & 1 \end{bmatrix}, \quad \alpha_{xx}(x) = \begin{bmatrix} 0 & 0 \\ 0 & 0 \end{bmatrix}, \quad S_1 = \begin{bmatrix} 0 & 0 \\ 0 & 0 \end{bmatrix}, \quad S_2 = \begin{bmatrix} 1 \\ 1 \end{bmatrix}$$

Correspondingly, the solutions of the differential equations in the statement of Theorem 12.1 can easily be computed and we find

$$P_1(t) = \begin{bmatrix} 0 & 0 \\ 0 & 0 \end{bmatrix}, \quad P_2(t) = \begin{bmatrix} 1 \\ 2 - t \end{bmatrix}, \quad P_6(t) = \frac{(t-2)^3 + 1}{3}$$

and conclude that the requirement of Theorem 12.2 on $P_6(\cdot)$ of Theorem 12.1 is satisfied.

Unlike in Problem 12.1, the Hamilton–Jacoby theory (see [15]) allows us to conclude that the solution which satisfies the local sufficient conditions is globally optimal.   ∎

12.3 $x(t_f) =$ Free, $t_f =$ Free

We consider optimal control problems defined by Eq. (12.1) where both the final state $x(t_f)$ and time t_f are free. For such problems we have the following result.

Theorem 12.3

Let $(x^o(\cdot), u^o(\cdot), t_f^o)$ be a triple which satisfies the necessary optimality conditions stated in Theorem 2.2 under the assumption that both $x(t_f)$ and t_f are free. Furthermore assume that $R(t) > 0, t \in [0, t_f^o]$ and there exist solutions of the differential equations

$$\dot{P}_1(t) = -P_1(t)A(t) - A'(t)P_1(t) + P_1(t)K(t)P_1(t) - Q(t)$$
$$\dot{P}_3(t) = -[A(t) - K(t)P_1(t)]' P_3(t)$$
$$\dot{P}_4(t) = P_3'(t)K(t)P_3(t)$$

with the boundary conditions

$$P_1(t_f^o) = S_1, \ P_3(t_f^o) = S_3, \ P_4(t_f^o) = S_4$$

such that $P_4(t) > 0, \ t \in [0, t_f^o]$. *Then* $(x^o(\cdot), u^o(\cdot), t_f^o)$ *is a locally optimal triple, in a weak sense.* ∎

Problem 12.3

Let

$$x(0) = \begin{bmatrix} 0 \\ 0 \end{bmatrix}, \quad x(t_f) = \text{free}, \ u(t) \in R, \ \forall t$$

$$J = \int_0^{t_f} l(x(t), u(t), t) \, dt + m(x(t_f), t_f), \ l(x, u, t) = t^k + \frac{u^2}{2}, \ m(x, t) = 2x_2 - x_1$$

where the final time is free and we have either $k = 0$ or $k = 2$.

The hamiltonian function and the H-minimizing control are

$$H = t^k + \frac{u^2}{2} + \lambda_1 x_2 + \lambda_2 u, \quad u_h = -\lambda_2$$

so that Eq. (2.5a) and its general solution over an interval beginning at time 0 are

$$\begin{aligned} \dot{\lambda}_1(t) &= 0 & \lambda_1(t) &= \lambda_{10} \\ \dot{\lambda}_2(t) &= -\lambda_1(t) & \lambda_2(t) &= \lambda_{20} - \lambda_{10} t \end{aligned}$$

The orthogonality condition at the final time is

$$\begin{bmatrix} \lambda_1(t_f) \\ \lambda_2(t_f) \end{bmatrix} - \frac{\partial m(x, t)}{\partial x} \bigg|_{x = x(t_f)}' = \begin{bmatrix} \lambda_{10} \\ \lambda_{20} - \lambda_{10} t_f \end{bmatrix} - \begin{bmatrix} -1 \\ 2 \end{bmatrix} = \begin{bmatrix} 0 \\ 0 \end{bmatrix}$$

and supplies

$$\lambda_{10} = -1, \quad \lambda_2(t_f) = 2, \quad \lambda_{20} = 2 - t_f$$

Thus we can easily compute the motion of the second state variable and find

$$x_2(t_f) = \frac{t_f^2}{2} - 2t_f$$

The transversality condition at the final time is

$$H(t_f) = t_f^k + \frac{u^2(t_f)}{2} + \lambda_1(t_f) x_2(t_f) + \lambda_2(t_f) u(t_f) =$$

$$= t_f^k - \frac{\lambda_2^2(t_f)}{2} - \frac{t_f^2}{2} + 2t_f = t_f^k - \frac{t_f^2}{2} + 2t_f - \frac{1}{2} = 0$$

so that

$$k = 0 \Rightarrow t_f = \begin{cases} t_f = t_{f1} := 3.41 \\ t_f = t_{f2} := 0.59 \end{cases}, \quad k = 2 \Rightarrow t_f = t_{f3} := 0.83$$

Therefore we have found two solutions which satisfy the NC when $k = 0$ and one solution when $k = 2$.

We now check whether these solutions verify the local sufficiency conditions as well. With reference to Eq. (12.4) we have

$$R(t) = 1, \quad A(t) = \begin{bmatrix} 0 & 1 \\ 0 & 0 \end{bmatrix}, \quad B(t) = \begin{bmatrix} 0 \\ 1 \end{bmatrix}, \quad Q(t) = \begin{bmatrix} 0 & 0 \\ 0 & 0 \end{bmatrix}$$

$$K(t) = \begin{bmatrix} 0 & 0 \\ 0 & 1 \end{bmatrix}, \quad S_1 = \begin{bmatrix} 0 & 0 \\ 0 & 0 \end{bmatrix}, \quad S_3 = \begin{bmatrix} 0 \\ -1 \end{bmatrix}, \quad S_4 = \begin{cases} 2 & \text{if } k = 0 \\ 3.66 & \text{if } k = 2 \end{cases}$$

Correspondingly, the solutions of the differential equations in the statement of Theorem 12.3 can easily be computed and we find

$$P_1(t) = \begin{bmatrix} 0 & 0 \\ 0 & 0 \end{bmatrix}, \quad P_3(t) = \begin{bmatrix} 0 \\ -1 \end{bmatrix}, \quad P_4(t) = P_4(0) + t$$

where

$$P_4(0) = \begin{cases} -1.41 & \text{if} \quad t_f = t_{f1} \\ 1.41 & \text{if} \quad t_f = t_{f2} \\ 2.83 & \text{if} \quad t_f = t_{f3} \end{cases}$$

We conclude that the requirement of Theorem 12.3 on $P_4(\cdot)$ are satisfied only when $t_f = t_{f2}$ or $t_f = t_{f3}$.

We can gain further insight into the significance of the above conclusion by noticing that it is possible to obtain a globally optimal solution of the problem where the final time t_f is given and the performance index is

$$I = \int_0^{t_f} l^*(x(t), u(t), t) \, dt + m(x(t_f), t_f), \quad l^*(x, u, t) = \frac{u^2}{2}, \quad m(x, t) = -x_1 + 2x_2$$

In fact a straightforward application of the Hamilton–Jacobi theory (see [15]) allows us to compute the (optimal) value $I^o(t_f)$ of the performance index as a function of t_f. It results

$$I^o(t_f) = \frac{(2 - t_f)^3 - 8}{6}$$

For any given t_f the (minimum) value $J(t_f)$ of the performance index of the original problem is $J(t_f) = t_f^k + I^o(t_f)$. The plots of $J(\cdot)$ are shown in Fig. 12.2. We note that the first derivative of this function vanishes at the values of the final time which satisfy the NC. We also note that the sign of the second derivative agrees with the

Fig. 12.2 Problem 12.3. The
function $J(t_f)$ when $k = 0$
and $k = 2$

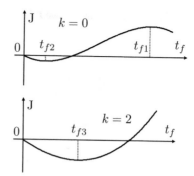

local sufficient conditions not being satisfied at t_{f1}. Finally, we observe that, when
$k = 0$, the original problem does not admit a solution even though we have been able
to compute a locally optimal triple $(x^o(\cdot), u^o(\cdot), t_f^o)$. On the contrary, when $k = 2$
the solution which satisfies the NC is also globally optimal. ∎

12.4 $x(t_f) =$ **Free**, $t_f =$ **Given**

We consider optimal control problems defined by Eq. (12.1) where the final state
$x(t_f)$ is free and the final time t_f is given. For such problems we have the following
result.

Theorem 12.4

*Let $(x^o(\cdot), u^o(\cdot))$ be a pair which satisfies the necessary optimality conditions stated
in Theorem 2.2 under the assumption that $x(t_f)$ is free and t_f is given. Furthermore
assume that $R(t) > 0$, $t \in [0, t_f]$ and there exists a solution of the differential
equation*

$$\dot{P}_1(t) = -P_1(t)A(t) - A'(t)P_1(t) + P_1(t)K(t)P_1(t) - Q(t)$$

with the boundary condition

$$P_1(t_f) = S_1$$

Then $(x^o(\cdot), u^o(\cdot))$ is a locally optimal pair, in a weak sense. ∎

Problem 12.4

Let

$$x(0) = \begin{bmatrix} 0 \\ 0 \end{bmatrix}, \quad x(t_f) = \text{free}, \ u(t) \in R, \ \forall t$$

$$J = \int_0^{t_f} l(x(t), u(t), t) \, dt, \ l(x, u, t) = \frac{u^2}{2} + x_1, \ t_f = 1$$

The hamiltonian function and the H-minimizing control are

$$H = \frac{u^2}{2} + x_1 + \lambda_1 x_2 + \lambda_2 u, \quad u_h = -\lambda_2$$

so that Eq. (2.5a) and its general solution over an interval beginning at time 0 are

$$\dot{\lambda}_1(t) = -1 \qquad\qquad \lambda_1(t) = \lambda_{10} - t$$
$$\dot{\lambda}_2(t) = -\lambda_1(t) \qquad\qquad \lambda_2(t) = \lambda_{20} - \lambda_{10}t + \frac{t^2}{2}$$

The orthogonality condition at the final time is

$$\begin{bmatrix} \lambda_1(t_f) \\ \lambda_2(t_f) \end{bmatrix} = \begin{bmatrix} \lambda_{10} - t_f \\ \lambda_{20} - \lambda_{10}t_f - \frac{t_f^2}{2} \end{bmatrix} = \begin{bmatrix} 0 \\ 0 \end{bmatrix}$$

and supplies

$$\lambda_{10} = 1, \qquad \lambda_{20} = \frac{1}{2}$$

Therefore we have found that

$$u(t) = t - \frac{1 + t^2}{2}$$

satisfies the NC.

We now check whether it verifies the local sufficiency conditions as well. With reference to Eq. (12.4) we have

$$R(t) = 1, \quad A(t) = \begin{bmatrix} 0 & 1 \\ 0 & 0 \end{bmatrix}, \quad B(t) = \begin{bmatrix} 0 \\ 1 \end{bmatrix}, \quad Q(t) = \begin{bmatrix} 0 & 0 \\ 0 & 0 \end{bmatrix}$$
$$K(t) = \begin{bmatrix} 0 & 0 \\ 0 & 1 \end{bmatrix}, \quad S_1 = \begin{bmatrix} 0 & 0 \\ 0 & 0 \end{bmatrix}$$

Correspondingly, the solution of the differential equation in the statement of Theorem 12.4 can easily be computed and we find

$$P_1(t) = \begin{bmatrix} 0 & 0 \\ 0 & 0 \end{bmatrix}$$

We conclude that Theorem 12.4 is satisfied.

As in Problem 12.2, the Hamilton–Jacoby theory (see [15]) allows us to conclude that the solution which satisfies the local sufficient conditions is globally optimal. ∎

Appendix

The impressive number of papers and books concerning optimal control theory which have appeared in the literature witnesses the interest raised by such a topic. Coherently with the scope of this book (a simple, self-consistent and effective presentation of the necessary conditions which constitute the founding results of this theory), we here follow the approach in [15] and mention some *classic* textbooks together with some significant and more recent contributions.

Undoubtedly, the corner stone is the book [17] where the *Maximum Principle* was stated, proved and brought to the knowledge of the international scientific community.

The Maximum Principle and first order variational methods are extensively dealt with in [1, 2, 7, 12, 13].

The particular issue of singular solutions can be more deeply tackled in [4, 5, 8], whereas minimum time problems are extensively illustrated in [2, 17].

Among the numerous recent contributions we mention [3, 6, 9, 18].

A classical presentation of the Hamilton–Jacoby theory, the basic tool for global sufficient conditions, is [11], whereas a more recent one is [15]. A simple derivation of the weak local sufficient conditions can be found in [14].

Finally, the textbooks [7, 8, 10, 16, 19] can be seen as a reference point for basic knowledge of computational methods, an issue which has not been taken into consideration in this book, although, obviously, of primary importance.

© Springer International Publishing Switzerland 2017
A. Locatelli, *Optimal Control of a Double Integrator*, Studies in Systems, Decision and Control 68, DOI 10.1007/978-3-319-42126-1

References

1. Alekseev, V.M., Tikhomirov, V.M., Fomin, S.V.: Optimal Control. Consultant Bureau (1987)
2. Athans, M., Falb, P.L.: Optimal Control. Mc Graw-Hill, New York (1966)
3. Athans, M., Falb, P.L.: Optimal Control: An Introduction to the Theory and Its Applications. Dover Books on Engineering, New York (2006)
4. Bell, D.J., Jacobson, D.H.: Singular Optimal Control Problems. Academic Press, New York (1975)
5. Berhnard, P.: Commande optimale, decentralization et jeux dynamiques. Dunod, Paris (1976)
6. Betts, J.T.: Practical Methods for Optimal Control Using Nonlinear Programming. SIAM Press, Philadelphia (2009)
7. Bryson, A.E., Ho, Y.C.: Applied Optimal Control. Hemisphere Publ. Co., New York (1975)
8. Burlisch, R., Kraft, D. (eds.): Computational Optimal control. Birkhauser, Basel (1994)
9. Cassel, K.W.: Variational Methods with Applications in Science and Engineering. Cambridge University Press, Cambridge (2013)
10. Gregory, J.: Constrained Optimization in the Calculus of Variations and Optimal Control Theory. Van Nostrand Reinhold, New York (1992)
11. Kalman, R.E., Falb, P.L., Arbib, M.A.: Topics in Mathematical System Theory. Mc Graw-Hill, New York (1969)
12. Kirk, D.E.: Optimal Control Theory: An Introduction. Dover Books on Electrical Engineering, New York (2004)
13. Knowles, G.: An Introduction to Applied Optimal Control. Academic Press, New York (1981)
14. Locatelli, A.: Elementi di controllo ottimo. Clup (1987). (in italian)
15. Locatelli, A.: Optimal Control: An Introduction. Birkhauser, Basel (2001)
16. Miele, A.: Gradient algorithms for the optimization of dynamic systems. In: Leondes, C.T. (ed.) Control and Dynamic Systems, vol. 16, pp. 1–52. Academic Press, New York (1980)
17. Pontryagin, L.S., Boltyanskii, V.G., Gamkrelidze, R.V., Mishchenko, E.F.: The Mathematical Theory of Optimal Processes. Interscience Publ, New York (1962)
18. Ross, I.M.: A Primer on Pontryagin's Principle in Optimal Control. Collegiate Publishers, San Francisco (2015)
19. Teo, K.L., Goh, C.J., Wong, K.H.: A Unified Computational Approach to Optimal Control Problems. Longman Scientific & Technical, New York (1991)

© Springer International Publishing Switzerland 2017
A. Locatelli, *Optimal Control of a Double Integrator*, Studies in Systems, Decision and Control 68, DOI 10.1007/978-3-319-42126-1

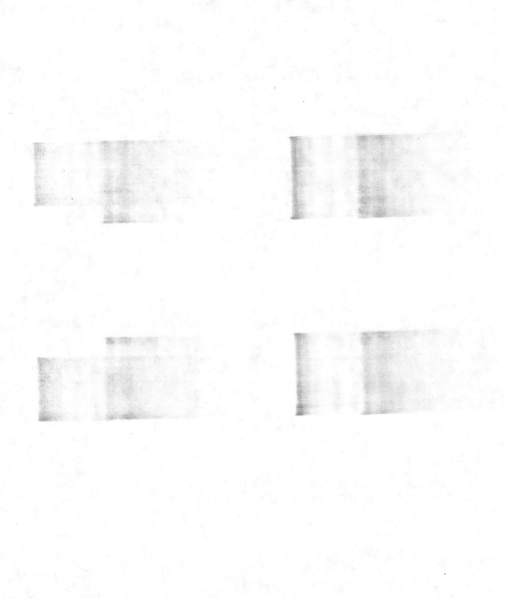

Printed in the United States
By Bookmasters